William Whewell

A Composite Portrait

edited by

MENACHEM FISCH

and

SIMON SCHAFFER

CLARENDON PRESS · OXFORD

1991

Oxford University Press, Walton Street, Oxford OX2 6DP
Oxford New York Toronto
Delhi Bombay Calcutta Madras Karachi
Petaling Jaya Singapore Hong Kong Tokyo
Nairobi Dar es Salaam Cape Town
Melbourne Auckland
and associated companies in
Berlin Ibadan

Oxford is a trade mark of Oxford University Press

Published in the United States
by Oxford University Press, New York

British Library Cataloguing in Publication Data
William Whewell: a composite portrait.
1. Great Britain. Social life, 1837–1901. Biographies
I. Fisch, Menachem II. Schaffer, Simon 1955–
941.081092
ISBN 0–19–824900–4

Library of Congress Cataloging in Publication Data
William Whewell, a composite portrait / edited by
Menachem Fisch and Simon Schaffer.
Includes bibliographical references.
1. Whewell, William, 1794–1866. 2. Philosophy, English—19th century. 3. Science—History—19th century.
4. Science—Philosophy—History—19th century.
5. Apologetics—History—19th century. I. Fisch, Menachem.
II. Schaffer, Simon, 1955– .
B1612.W54W55 1991 170'.92—dc20 90–41928
ISBN 0–19–824900–4

Set by Hope Services (Abingdon) Ltd.
Printed in Great Britain by
Biddles Ltd.
Guildford & King's Lynn

Preface

> I remember praising Whewell's *History of the Inductive Sciences* to [Robert Brown], and he answered 'Yes, I suppose that he has read the prefaces of very many books.'
>
> Charles Darwin, 1876[1]

THE achievement of William Whewell (1794–1866) poses an immediate, if superficial, problem to his interpreters. His celebrated 'omniscience' is hard to deal with, especially given the constricting parochialism of current intellectual fields. Former studies of Whewell have typically compartmentalized their subject. We have been presented with Whewell the Kantian, Whewell the neologist, Whewell the writer of student textbooks, Whewell the sounding-board for John Stuart Mill, Whewell the exemplary 'gentleman of science'. He was obviously all of these at once, hence the salient problem of coping with a polymath. Even the few more holistic treatments of his *œuvre* have been strangely one-sided. The great division between externalist and internalist approaches to intellectual history has been all too obvious here. None of the excellent studies of the public culture in which Whewell led his life have, thus far, established connections with the content of his philosophical, educational, and moral writings. Similarly, critical analyses of his epistemology haul this work up before the court of late twentieth-century philosophy of science, in a manner unrelated to the culture in which he plied his trade and which prompted the great questions to which he sought decisive answers. As a result, the problems of determining the way in which this culture made demands upon his philosophical work, and vice versa, have not been addressed. This hiatus is a result of the schism between the best recent studies of the social world of early Victorian science and recent authoritative studies of Whewell's philosophy. Whewell's work provides plentiful resources for both approaches. His career as a pillar of Victorian society and a 'gentleman of science' gives ample scope for a socio-historical reading.[2] His massive literary output, especially in epistemology, leaves rolling fields ripe for the philosopher's scythe.[3] This apparent division offers a chance to apply the best contemporary historiography to an exemplary case.

Initially the editors of this volume had a more modest aim. We intended that a series of specific, local studies might produce a picture of Whewell designed by

[1] C. Darwin (1974), 61.

[2] Morrell and Thackray (1981).

[3] For some previous philosophical analyses of Whewell's achievement, see Butts (1968); L. Laudan (1971); Marcucci (1963); Blanché (1935); Ducasse (1966); Stoll (1929).

committee, as it were. We planned a patchwork evocation of the full scope of Whewell's intellectual world. In a darkened Oxford pub we sketched out on a paper napkin a list of chapter headings such as 'Whewell the Teacher', 'Whewell the Priest', 'Whewell the Historian', and so on. However, as this volume emerged, these carefully constructed compartments dissolved. Pleasingly, contributors produced accounts which simultaneously confronted contextual, cognitive, and prospective issues. We, the editors, welcomed and encouraged this process. As the dykes burst under a deluge of cross-reference and debate, we attempted to preserve and encourage a measure of fruitful argumentative overlap between chapters.

Much recent historical attention has been paid to the nature of that 'Second Scientific Revolution' which some have seen as marking a dramatic transformation in the realm of natural knowledge in the second quarter of the nineteenth century. Chief amongst such studies have been those works devoted to the newly constituted sciences of geology and biology. The work of the London Geological Society, the nascent institutions of the metropolitan medical communities, the new research institutions in Germany and France, the formation of debate among palaeontologists, naturalists, and geographers, the troubling category of 'Humboldtian science', and, not least, the Darwin industry itself, have all been subject to exemplary social and historical treatment.[4] The study of natural theology's cultural role has also been pursued, with an especial focus on the celebrated Bridgewater Treatises, that group of eight essays by several hands, Whewell among them, produced in the 1830s with the aim of establishing anew the links between revealed and natural knowledge. There were, on the other hand, those radical and sceptical contemporaries for whom the Bridgewater collection was just 'Bilgewater'. So, interactions, both amicable and hostile, between élite savants and plebeian activists have been recovered and restored to their rightful place in our picture of the map of knowledge and power.[5] Such has not been the case with the mathematical and physical sciences of this epoch. Nor have historians of philosophy been conspicuous in this move. From a Whewellian perspective, this situation has fascinating resonances. Whewell was a massive presence in the British mathematical and philosophical scene. Comparative historical study of his career may help extend this historiography to these relatively uncharted territories. Furthermore, Whewell himself distinguished between 'pure' and 'mechanical' sciences on the one hand, and 'palaetiological' sciences on the other. The latter described the processes of progress in explicitly historical terms. It is these which have been subjected to socio-historical treatment in our times. This is not surprising, since here the forces of the historical and the progressive are evident. But Whewell's analysis of the pure and mechanical sciences was equally palaetiological. He gives us excellent clues as to how we might bring

[4] Rudwick (1985); Secord (1986); Desmond (1982); S. Cannon (1978); Kohn (1985); Lenoir (1982); Lesch (1984); Rehbock (1983); Winsor (1976); Engelhardt (1979).

[5] Desmond (1989); Cooter (1984); Barrow (1986); Brooke (1974).

these commanding heights of the Victorian scientific empire within the ambit of a thoroughgoing historicism.[6]

Whewell's historico-philosophical perspective thus not only provides a fascinating case-study for contemporary historiography but also bears reflexively upon contemporary historiographical debates. In 1936 George Sarton complained that 'nothing illustrates better the backwardness of our studies than the fact that Whewell's book was still commanding the respect of many thoughtful readers at the beginning of the century.'[7] In fact, nothing now illustrates better the advance in our studies than the fact that in this reassessment of William Whewell we have been able to draw upon the work of a number of leading scholars in history and philosophy of science, contributions so marked by a Whewellian attitude to their own field. However, Whewell's relevance is by no means limited to historiography. His philosophy (when read within the context of his own problematic) is found to be profoundly relevant to issues of central contemporary philosophical concern. For Elkana, for example, 'Whewell's thought, astonishingly in advance of its time, strikes us now as a lively refutation of that twentieth-century mode of philosophy' known as logical positivism.[8] It must be stressed that Whewell's actual role in our own philosophies can only be rendered through a detailed contextualization of the emergence and purpose of his philosophical project. When read this way, Whewell does not challenge a latter-day consensus; rather, he participates in a latter-day debate. Once again, we can learn a reflexive lesson from Whewell himself. No other writer of his generation understood so well the right means by which properly contextualized philosophies of the past could be deployed in contemporary debate. This process of deployment, that of a 'constitutional conservative' innately opposed to the heresies of the 'new Whigs', avoided the pitfalls of Whiggery which have so exercised the historians of philosophy of our own generation.[9]

This mutually supportive historical and philosophical aim is reflected both in the structure and in much of the content of this book. In this respect, the volume intriguingly recapitulates the lay-out of the *History* and *Philosophy of the Inductive Sciences*. However, the task of this collection of papers is even more formidable than that which Whewell undertook. Whewell took all the sciences as his province. His project ran into difficulties through the width of its scope. It is highly misleading to use the term 'polymath' for Whewell, since he was precisely in search of a means of synthesizing a vast range of allegedly disparate material. He was after the underlying unity in the stressful diversity of the sciences of his age, and hence in pursuit of means of extending these

[6] For the historiography of early nineteenth-century mathematical physics, see C. Smith (1976); Grattan-Guinness (1985); D. Wilson (1985); D. Miller (1986). For exemplary treatments of mathematical physics in Germany, see Stichweh (1984); McCormmach and Jungnickel (1986).

[7] Sarton (1936), 65.

[8] Elkana (1984), p. xvii.

[9] For problems in the historiography of philosophy see Skinner (1969); Rorty, Schneewind, and Skinner (1984); for histories of the sciences see Graham, Lepenies, and Weingart (1983). For Whewell on the 'new Whigs' see Todhunter (1876), ii. 318.

insights to other realms of knowledge: morality and aesthetics, economics and jurisprudence. We try to make sense of this project through analysis of both context and content. We have signalled the need to recover Whewell's own sense of his purpose. Our contributors vary between stressing that the philosophical project of the 1830s was driven principally by Whewell's personal view of his problematic as it developed in that period, or in portraying it as a principled response to the fierce pressures of the decade of Reform. Furthermore, some contributions demonstrate that, after 1840, Whewell never reneged on the fundamental doctrines which he had established in his *History* and *Philosophy*. Others show that in interaction with the aims and endeavours of other workers, Whewell's account of the sciences, natural and moral, underwent significant modification. Once his principles were in place, Whewell single-mindedly committed himself to a generalized picture of the antithetical structure of science construed as a unified system. He also sought to map and maintain some key boundaries within the sciences. Details of this exercise were modified, debated, or even abandoned as his career progressed. We adopt an approach which renders Whewell's venture in his own terms, while simultaneously understanding the aims of contemporary admirers, targets, and critics.

So this book has much to say on the issues of historiography and philosophy of the sciences, on Whewell's own endeavours, and on the culture of his time. His work began as a relatively simple effort at understanding the monumentality of excellent science; it became a grand scheme for interpreting the emergence and structure of the intellectual world. So our book began as a simple-minded effort to understand a Victorian monument; it has ended, we hope, as a bold effort to exemplify the right way of studying the context and content of intellectual life. The chapters in this book follow this sequence: diachronic historical accounts of the emergence of various aspects of Whewell's activity are succeeded by essays which take a more synthetic view, linking such themes as theology and epistemology, history and linguistics, or physical and life sciences. The final chapters present philosophically-oriented critiques of Whewell's work, while maintaining the thoroughgoing historicism which is a precondition for a satisfactory reading of these texts.

Harvey Becher's introductory chapter offers an integrated overview of Whewell's achievements and career. The next group of chapters analyses the historical emergence of Whewell's multi-faceted project. Menachem Fisch charts the developing relationship between Whewell's work as a teacher in Cambridge and the problems which he confronted as he strove to develop his philosophy from his initial involvement with the Analytical Society to the critical conjuncture of the earlier 1830s. Geoffrey Cantor provides an account of Whewell's work as a major historian. The tradition of 'Liberal Anglican' historiography emerges as a cultural context in which to place the analysis of progress and recurrence which characterized this work and that of Whewell's immediate sources and colleagues. Following Michael Ruse's synthetic treatment of Whewell's various accomplishments in the natural sciences, including tidology, political economy, and mineralogy, Perry Williams charts the political context in which Whewell's work as educational reformer developed. Williams emphasizes the struggle which Whewell led against what he perceived as a Benthamite, utilitarian threat to the ideals of 'liberal education'. Williams divines important links between Whewell's definition of the liberal curriculum and both his epistemology and political view of the university's civilizing mission.

John Brooke bases his reading of Whewell's theological outlook on a detailed study of his sermons. These reveal a priest engaged in developing a theodicy directed against modish irreligion and deism. But while Whewell certainly drew on highly traditional apologetic resources of natural theology, Brooke shows convincingly that he saw the deficiencies of purely rational physico-theology, and, in consequence, also worked hard to construct a Pauline account of conscience and of redemption which owed little to the received wisdom of Paleyite priestcraft. Richard Yeo's chapter on Whewell's philosophy shares Brooke's interest in the theological complexity of this project. Yeo discusses Whewell's celebrated stress on the 'unity of the sciences'. This allowed a coherent place for the moral sciences. The laws of nature which sciences rendered were not to be lumped together. It was in their emergence, Yeo argues, that the sciences, natural and moral, displayed their progressive and their unitary quality. Not all Whewell's audiences welcomed this manœuvre. Yeo examines the problems Whewell encountered in dealing with these dissenters.

Chapters by Simon Schaffer, David Wilson, and Jon Hodge all deal with the relation between the historical development of Whewell's career and themes which have drawn much attention from subsequent commentators. They pursue Yeo's concern with the relation between Whewell's aims and the response of his contemporaries in fields such as language, physics, and evolution theory. Schaffer seeks to understand the relationship between the *History* and the *Philosophy* through a close reading of Whewell's language and of his theory of language. Schaffer follows Cantor in arguing that apparent dichotomies between history and philosophy, induction and deduction, permanent and progressive sciences, provided Whewell with problems and resources with which to make a model of scientific language of great local power. Whewell's famous coinages, including the term 'scientist', reveal themselves as marks of the emergence of his synthetic, and antithetical, epistemological scheme. Schaffer uses the exchanges with heroic discoverers such as Faraday to uncover Whewell's notion of the right management of the progressive sciences. Wilson also examines Whewell's dealings with the leading natural philosophers of his time. He uses this material to see the way in which Whewell modified and developed his account of 'fundamental ideas' and of the demarcation of the sciences in the light of the best available scientific inquiry to which he had direct access. Hodge confirms this sense of Whewell as 'judicious critic and cautious reformer', especially in his close dealings with the actualism and transmutationism debated with Charles Lyell, and the more complex and tortuous interactions with Charles Darwin and the author of *Vestiges*. Hodge judges Whewell's own account of the boundaries and content of the 'palaetiological sciences', especially geology, noting that for Whewell no real contradiction existed between the stress of divine action through natural law and the discontinuity of law-like series in the histories of this planet, life, and intelligence. At the same time, Hodge scans Ruse's 'neo-Whewellian' notion that scientists, notably Darwin, have great philosophers lurking behind them. Whewell's profound historicism, and the 'neo-Whewellian' historicism with which his work must be read, emerge quite clearly from this story.

Furnished with an account of the *emergence* of Whewell's philosophy of science, Fisch's second contribution endeavours to explicate and critically assess Whewell's theory of scientific truth. Against the claims of earlier critics, Whewell's peculiar notion of 'necessity' emerges as both novel and viable. By contrast, on the showing of Whewell's own philosophy, his notion of 'empirical' truth is seen by Fisch to be misguided. Gerd Buchdahl's contribution is a much enlarged version of a classic paper which first appeared in 1971. Buchdal develops a quintuple categorization of modes of appraisal of hypotheses. He uses this scheme to analyse and chart the respective methodologies of Whewell and of his opponent J. S. Mill. Finally, John Wettersten and Joseph Agassi analyse Whewell's 'problematic heritage'. They point to the suppression of Whewell's achievement in the philosophical writings of John Herschel, J. S. Mill, Ernst Mach, C. S. Peirce, and Pierre Duhem.

Wide as is the scope of this book, it also points towards some key areas for further enquiry. Notable amongst these are, first, the profound significance of Whewell's engagement with the mathematics of his time. Study of the problematic trajectory from the Cambridge Analytical Society through figures such as Augustus de Morgan to the accomplishment of George Boole and his contemporaries could be illuminated from this perspective. The new zones of mathematical physics which appeared under the aegis of Cambridge wranglers and their associates in the mid-nineteenth century should be studied from the perspective of Whewell's epistemological analysis. Here the links between Whewell, William Rowan Hamilton, and James Clerk Maxwell would be a fine site for research.[10] Second, considerable scholarship has been devoted to the development of the science of political economy at this epoch. Some recent studies have begun to show us how exponents of political economy met radical challenges from so-called 'Baconian' critics in Cambridge and elsewhere. Whewell and his close friend Richard Jones can now take their rightful place within this development, and may well emerge as crucial sources for the debate on application of economic models in natural sciences, and for the 'marginal revolution' which characterized the science of economics in the third quarter of the nineteenth century.[11] Finally, this volume has skirted Whewell's massive

[10] Hankins (1980); Hendry (1986); Becher (1981); Harman (1985); Wise (1982).

[11] Hollander (1983); Corsi (1987); Rashid (1979); Checkland (1951).

accomplishments in the area of literary and moral sciences. We have debated his relation with the Coleridgean philologists and with the consequentialist moralists, but evidently revision of some received views of the growth of the moral sciences is now needed. As the connections with William Rowan Hamilton, Julius Hare, and S. T. Coleridge suggest, the time is ripe for a sustained investigation of the relation between this work and the development of literary Romanticism after 1815.[12] It is a tribute to the fertility of Whewell's intellectual world that so much of such importance now rates so high on our agenda as historians, philosophers, and cultivators of the sciences.

Menachem Fisch
Simon Schaffer

Cambridge
June 1988

[12] Abrams (1958); Levere (1981); Knight (1970); Shaffer (1974).

Contents

1. *William Whewell's Odyssey: From Mathematics to Moral Philosophy*

by HARVEY W. BECHER

WILLIAM Whewell was born on 24 May 1794, and died on 6 March 1866. What the 'ruddy strapping divine' crowded into his life brought forth from Sidney Smith, the satirist, a breakfast-party quip: 'Science is his forte and omniscience is his foible.'[1] Smith's epigram was clever, but it provided little insight into the object of his barb. Even today the name Whewell is ubiquitous in historical literature while the man himself remains an unresolved nebula, a testimonial to the difficulty of placing Whewell's wide-ranging interests into some kind of fathomable pattern. Since Whewell was both a significant influence, and a reliable mirror, on his age, bringing him into focus is of historical importance, and the direction of this essay is towards that end.

During his fifty-four years at Trinity College in Cambridge University, in an age when knowledge reverberated throughout an intellectual world unencumbered by barriers erected by disciplines narrowly defined as means and ends in and of themselves, Whewell incessantly studied and promoted the science, literature, and pedagogy which engulfed him. His forte was not scientific research, but he had an uncanny talent for assimilating, dissecting, exposing, and developing the foundations of the research of others. With this extraordinary ability, with a philosophy of knowledge that expeditiously guided his enquiries, and with a seemingly infinite store of energy, Whewell acquired not omniscience, but rather a medley of information from which he composed a cerebral polyphony resonant with the intellectual tones of his times.

A list of errata correcting what has been written about Whewell would require a book. An erratum on that list would be that Whewell did not

Without the financial aid provided by the Northern Arizona Organized Research Committee, I would not have been able to complete this essay. I am grateful to friendly and helpful librarians at the Trinity College Library, the Cambridge University Library, The Royal Society of London, the British Museum, Yale University, and Northern Arizona University without whose help my task would have been much more difficult. I would like to thank the Master and Fellows of Trinity College for the use of the Library and for their permission to quote from the Whewell Papers. I would also like to thank the Bursar of Trinity College for making the Account Books available to me.

[1] Todhunter (1876), i. 410.

contribute to the one and only issue of the *Memoirs of the Analytical Society* (1813).[2] When Whewell came up to Trinity in 1812, Cambridge was in the throes of the analytic revolution which brought contemporary mathematics into its curriculum.[3] Whewell probably was too immature a mathematician to make a contribution to the *Memoirs*, which forwarded the revolution. Even more, he never exhibited the zeal for pure mathematics displayed by the undergraduate leaders of the revolution—George Peacock, John F. W. Herschel, and Charles Babbage. For Whewell, mathematics was and remained only a means to an end and not an end in itself.

Initially mathematics enabled Whewell to escape from his father's building and joining business in Lancaster. Mathematical talent—if discovered—opened doors at Cambridge for those not born to open doors. When the Lancastrian élite with connections in Haversham and Cambridge discovered the underprivileged 'prodigy' in their midst, Whewell was soon on his way to Haversham Grammar School for two years, and then to Trinity as a sub-sizar with an exhibition that a Haversham squire held in his gift.[4] Whewell pursued his studies with the vigour that was to mark his activities for a lifetime. He led his class at the annual college examinations, and in 1813 he earned the honour of reciting a Latin declamation on Caesar and Brutus in the college chapel. He gained a sizarship, a scholarship, the Chancellor's Medal for the best English poem on the subject of Boadicea, and the third English declamation prize. The medal did not enitrely please Whewell's former schoolmaster who feared that the young scholar might be neglecting mathematics. Indeed, in the summer of 1815, when he should have been drilling for the Tripos, Whewell found it difficult to limit his readings to mathematics. He conceded that 'some people bid us beware of the Demon of universal knowledge, and I suppose some people are wise', but he savoured the idea that he had access to almost every book ever written. Four months later, Whewell provided grist for Smith's mill by admitting that he yearned for 'something like omniscience'. Despite continuing in his 'all-reading, all-learning mood', Whewell graduated second wrangler and second Smith's Prizeman in 1816.[5]

The second wrangler of 1816 had neither the means nor the family connections to pursue a career in either the Church or the law, so like other high wranglers

[2] Without documentation, Martha McMackin Garland states that Whewell contributed to the 'journal of the Analytical Society' which was 'published periodically' (Garland (1980), 29). The only 'journal' of which I know was the *Memoirs*. It had one issue, and Whewell did not contribute to it.

[3] Dubbey (1963); Becher (1981); Enros (1983).

[4] Thomas Satterthwaite to John William Whittaker, 15 Nov. 1809 and 12 Oct. 1812, Lancaster Public Library, MSS 6695 and 6696; Whewell to his father, 5 Sept. 1811, Whewell Papers, Trinity College Library, Cambridge University, Add. MS c. 191^1; Humber [1968?], 28; Stair-Douglas (1881), 2–6.

[5] Pryme (1870), 332; Todhunter (1876), i. 3–6; Satterthwaite to Whittaker, 22 Nov. 1812 and 24 July 1814, Lancaster Public Library, MSS 6697 and 6702; Whewell to his aunt and sisters, 22 Dec. 1813 and 17 Dec. 1814, Whewell Papers, Add. MSS c. $191^{7,67}$; Whewell to George Morland, 10 Aug. [1815?] and 15 Dec. 1815, Todhunter (1876), ii. 6–8, 10–12; Stair-Douglas (1881), 39–40.

before and after him who had to make their own way in the world, he took private students. More importantly, since an academic career for a poor boy depended upon winning a Fellowship, Whewell studied for the Fellowship examinations. In 1817 he won a Fellowship, a 'Freehold for life', as the Professor of Geology, Adam Sedgwick, would later define it.[6] The following year Whewell became an assistant mathematics tutor at Trinity, thus beginning a career that would culminate in the Mastership and end only with his death in the Lodge.

In 1819 Whewell published the first and most successful of his textbooks, *An Elementary Treatise on Mechanics*. The *Elementary Treatise* went through seven editions by 1847 and was translated into German and Chinese. In 1823 Whewell became a head tutor at Trinity and published *A Treatise on Dynamics*, which ran through three editions, as a sequel to his *Elementary Treatise*. Through these and his other textbooks, through his tutorial lectures, through the publication of pedagogical tracts, and through hard work on innumerable syndicates, Whewell conditioned the reform of the Cambridge mathematics curriculum in line with his own views.[7] His textbooks also served for propagating pieces of his philosophy of science, and, of course, for earning royalties.

Although he published papers on mathematics, reduced more abstruse treatises to simpler formulations, and published textbooks, Whewell did not push forward the frontiers of mathematics. But because of his mathematical literacy, he could overcome the only real disciplinary barrier in early nineteenth-century Britain. Also because of this literacy, and because of the prestige accorded wranglers, Whewell enjoyed a special status at Cambridge and in the English academic community at large. Within England, Whewell, like other wranglers, participated in the forming and running of various scientific organizations. Within Cambridge, wranglers held claim to all ten university science professorships. Whewell decided to procure one, and in 1826 considered the Lucasian Professorship of Mathematics, but abstained from canvassing for the position because a tutor could not hold it; however in 1828 he obtained the Professorship of Mineralogy which had become vacant.[8]

Whewell's career in mineralogy illustrated his activities elsewhere. He early studied mineralogy, as witnessed by his attendance at the Professor of Mineralogy's lectures as an undergraduate. In 1818 he undertook the history of mineralogy, and in 1821 he studied geology and took a geological tour with Sedgwick. In 1825 he rounded off his education with visits to mineralogical museums in Berlin and Vienna and with three months under Friedrich Mohs at Freiberg and Vienna studying crystallography, a subject upon which he had

[6] 'The Commissioners and the Colleges' [cut-out from *Cambridge Chronicle*], Cambridge University Library, Add. MS 725[c], 96.

[7] Schnuse (trans.) (1841); Atkins and Shan-lan Li (trans.) (1867); Becher (1981), 14–48.

[8] Whewell to Richard Jones, 13 and [n.d.] Oct. 1826, Todhunter (1876), ii. 71–2, and Whewell Papers, Add. MS c. 51[30].

already presented papers to the Royal Society of London and the Cambridge Philosophical Society.[9]

Whewell went straight to the foundations of his subject. In his lectures he set out to bring together the natural historians who classified minerals by their external characteristics and the mathematically and physically oriented who classified minerals on the basis of their chemical composition. In his publications he provided a nomenclature, a taxonomy, and a simple mathematical foundation for crystallography. Having thus set the discipline on what he considered to be its proper path, having no desire to pursue the nitty-gritty of the subject or make it a lifetime career, and having found a successor in his protégé William Hallowes Miller, Whewell resigned the Chair in 1832, four years after he had assumed it. Intellectually, Whewell achieved his goal: his work and that of Miller laid the foundations for British crystallography for the rest of the century.[10]

A professorial chair provided political clout in the university because of the prestige associated with it and because the professor was a logical appointee to any syndicate dealing with matters impacting his discipline. As Professor of Mineralogy, Whewell had served on syndicates that reformed the medical curriculum and fought for the construction of a central university complex which would have included up-to-date teaching and research facilities for all the science professors. By 1832 the new medical curriculum was in place, plans for building the university complex had fallen through, and Whewell had served on syndicates that had secured new facilities for the Professors of Anatomy and Chemistry.[11] Little more could be done, and Whewell had no further use for his professorship, especially with Miller waiting in the wings.

Whewell resigned his chair also because he had outlined a decade of work in areas other than mineralogy and wanted to get on with it. In 1830 as Professor of Mineralogy he had published *Architectural Notes on German Churches, with Remarks on the Origin of Gothic Architecture*. Whewell's education in architecture had begun in boyhood. His father had been reluctant to give him up to education because the lad knew more about some aspects of building than he. Whewell's father had sent some of his son's drawings to a London architect; the architect had foreseen the possibility of another Inigo Jones or Christopher Wren in the drawings. Whewell learned about building by doing, and when he provided imagery for force in his mechanics textbooks, he was writing from experience: a footlathe's pressure on the foot, a winch thrusting against the hand, a horse pulling on its traces, and steam pushing against a piston. Likewise, Whewell knew first-hand about friction, the essence of architecture,

[9] Whewell to H. J. Rose, Whewell Papers, R2 99[25]; Whewell (1822); 'On the Classification of Crystalline Forms, Particularly with Reference to the Systems of Weiss of Berlin and Mohs of Freyburg' (read 16 May before the Cambridge Philosophical Society), cited in Clark (1889–92), p. xxvii; Whewell (1825).

[10] Whewell 'Mineralogy Lecture Notes' [1828?], Whewell Papers, Add. MS a. 61[34]; Deas (1959).

[11] Becher (1986), 74–8.

and he mathematized it, together with arches and other architectural devices in his mechanics textbooks.[12]

Whewell also studied the history of architecture, in published works and in the archaeological record. His intellectual inspiration was Thomas Rickman, a Birmingham architect and the foremost historian of English medieval buildings. As early as 1818 Whewell became acquainted with Rickman's universally acclaimed *An Attempt to Discriminate the Styles of English Architecture, from the Conquest to the Reformation*.[13] In his 1817 publication, Rickman defined and classified English buildings as Early (1200–1300), Decorated (1300–1400), or Perpendicular (after 1400) with such force that all subsequent historians used his taxonomic devices without questioning their validity. Architectural history had become a science rather than an inaccurate collection of antiquarian curiosities.

In 1821, in the *Journal of the Royal Institution*, Whewell published his translation of an article by M. F. de Lassaulx, the Prussian architectural historian, on a means of erecting light vaults over churches and similar buildings. In 1823 Whewell surveyed ecclesiastical buildings in Normandy, making notes, drawings, and tables to integrate the data and to make visual and immediate the distinguishing features of the different orders according to Rickman's taxonomy. In 1824 he studied cathedrals and churches in Cumberland, and in his notes he included cross-references to Rickman's publication. Whewell also scrutinized architecture on the continent. In 1829 he found the Netherlands disappointing, but became so enthralled with German churches that he decided against travelling on to Switzerland and neglected his study of mineralogy. What caught Whewell's attention were the beautiful specimens of Gothic architecture and the circumstance that Germans had made no attempt to theorize about them.[14]

In 1830, the same year he published the *Architectural Notes*, Whewell toured Devonshire and Cornwall with Rickman, studying 'some dozens of churches'. Two years later the duo trekked through Picardy and Normandy, examining four Roman constructions, fourteen cathedrals or collegiate churches, forty-three larger churches in towns, fifty smaller churches in towns and villages, fourteen domestic and civil buildings, and six smaller edifices. The tour resulted in Whewell publishing in 1832 a second edition of his *Architectural Notes* in which he added 143 pages of 'notes written during an Architectural Tour in Picardy and Normandy' to the first edition of 118 pages. Seven years later he published the third edition, this time adding 87 pages of 'notes on the

12 W. Grayson to John Whewell [n.d.], Whewell Papers, Add. MS c. 193[192]; Whewell (1819), 240–56; id. (1841*b*), 71, *passim*; id. (1833*d*), 115 ff.; id. (1841*c*), 84–111.

13 Whewell to Jones [1818?], Whewell Papers, Add. MS c. 51[8]; before 1817 Rickman published an outline of his work, Rickman (1835), i. p. iii.

14 Stair-Douglas (1881), 37–8, Whewell to his sister, 29 Oct. 1829, Whewell Papers, Add. MS c. 192[107]; Whewell, 'Sketch Books, 1821–53', ibid., R6 12[5–7, 11, 16, 25]; Whewell to Jones, 31 July 1829, Todhunter (1876), ii. 99–101.

Churches of the Rhine', a translation of a treatise by de Lassaulx, with whom Whewell had formed a cordial intellectual relationship.[15]

In the *Architectural Notes* Whewell used the classificatory system devised by Rickman and displayed an appreciation similar to his for the whole of each style. He relied also, however, upon current German works. His purpose was twofold: to point out and interpret the form and the evolution of the Gothic, and to suggest a theory of why Gothic (pointed) arches came into being. Whewell achieved success with both endeavours, but in dissimilar ways.

In regard to his first aim, Whewell detailed how German architects, independently of the English and often independently of one another, stumbled along from the Romanesque through a mixture of styles towards the Gothic until they distinctly envisioned the basic concept of the dominance of the vertical (grasped what in his philosophy of science Whewell would label the fundamental idea). With this concept firmly in mind, the Germans constructed fully Gothic buildings after and not when the pointed arch had come into being. On the basis of this chronology, Whewell rejected the prevailing belief that the pointed arch was the salient feature of Gothic architecture. In these archaeological reconstructions, Whewell's research was exemplary. When he contrasted the distinct elements in the designs of the German Late Romanesque with those of the Early English by a painstaking analysis of what in each of them was still Romanesque and what was Gothic, Whewell manifested 'an understanding for the totality of these motifs as a specific style such as had not been shown by any archaeologist, not even in Germany'.[16]

In his theoretical reconstructions, Whewell suggested that the pointed arch arose out of mechanical needs for vaulting. Rickman agreed, but Whewell's cohort, Robert Willis, the Jacksonian Professor of Mechanism and Experimental Philosophy, showed that Whewell's explanation of the origin of the pointed arch was erroneous, and challenged his view that the concept of vertical predominated over other dimensional concerns in the fully Gothic works. Although these were Whewell's major theoretical points, he appeared unperturbed by Willis's determinations.[17]

To understand Whewell's rather placid acceptance of Willis's confutation of his theories and to understand nevertheless why the theoretical side of Whewell's work was a success in his own eyes, it is necessary to understand his view of the progress of knowledge. For Whewell, the accumulation of masses of uninterpreted data led only to confusion; research based upon theory, even an incorrect or inadequate one, made clear facts and relations 'which would

[15] Whewell to his aunt and sisters, 27 Aug. 1830, to his sisters, 3 Aug. and 9 Oct. 1832, and to Vernon Harcourt, 12 Sept. 1832, Whewell Papers, Add. MSS c. 191[112, 123–4, 195] and O15 47[104a]; Rickman (1835), i. 309; Whewell (1842), 204 n.; Todhunter (1876), i. 43.

[16] Whewell (1842), ix–46, 305–6, 312–29, *passim*; Pevsner (1970), 6.

[17] Rickman to Whewell 12 Mar. 1830, Whewell Papers, R6 11[30]; Willis, 'On the Construction of Vaults in the Middle Ages' [*Transactions of the Institute of British Architects*], Cambridge University Libarary, Cam. a. 500. 7, pp. 25–69; Willis (1830), 115, 120, 130, *passim*; Whewell (1842), pp. xi–xii, 16.

otherwise seem vague and unclear', and provoked and defined questions which stimulated discussion which led to more theory-directed research which led to truth. Thus we can take at face value what Whewell wrote in the *Architectural Notes*: 'No one will rejoice more than myself to find what I have put forward superseded by the complete labours and sounder views of persons who may bring . . . more leisure, talent, and research.'[18]

Whewell's work, along with Rickman's and Willis's, brought previously obscure facts and relations clearly into view, and Whewell's *Architectural Notes* stimulated research, especially since the conflict between the advocates of the Grecian revival and those of the Gothic revival reached its peak in Britain just as Whewell was publishing the first edition. The forceful Grecian advocate, C. R. Cockerell, who blended together Byron-like romanticism, philhellenism, liberalism, and nationalism, in 1829 acknowledged that a 'temporary celebrity' had been given the Gothic 'of late years', but predicted that it would 'again probably sink into disrepute'.[19] Cockerell was wrong; the unfinished National Monument on Calton Hill in Edinburgh (1820–1829), which Cockerell designed, was almost the last gasp of this Grecian revival, and the Gothic revival had not yet reached its peak. Rickman, Whewell, and Willis kept the Gothic intellectual revival alive until the Ruskinian onslaught buried the Greeks.

Although Whewell was enamoured of the Gothic, and though Ruskin was the main drive behind the Gothic triumph, Whewell was less than enthusiastic about Ruskin's *Seven Lamps of Architecture*, published in 1849. What Rickman, Whewell, and Willis brought to the study of medieval architecture was a structural-analytical approach reinforced by geometrical descriptions. Ruskin was an art critic with little understanding of the technical aspects of building, engineering, or architecture. When Whewell wrote a review of the *Seven Lamps*, this difference stood out. Whewell began the review by defending the contributions made to the 'art of architecture' by 'unprofessional writers' on the grounds that the labours of Rickman, himself, and Willis had had a positive influence upon archaeological research, upon architectural criticism, and upon architecture itself. Whewell found Ruskin's writing elegant, even poetical, but also indistinct, and he spent much of the review redefining Ruskin's seven lamps. Also, Whewell used his mineralogical knowledge to undermine Ruskin's argument that beauty is only achieved in the imitation of natural forms, invoked his engineering expertise to destroy Ruskin's contention that the King's College Chapel defied the 'lamp of truth' by concealing the support for its vault, and employed his archaeological knowledge to illustrate why 'those among us whose architectural views have been formed by a comprehensive study of English, French and German medieval architecture' could not share

[18] Whewell (1833); Whewell to L. A. J. Quetelet, 2 Oct. 1835, and to George Biddell Airy, 18 Jan. 1843, Todhunter (1876), ii. 227–9, 306–8; Whewell (1842), pp. xi, 231.

[19] C. R. Cockerell to the Honourable Vice-Chancellor and the syndics appointed to consult respecting the application of the premises lately purchased by the university, for the arrangement and improvement of the Public Library, the erection of schools, lecture rooms, Museums etc., 31 Oct. 1829, Cambridge University Library, Add. MS 6630, ff1.

Ruskin's 'unmingled admiration' of Italian constructions. For the down-to-earth Whewell, Ruskin was somewhat ineffectual.[20]

Returning to the question of omniscience, the disciplines of mineralogy and history (the *Architectural Notes* was Whewell's first publication in history) are disparate. Yet there was a common approach to them on Whewell's part, an approach that pervaded his work in all areas. He had a long-standing interest in both fields (his last publication in architecture being an analysis of the decline of the Gothic presented to the Royal Institute of British Architects in 1863). In both he studied not only the best works, but also had access to the best minds. He clearly grasped what had been done and what needed doing. He then did his own research (when he resigned his professorship, Whewell donated approximately one thousand mineralogical specimens to the university).[21] Finally, he laid down the channels through which he believed the discipline should flow and turned further research over to other men as competent, or more so, than he. Whewell desired to assimilate and organize knowledge, to get the big picture. He had no urge to investigate every nook and cranny not yet explored.

What also stands out is Whewell's rationality, an almost clinical approach to his subject. At the time of his professorship, a rift within mineralogy had separated those who were enraptured with dilettantish field collecting and those who argued that the future of mineralogy lay in a mathematical-experimental approach. Whewell clearly belonged in the latter camp. In his historical work, Whewell's position is even more obvious. Prior to Rickman's opus, the study of Gothic architecture was superficial and wrapped in romanticism. Rickman, Whewell, and Willis brought a scientific attitude to bear on the subject. To some degree, Ruskin represented a return to the earlier period. That is why Whewell viewed Ruskin's *Seven Lamps* to be as much mysticism as science.[22]

Another unifying factor in Whewell's work was his philosophy of science which, though he did not publish until later, he consciously practised. On the basis of his philosophy, Whewell knew what had to be done and where to go. He did not waste time blundering around aimlessly. In general, the procedure was to establish a classificatory system for organizing data, to derive laws from the organized data by the imposition of conceptions provided by the mind, and then to propose theories (causes) for the laws. Whewell tended to leap into areas where some progress in this direction had been made. His excursions into other realms, for instance his 1831 articles on chemical notation and chemical isomorphism, were usually tied to some of his other work (in the case of chemistry, mineralogy), and were short-lived (he published nothing else in chemistry other than a reply to Berzelius's criticisms of his article on notation).[23] In both mineralogy and the history of medieval architecture, foundations had

[20] [Whewell] (1850*a*), 151–9.

[21] Todhunter (1876), i. 58–9, 238.

[22] Becher (1986); Anon. (1821), 13–47, and (1842), 11–149.

[23] Whewell (1831*c*); id. (1831*e*); id. (1834*h*), 9–10.

been laid in the classification and organization of specifics. Whewell made contributions to both disciplines at this level, and he did not procrastinate in theorizing. His philosophy, and his ability to envision the entire landscape of disciplines in order to pick out those plants ready for harvesting, expedited his march through knowledge.

Whewell's willingness to theorize contrasted sharply with the temperament of his friend, the political economist Richard Jones. When he saw how 'oppressive' was Jones's fear to publish, he became 'less angry' with his own 'fiddle faddle way of swimming cork boats in a gutter'.[24] As Whewell learned medieval architecture from Rickman, he learned political economy from Jones. However, in his relationship with Jones, Whewell exhibited another trait not illustrated as clearly elsewhere: his fervent commitment to the advancement of knowledge. Over more than a decade, he urged, cajoled, threatened, and soothed Jones, in hopes of provoking him into publishing. Whewell informed Jones about pertinent publications, reported to him his observations of the farming practices and the local economic idiosyncracies he encountered in his travels, and acquired for him (from John William Lubbock, his former student) data on mortality rates and on the relationship between the average age of couples at the time of marriage and the average number of offspring produced. In addition, the Professor of Mineralogy became the midwife for Jones's only major publication, *An Essay on the Distribution of Wealth and on the Sources of Taxation: Part I, Rent* (London, 1831). Whewell wrote articles that helped to pave the way for Jones's book, extricated piecemeal from the procrastinating author the manuscript of the *Essay*, sought from the University Press funds to defray the costs of its publication, corrected page-proofs, and published perhaps its only fully favourable review. Whewell thereafter continued to prod Jones for the second part of the contemplated multi-volumed *Political Economy of Nations* (unsuccessfully as it turned out), published more on economics, and after Jones's death, wrote a preface for a collage of his work printed in 1859.[25]

In the articles he published, Whewell transformed some of the leading principles of classical economics into mathematical equations, and with simple algebra deduced from them consequences that showed the principles to be

[24] Whewell to Jones, 4 Feb. 1829, Todhunter (1876), ii. 97.

[25] Whewell to Jones [n.d.] Nov. 1822, 24 May 1825, 28 June 1829, [n.d.] Feb. 1831, [n.d.] Feb. 1831, 8 Nov. 1831, [n.d.] May 1831; to John Herschel, 8 June 1831; and to Rose, 29 Dec. 1833, Whewell Papers, Add. MSS c. 51[14, 22, 67, 98, 99, 118, 119], O15 97[129], and O15 47[405]; Whewell to Jones, [n.d.] Oct. 1825, 18 Oct. and 10 Dec. 1826, 21 May, 30 May, 9 Sept. and 18 Nov. 1828, 27 Aug. 1829, 28 Feb. and 21 May 1830, 15 July 1831, 23 Aug. 1832, 13 Nov. 1833, 19 Dec. 1834, 26 May, 20 July, and 1 Sept. 1835, 14 July 1839, 14 Feb. and 6 Oct. 1843, 18 July 1845, 30 Apr. 1848, 23 Apr. 1849, 29 Oct. 1851, 20 Jan. 1852, Todhunter (1876) ii. 61–2, 74–5, 79–80, 91–6, 102–5, 106–8, 122–4, 143–6, 172–3, 198–9, 213–15, 219–20, 223–5, 280–2, 312–13, 318–19, 326–7, 345–6, 352–3, 369–70, 371–2; J. W. Lubbock to Whewell, 22 and 25 June 1828, and 28 Oct. 1829, Royal Society of London, Lub. W 248 and Whewell Papers, Add. MSS a. 208[76–7]; Airy (ed.) (1896), 69; Whewell (1830); id. (1831*b*); id. (1850); id. (1850*c*); id. (1831*b*); id. (ed.) (1964 [1859]).

either erroneous or lacking in universality. What was needed in economics, Whewell concluded, was induction from facts and not deductions from inadequate principles as the Ricardians were wont to do. Again, mathematics for Whewell was a means to an end. As he declared to Jones, 'people will not mind me when I write Political Economy, except indeed I join it to mathematics, and make nonsense out of it'.[26] And that is exactly what Whewell wanted to do: make nonsense out of classical economics.

Whewell has been lauded by economic historians as the first mathematizer of the Ricardian model.[27] He has also been charged with being self-serving, with writing to protect the interests of landlords because as a Fellow and Master of Trinity he was a landlord. S. G. Checkland originally made this accusation with no substantiating documentation other than a vague time correlation between political-economic events in England and Whewell's publications.[28] Others have reiterated this line without attempting to verify it, probably because verification is impossible.[29]

To consider that element of Whewell's motivation for attacking Ricardian economics which can be substantiated is to glimpse the very font of his being. Whewell entered Trinity imbued with the deep religious feelings of his birthplace. That an education at Cambridge could only reinforce that feeling was attested to by many, including Whewell's undergraduate classmate, J. M. F. Wright, who 'began to see and feel the reasonableness and truth of Christianity' after exposure to natural theology at Cambridge. As all undergraduates, Whewell and Jones (who also took his degree in 1816) studied natural theology and moral philosophy. These subjects constituted the context in which they talked 'day by day about Malthus and Ricardo'.[30]

To retain most college Fellowships at Cambridge, ordination became mandatory after a certain number of years. In 1826 Whewell became a priest, and by December he was preparing a course of sermons he would give the following February. Upon seeing Whewell's outline for the sermons, Jones became concerned that Whewell was going to unfurl some of his material.[31] Whewell assured him that:

> I never intended to publish, and I do not think that I shall preach, any thing which will even brush the most delicate novelty off your plums. . . . and . . . though in my sketch I have talked of taking more than one example of the false philosophy of the irreligious school, I shall not in fact have . . . room for more than a very short attack on . . . the false induction . . . generalizing the impulse to increase, and not generalizing . . . the moderating and controlling influences which the nature of society and of man contain . . . . I can either . . . leave out the moral part of the subject for another time, when I can set

[26] Whewell to Jones, 23 Apr. 1849, Todhunter (1876), ii. 352–3.
[27] Cochrane (1970); Caravale (1985), 164–6.
[28] Checkland (1951), 53.
[29] Rashid (1980).
[30] [Wright] (1827), i. 218–9; Whewell to Jones 30 Apr. 1848, Todhunter (1876), ii. 345–6.
[31] Whewell to Jones [1826], Whewell Papers, Add. MS c. 51[31].

my back against your book and fight out the battle; or I can take something which does not exactly come your way.[32]

Whewell omitted those portions to which Jones had objected. None the less, those suppressed passages reveal what lay at the bottom of his economics:

It has been maintained . . . that the fiat of His will by which the Creator ordained the increase and multiplication of men, impelled them . . . to want and degradation, to vice and misery. It has been passed on . . . that the tendency of mankind to replenish the earth ever pushes them on till the sharp discipline of pain, the iron hand of want . . . drive them back. . . . that . . . those portions of society that win their bread by the labour of their hands and eat in the sweat of their brows, are destined to eternal and irredeemable degradation . . . to become more numerous *lest* they should become more happy [an allusion to the fundamental tenet of utilitarianism]. . . . That the proclamation of such a doctrine . . . shook and startled the minds of pious and benevolent men, and seemed like an oppressive and disquieting thought forced in among their belief and trust in God's goodness, . . . many who bear in mind the youth and first appearance of these doctrines . . . will still recollect.[33]

Whewell lumped together Ricardians, 'ultra Malthusians', and utilitarians as the 'irreligious school' bent on representing the labouring classes as naked apes. *Knowing* that man was created in God's image, Whewell applied the cutting edge of mathematics to show the inadequacies of the assumptions upon which the irreligious school based its interpretations. He hoped that Jones, through the inductive method of collecting facts and deriving from them laws supported by theories based upon human actions reflecting God-given morality and ethics, would supply a better system of economics. 'To other times', Whewell wrote in the draft of his sermon, 'to other hands must be left the labour, the high and enduring honour, of combating in particulars those errors from which foolish and unhappy men have drawn inferences which darken the face of moral nature and divine truth.'[34] Only then would the 'oppressive and disquieting thought' be thrust out of Christian certainty.

As a mathematician Whewell did what he could to show the inadequacies of classical economics, and Jones added empirical data which contradicted conclusions of the Ricardians. As Whewell wrote in his preface to Jones's *Literary Remains* in 1859, interested parties could read about 'rents that are actually paid in every country, or the Ricardian rents which are *not* those paid in *any* country'.[35] And that was the line Whewell followed when he tutored the Prince of Wales in economics in the Lodge in 1861.[36] But for Whewell, the task of establishing a 'true' economics required the creation of a system that would incorporate any truths that lay within the Ricardian formulations and, more importantly, would depend upon causes embedded in the God-given social conditions and morality of man. This was the task of what Whewell called the

32 Ibid., Add. MS c. 51^{33}.
33 Whewell, 'Drafts of Sermons [1827]', Whewell Papers, R6 17^{22-6}.
34 Ibid. 35 Whewell (1859), p. xx. 36 Id. (1862).

'ethical' economists, and this was the task towards which Whewell tried to push Jones for two decades. Jones failed in this 'employment of higher philosophical dignity and importance than any office to which the Mathematician . . . [could] aspire'.[37]

If Whewell suppressed the economics portion of his sermons in 1827, he did preach natural theology in the context of the natural philosophy. And while still Professor of Mineralogy, he gained an opportunity to reach a wider audience when named to write one of the Bridgewater Treatises, conceived of and funded by the Revd Francis Henry, Earl of Bridgewater, who had left £8,000 for the writing, printing, and publishing of 1,000 copies of a work or works 'On the Power, Wisdom and Goodness of God, as manifested in the Creation'. Whewell relished his assignment. Quickly finishing with the drudgery of revising his mechanics textbooks, and having resigned his professorship, he immersed himself so enthusiastically that he was the first of the original eight selected authors to publish.[38]

In his *Astronomy and General Physics Considered with Reference to Natural Theology* (1833), the Cambridge mathematician outlined the mechanism of the heavens to show intelligent design. However, Whewell also devoted much space to meteorology, a subject he researched with great vigour as a promising field both for his philosophy of science and his natural theology. Whewell's central point was that the correspondence between the astronomical cycles and the physiological cycles of life on earth, the constitution of the atmosphere, and the laws governing gases and liquids, fit so intricately the needs of the earth's flora and fauna (including man) that mere chance could not have brought them into being. He also displayed that there was no physical necessity for these conditions and, therefore, that a beneficent God had created the flora and fauna and their environment for one another. In a condescending manner, he dismissed as a 'baseless assumption . . . useless for the purposes of explanation' the 'supposition' that the structure of organisms came into being through the action of the environment and that those that could not adapt became extinct.[39]

In general, Whewell's *Astronomy* was well received: it went through four editions by 1837 and a sixth by 1864. However, Whewell's friend from undergraduate days, the scholar, theologian, and editor of the *British Magazine*, Hugh James Rose, who feared that science could not accord with religion, raised an important and valid point. 'I see that you have contented yourself . . . with asserting the strength of proof from contrivance to a contriver', Rose wrote, 'without arguing against the denial of that truth. Will you tell me where

[37] Whewell to Jones, 16 Aug. 1822, Whewell Papers, Add. MS c. 51^{14}; Whewell (1830), 156.

[38] Whewell to Henry Wilkinson [n.d.] Mar. 1832, Whewell Papers, O15 46^{65}; Whewell to his sister, 13 Mar. 1832, 9 Oct. 1832, 16 Nov. 1832, and 2 Mar. 1833, Stair-Douglas (1881), 143; Whewell Papers, Add. MSS c. 191^{125-7}.

[39] Whewell to Jones, 23 July 1831 and to J. D. Forbes, Todhunter (1876), ii. 124–6, 136–8; Whewell (1839*a*), 37.

that is best done?'[40] Whewell had no answer. He had already done the best he could when the shoe had been on somebody else's foot, that of Charles Lyell. In his reviews of Lyell's *Principles of Geology*, Whewell found what he termed the 'uniformitarian theory' implausible, indeed a hypothesis to hold up to ridicule. And Whewell exulted that Lyell had concluded 'that the appearance of new species, at successive epochs, which we learn from irresistible geological evidence to have repeatedly occurred, is a fact *not* belonging to the operation of that tendency to change in organized beings, which we still see brought into play'. From this admission, Whewell chortled, ensued a fundamental contradiction in Lyell's reasoning because Lyell's inchmeal geology depended upon the gradual introduction of species. Until someone could provide a cause for the gradual evolution of species, Whewell could rest secure in what he named catastrophism, which, of course, fitted neatly into natural theology.[41]

Having finished off mineralogy, architecture, political economy, geology, and natural theology, at least temporarily, in 1833 Whewell decided 'that he was going to do something about tides',[42] and, in conjunction with Lubbock, do something he did. Enlisting the aid of the Admiralty, the East India Company, the British Association for the Advancement of Science, foreign governments, friends such as Herschel (while he was the Cape Astronomer), and missionaries enticed with promises of contributions, Whewell amassed a body of data which placed tidal studies on a new footing. On tides, Whewell published fourteen memoirs in the *Philosophical Transactions of the Royal Society of London* between 1833 and 1850, printed eight occasional papers, outlined practical instructions for making tidal observations in an article in the Admiralty's *Manual of Scientific Inquiry*, presented reports to the Royal Society, to the BAAS, and to the Cambridge Philosophical Society, wrote up an account of his researches to 1837 for the *United Service Journal*, published an article on tidal observations at the Cape of Good Hope in the *South African Quarterly Journal*, defended his use of a level line in the *Philosophical Magazine*, and directed tidal observations made around the world.[43] Overall, his accumulation of data and his mathematical analysis of them were sound, and he tried desperately to establish a global map of cotidal lines, lines passing through all points where high water occurs at the same time. Whewell managed to work out his cotidal lines for restricted areas, and to establish laws for the diurnal inequality. However, Whewell and Lubbock relied upon the eighteenth-century equilibrium theory rather than the dynamical theory of Laplace, for which they met with the disparagement of George Biddell Airy, the Lucasian Professor of Mathematics cum Plumian Professor of Astronomy and Experimental Philosophy cum

[40] Todhunter (1876), i. 74.

[41] [Whewell] (1831); id. (1832).

[42] Whewell to Herschel, 14 Jan. 1833, Herschel Papers, Royal Society of London, H18 184.

[43] Whewell Papers, R6 20$^{43–190}$, O15 46$^{51, 66–7}$, O15 47$^{48, 51c, 185–8, 193, 196, 200–28, 232, 298–9, 301, 355}$, Add. MS c. 51$^{167, 169}$. Todhunter (1876), i. 75–88; Whewell to his sister, 19 Dec. 1834, 1 Mar. and 2 June 1835, Stair-Douglas (1881), 162, 171, 174; Farquharson (1953).

Astronomer Royal. In print Airy called Whewell to task for employing the 'miserable' and 'contemptible' equilibrium theory.[44]

Whewell won the Royal Society's Royal Medal in 1837, but despite all his efforts, and though he practised his philosophy diligently, Whewell did not achieve all that he had hoped. When he began his study, Whewell had imagined that in a few years he would discover the law of cotidal lines, and that maps of them would be published for the entire world. In 1840 Whewell termed his failure to uncover the law of cotidal lines a 'blot' on the theory of gravitation. In 1851, in 'On our Ignorance of Tides', he outlined problems still unresolved. By 1856 he concluded that there was no theoretical way to establish cotidal lines for oceanic tides.[45] Whewell's investigation of the tides taught him the real difficulties involved in doing science.

Despite his failure to establish universal cotidal lines, not all of Whewell's theoretical work in this respect was in vain. At the beginning of the twentieth century, his interpretation of cotidal lines as being the result of rotating tides proved to be far in advance of Airy's alternative explanation which came to dominate British tidal studies in the interim.[46] And Airy's denunciation of the use of the equilibrium theory was not quite fair. Airy determined that Whewell's findings were 'practicably useful'.[47] Whewell himself realized that the equilibrium theory was not a faithful representation of reality. In an article on economics in 1833, to develop by analogy the inadequacies of the Ricardians, Whewell had contrasted the inexact equilibrium theory with the 'true' theory of Laplace, but also had noted that the former theory explained 'rightly the main features of the phenomenon'. And, by 1835 at least, Whewell saw through the analytic equations of Laplace to the inadequate physical foundation upon which the equations rested.[48] Moreover, Lubbock had begun the research, and accepting that neither the equilibrium theory nor the dynamical sufficed to explain all the phenomena, he had opted for equilibrium. Lubbock was not a sophisticated mathematician. That is why he had graduated a senior optime rather than a wrangler. Thus, Whewell's answer to Airy that he saw no 'superiority in his [Laplace's] theory, except in his having formed differential equations which we cannot integrate' was appropriate in general and even more so specifically.[49] It was left to Airy, a better mathematician than either Whewell or Lubbock, to struggle with Laplace's equations.[50] A better mathematician still, George Gabriel Stokes, used Whewell's data and his mathematical equations. When William Thomson, later Lord Kelvin, took up the study of tides, he relied upon the equilibrium theory where appropriate, ascertained that in some cases Laplace's theory required 'radical amendment', and recalled warily

[44] Airy (1845), 259, 381, 390. Airy completed his article in 1842, see Grattan-Guinness (1985), 87–8.

[45] Whewell (1847*a*), 160; Whewell to Forbes, 23 Oct. 1856, Todhunter (1876), ii. 408–9.

[46] Deacon (1967), 269–70.

[47] Airy to S. H. Christie, 11 Feb. 1850, Royal Society of London, RR2 262.

[48] Whewell (1831*b*), 167; Whewell to Charles Lyell, 5 Mar. 1835, Todhunter (1876), ii. 310–12.

[49] Whewell to Airy, 2 Mar. 1833, ibid.

[50] Airy (1845).

how Airy had called the equilibrium theory 'contemptible except when . . . [he called] it miserable'.[51]

While engaged in his research on tides, Whewell undertook another task which 'obliged' him to 'contract' his 'efforts in other quarters'.[52] That undertaking was writing the *History of the Inductive Sciences* (1837) which formed the foundation of his *opus magnum*, *The Philosophy of the Inductive Sciences, Founded upon their History* (1840). By the time he began writing the history in 1833, Whewell was almost 40 years old, certain that writing the two books was his last chance to produce something extraordinary, and the best-prepared man in England for the enterprise.[53] Whewell's *History*, unlike the Bridgewater Treatise, was written for those with a knowledge of the sciences, and with the exception of a somewhat paranoid review by the Scottish physicist, David Brewster, in which he continued long-standing feuds with Whewell, the *History* was well received.[54] It was a different story with the *Philosophy* which elicited widespread criticism.

Whewell was disappointed in the almost universal rejection of his Kant-like fundamental ideas (that element of his philosophy which Whewell considered most crucial), but he had anticipated trouble. In his textbooks beginning in 1819 and carried through the 1820s into the 1830s, in his review in 1831 of Herschel's *A Preliminary Discourse on the Study of Natural Philosophy*, in his Bridgewater Treatise, in his *Thoughts on the Study of Mathematics as Part of a Liberal Education* (1835), in the first full exposition of his pedagogical beliefs in 1837 (*On the Principles of English University Education*), and in the *History* where his main objective was to prove the existence of fundamental ideas in mechanics, Whewell revealed his philosophy bit by bit in hopes of preparing his audience to accept his full views without generating alarm.[55] But he could not bring the mass of his contemporaries to accept his full philosophy because it incorporated certain tenets of Scottish Common Sense philosophy, and because his fundamental ideas owed their origin to Kant. To nineteenth-century Englishmen, Scots were 'obscure and wearisome' while German philosophy was 'madness . . . mere jargon of unmeaning words'.[56] In this tempest, John Stuart Mill easily drowned Whewell's philosophy in a deluge of refined empiricism. Leslie Stephen, Cambridge don, agnostic, and historian of the English utilitarians, would 'not attempt to explain a theory [Whewell's] which seems to be radically incoherent, and which made no converts'.[57] In 1876, Henry Sidgwick, Lecturer in Moral Philosophy, noted that John Stuart

[51] G. G. Stokes, 'Notes: Tides in the Panama Canal', W. Thomson to Stokes, 29 Oct. 1861 and 4 Apr. 1862, Cambridge University Library, Stokes Collection, Add. MSS 7656 PA186, 7656 K130, and 7656 K135.

[52] Whewell to Henry Wilkinson, 5 June [1837], Whewell Papers, O15 46[68].

[53] Whewell to Jones, 21 Oct. 1833, Whewell Papers, Add. MS c. 51[158].

[54] Brewster (1837), 110–51.

[55] Whewell (1831*a*); Whewell to Jones, [n.d.] Feb. 1831, and 14 July 1839, Todhunter, (1876) ii. 115–16, 280–2; Whewell to Jones [1833], 26 May 1835, and [1835], Whewell Papers, Add. MSS c. 51[143, 183, 191].

[56] Anon. (1847), 6–8.

[57] Stephen (1950), iii. 143.

Mill had had a greater influence on Cambridge than any other thinker since Newton. If Sidgwick was right, Whewell's philosophy of the progress of knowledge had been fulfilled with a vengeance, because Sidgwick also cited Whewell as being the primary stimulus for the revival of philosophy in the university.[58] Yet there were positive results of Whewell's philosophy, among them James Challis, the Plumian Professor, who adopted Whewell's notions, the short-lived but brilliant mathematicians and scholars, Duncan F. Gregory and Robert L. Ellis, who found that aspects of Whewell's philosophy anticipated their own, James Clerk Maxwell, who not only sought Whewell's 'appropriate ideas' in his physics but also accepted philosophically Whewell's fundamental ideas and other of Whewell's tenents, and present-day historians of philosophy, who find much of interest in Whewell's idealistic position.[59] The latter group has fulfilled Whewell's bitter-sweet premonition. In 1838 he wrote that his philosophy carried him 'into regions where I can hardly expect any contemporary friends to follow me—at least for a long time; and if a younger generation learn to sympathize with me (of which I hope very boldly), they cannot quite replace my *aequales*'.[60]

Capped by the *Philosophy*, the 1830s were the years of Whewell's greatest achievements in areas related to science. They were also the years of Whewell's emigration from natural philosophy to moral philosophy, a discipline which in Whewell's time was composed of intertwined portions of theology, law, political economy, ethics, and morality. This transition was facilitated to some extent by chance, but certainly was not unplanned.

When the Lowndean Professorship of Astronomy and Geometry fell vacant in 1836, Whewell sought it, only to be sorely disappointed when Lord Monteagle, a member of the government, voted out of friendship for Peacock, thereby barring Whewell from the position. What Whewell would have done in the chair is not certain, except that he would not have held it as a sinecure as, after a few years, Peacock did.[61] Whewell might have become a leading astronomer in his later years as did Charles Pritchard, the fourth wrangler of 1830, after he became Professor of Astronomy at Oxford in 1870.[62] Or he might have used the chair as he planned to use the Lucasian Professorship in 1826, 'to make very grand lectures on the principles of induction'. Or perhaps he would have made, as he said he might, his professorial 'business' the promotion of 'Tidology' and the 'Reform of University Education'. Regardless, the post went to Peacock, and in 1838 the Professorship of Moral Philosophy

[58] Sidgwick (1876), 241–2, 244.

[59] Challis (1873), 102; id. (1869), xix–xx; id. (1863), 285–6; Ellis, Diary, Whewell Papers, Add. MS c. 82[1], entries for 4 Apr. 1840 and 14, 15, 18, 24 June 1840; J. C. Maxwell to R. B. Litchfield, 6 June 1855, 4 July 1856, and 29 May 1857, Cambridge University Library, Maxwell Papers, Add. MSS c. 1[80, 83, 86]; Harman (1985), 221–3. For an example of the present day interest in Whewell's philosophy with citations of some of the vast literature on the subject, see Fisch (1986).

[60] Whewell to J. C. Hare, 7 Jan. 1838, Todhunter (1876), ii. 267–8.

[61] Whewell to Jones, 18 Nov. 1836, Whewell Papers, Add. MS c. 51[204].

[62] Pritchard (1897), especially 236–316.

became available. Whewell seized the chair and vivified what had been a sinecure.[63]

Whewell had always believed that no philosophy could 'be at all satisfactory to thoughtful men, which has not a theological bearing'. The lack of such a 'bearing' was what, for Whewell, vitiated Alexander von Humboldt's *Cosmos*, and, indeed, the convergence of science and religion in his *Philosophy* brought accolades from his contemporaries. However, Whewell always hoped that the ethics and morality of man could also be made into a science. The problem was, that until the mid-1830s he was perplexed as to how to accomplish this.[64]

If not before, Whewell studied moral philosophy as part of his undergraduate readings, the major portion of which was William Paley's 'proof' of utilitarian principles and his derivation of a system of morality from them. Whewell and most of his friends detested Paley's moral philosophy, but could not dislodge it from the curriculum because there was no systematic set of readings to replace it. Prior to 1818, Whewell read Joseph Butler's views on human nature and virtue as a possible antidote for Paley. However, he found Butler's effort 'a book of negatives, its object being not to prove but to remove the presumptions against natural and revealed religion'. Butler attacked '*dis*believers, but had little to say to mere *un*believers', those who denied the reality of moral distinctions. Thus, Butler had a 'sceptical effect', illustrating that with reason 'you may disprove with much more certainty than you can prove'.[65] That Whewell did not get beyond Butler in his Bridgewater Treatise wherein he attacked those who argued for physical necessity or random chance but had little to say to unbelievers (those who in the second half of the nineteenth century would label themselves agnostics), Rose disclosed in the question he raised.

Whewell barely touched upon moral philosophy in the Bridgewater Treatise. He would have included more, but he deemed himself unequal to the task. Nevertheless, he did write that the 'mode in which the belief of God's government of the physical world is important and interesting to man, is . . . through the connexion which this belief has with the conviction of God's government of the moral world.'[66] And the trepidations which engulfed his hopes for systematizing morality faded in the mid–1830s. In 1833 Thomas Rawson Birks, Whewell's student (and 1834 second wrangler) inspired his tutor. In an *Oration on the Analogy of Mathematics and Moral Certainty Delivered in the Trinity College Chapel*, which won him the chief English declamation

[63] Whewell to Jones, 13 Oct. 1826 and 6 Nov. 1836, and to his sister, 9 June 1838 and 6 May 1839, Todhunter (1876), ii. 71–2, Whewell Papers, Add. MSS c. 51[203] and c. 191[183, 194].

[64] Whewell to Herschel, 1 Nov. 1818, Royal Society of London, Herschel Papers, HS18 160; Whewell (1858*b*), and id. (1860), 396–7; Anon. (1841), 555–72, especially on 563–4; Whewell to Forbes, 29 Feb. 1860 and to Jones, [n.d.] Feb. and 23 July 1831, Todhunter (1876), ii. 418–20, and Whewell Papers, Add. MSS c. 51[99, 110].

[65] Whewell to Rose, 7 Jan. [1834] and 24 June 1818, Whewell Papers R2 99[39] and O15 47[379]; Whewell, Diary, R18 19[12], 14–15, entry for 29 Feb. 1825.

[66] Whewell (1839*a*), 280–1.

prize, Birks argued that there existed axioms for moral philosophy as certain as those for geometry. What plagued moral philosophers, Birks argued, was that they, unlike geometers, did not build upon their predecessors' work, but rather, for no good reason, felt compelled to destroy earlier elaborations before beginning anew with their own.[67]

In 1834 Whewell told Jones of his plans 'to discover the nature of the analogy' between the physical sciences and the non-physical sciences, including morality, after he had finished the *History* and the *Philosophy*. He proposed that he would trace the history of morals as he did the history of science to uncover axiomatic principles and ideal relations upon which to build a moral philosophy. When he had shown that the relations of morality were 'as fundamental a part of man's thoughts as the relations of time and space', he believed that he would have 'assigned man a moral and social constitution on firm grounds'. However, in morality, unlike in physical sciences, Whewell thought that it was 'neither possible nor necessary' that 'every word . . . be founded on close examination and proved by reference'. In morality he would take advantage of his own philosophy which posited 'that all knowledge comes by induction and that induction is guessing, [which] allows us to publish guesses, acknowledgedly imperfect, as contributions to knowledge'.[68]

In 1835 Whewell found what he had 'long been wishing and struggling for, of the inductive history of ethics'. What he had come across was James Mackintosh's *A Dissertation Exhibiting a General View of the Progress of Ethical Philosophy, Chiefly During the Seventeenth and Eighteenth Centuries*, a review and criticism of the thoughts of earlier moral philosophers. Best of all Whewell found that Mackintosh indicated that moral judgements stemmed from some faculty and could not be resolved into 'mere pleasure and pain'. Whewell was so impressed with this historical survey that he wrote a preface for it, edited it, had it published, and introduced it as a reading for Trinity students.[69]

In 1837 the Reverend Whewell preached four sermons on morals. He dedicated these sermons, in which he adverted to 'guesses at truth', to his friend from undergraduate days, J. C. Hare, Rector of Hurstmonceux, disciple of Coleridge, Broad-Churchman, co-translator of Barthold Niebuhr, and co-author of *Guesses at Truth* (1827), an unconnected romantic rejection of the reign of reason and an example of the 'liberal Anglican' philosophy of history in which the intellectual and moral faculties of man evolve unevenly to the present. The last of his sermons Whewell devoted to attacking Paley.[70]

[67] Whewell to J. C. Hare, 25 Dec. 1833, to his sister, 14 Jan. and 27 Mar. 1834, and to William Wordsworth, 14 May [n.y.], Whewell Papers, Add. MSS a. 215[29], c. 191[130], 193 and c. 91[107]; Birks (1855).

[68] Whewell to Jones, 27 July, 5, and 21 Aug. 1834, and 6 Oct. 1843, Whewell Papers, Add. MSS c. 51[159, 174, 175] and Todhunter (1876), ii. 318–19.

[69] Whewell to Jones, 9 and 26 May, 21 Aug., and 1 Sept. 1835, Todhunter (1876), ii. 210–11, 213–15, 222–5; Mackintosh (1872).

[70] Whewell (1837*d*), Forbes (1952), 64–5.

Obviously, when Whewell became Professor of Moral Philosophy, he was primed. He was anxious, as he said, to take off his 'mathematical boxing gloves, and go on with arms of wider range'.[71] In this 'employemnt of higher philosophical dignity and importance than any office to which the mathematician could aspire', Whewell again applied his philosophy of the progress of knowledge, and also his precept that truths in earlier elaborations were necessarily incorporated in those which superseded them. In his lectures, Whewell traced the history of moral philosophy in England, criticizing broadly and zealously, especially the utilitarians and specifically Jeremy Bentham, while at the same time extracting from his predecessors everything which he considered valid. He tried to establish an analogy between the natural sciences and the moral sciences, and that derivations from his system of morality included all the truths previously established and more.[72]

In 1845 Whewell published the *Elements of Morality Including Polity* which reached a fourth edition in 1864. In the preface to the first edition, Whewell maintained that there existed axioms in moral philosophy as certain as those in geometry. In 1830, for his application of mathematics to economics, Whewell had been ridiculed for applying mathematics to moral science. In 1845, nothing had changed. The reaction to Whewell's assertions was negative, and he withdrew them from the preface in the second edition. That does not mean that he gave up the idea. In fact, what he really wanted was to draw an analogy between morality and mechanics, but he feared that this would alienate his readers even more. His aim was not to establish axioms inductively—he had concluded that this was impossible in moral philosophy—but rather to 'guess' at the axioms and show their validity deductively, a remarkable reversal of what he had hoped Jones would accomplish.[73]

Whewell's primary concern was to create a science of morality, independent of immediate socio-economic conditions, on the basis of the supremacy of conscience. What he abhorred in the utilitarians was their conclusion that each individual's conscience dictated what was right and what was wrong. Whewell imposed fundamental ideas as the source of primary moral conceptions which derived neither from a physiological faculty nor from social conditioning. In this context, Whewell accepted Paley's contention that if a moral sense 'were implanted in man, it could not subsist except there were implanted also the ideas which it includes; and thus we are led to innate ideas'. In this context also, when Whewell argued that 'the true guide of man is Conscience, only so long as the guide of Conscience is Reason', the reaction of his contemporaries paralleled that of their reaction to his philosophy of science. Only when Whewell turned to show the correlation between moral science and theology,

[71] Whewell to Henry Wilkinson, 26 December 1837, Todhunter (1876), ii. 265–6.

[72] Whewell ['Lecture Notes'], Whewell Papers, R6 15[14], and R6 16[15].

[73] Anon. (1830); Blakey (1850), iv. 71–2; F. Myers to Whewell 31 July 1845, Myers to Whewell [n.d.], and Whewell to Myers, 6 Sept. 1845, Whewell Papers, Add. MSS a. 209[137, 139] and a. 214[57].

when he wrote that religion was 'the necessary supplement to a mere rational Morality', and that 'moral law was the law of God and man's conscience was the voice of God', did Whewell win sympathy.[74]

By the 1840s Whewell had introduced not only his editions of Macintosh and Butler, but also his *Elements of Morality*, and his *History* and *Philosophy* into various Trinity examinations; and Paley's *Moral Philosophy* had been deleted from his college's examinations.[75] A Professor of Morality, Whewell played a leading role in the creation of the Moral Sciences Tripos (first given in 1851), and his *History*, his *Philosophy*, his publications on moral philosophy, and his editions of Mackintosh and Butler (1849), took places in it.[76] Whewell accomplished what he had set out to do, and as with his Mineralogy Professorship, having found a sympathetic successor in the Trinity scholar, John Grote, he resigned his second professorship. Grote, like Whewell, assumed the existence of moral sense and rejected Paley's principle of expedience even though he could not disprove it. George Pryme, the Professor of Political Economy, asked both Whewell and Grote for references which disproved Paley's moral philosophy, and he received no help from either.[77] Little wonder then that Paley's moral philosophy remained a part of the Moral Sciences Tripos; it did not, as reported recently, get pushed 'out of the honours curriculum'.[78]

After his resignation, Whewell developed a subject which he had begun in his lectures by translating, editing, annotating (in some places rewriting), and publishing three volumes of *Platonic Dialogues*, not for classicists, but for the general English reader. Whewell revealed his link to Grote by arguing for the latter's interpretation that the Sophists were not the unethical rhetoricians that historians had made them out to be. Grote returned the favour by defending this aspect of Whewell's translation, but he could not go along with Whewell's minimizing of some of Plato's work. Likewise, Whewell rejected some of Grote's interpretations.[79] Here, as in other areas, Whewell was indebted to a colleague, but he also developed his own thoughts.

Despite his involvement with moral philosophy, Whewell kept an eye on what was happening in the sciences. In 1838, in his presidential address to the Geological Society, Whewell congratulated his fellow Lancastrian and anatomist, Richard Owen, for being a pious scientist and for winning the Wollaston

[74] Whewell (1862*a*), 170–2; id. (1864*a*), 17, 231.

[75] Trinity College Library, *Trinity College Conclusions Book* (1811–1886), 102, 115, 173, 192, 215; Whewell to J. C. Hare, 13 Oct. [1840] and 15 Dec. 1840, Whewell Papers, Add. MSS a. 215[49, 53].

[76] [Whewell] 'Draft of Books Bearing on the Moral Science Tripos' [1860?], Whewell Papers, Add. MS a. 62[47]; Whewell (ed.) (1855*c*).

[77] Pryme (1870), 332–3.

[78] Garland (1980), 68–9. See, however, Goodwin (1845), 6, 14; '[For the members of the Board of Moral Sciences]', Whewell Papers, Add. MS a. 62[54]; 'Draft List of Books Bearing upon the Moral Sciences Tripos', 'List of Books Published by the Board of Moral Sciences Studies, May 1860', ibid., Add. MSS a. 62[47, 48, 75, 78].

[79] Grote (1861), Cambridge University Library, LO22 12[4]; Whewell (1866).

Medal, and Charles Darwin for his speculations founded upon materials gathered on the voyage of the *Beagle*. He also reported the discovery of monkey remains in geological strata that previously had been thought not to contain 'traces of those creatures which most nearly imitate the human form'. Obviously disturbed, he exhorted geologists seeking the origins of man 'to close his own volume, and open one which has man's moral and religious nature for its subject'.[80] Whewell followed his own advice, but not all in attendance did. Darwin, in commenting upon Whewell's speech in his notebook, wrote that 'if all other animals have been so formed [by evolution], then man may be a miracle, but induction leads to other views'.[81]

During his professorship, Whewell continued to incorporate new discoveries into subsequent editions of his *History*, which Darwin found helpful.[82] Also, in reaction to the evolution-promoting *Vestiges of the Natural History of Creation* (1845), he published *Indications of the Creator* in which he used Owen's discovery of the suckling mechanism of the kangaroo as evidence of design in addition to other materials he had gathered together from his previous publications. Robert Chambers, the anonymous author of the *Vestiges*, replied that Whewell's 'professional position' necessitated 'pre-possessions' which determined his ideas, and criticized Whewell for taking refuge in the ignorance of the processes of life, an ignorance that also precluded knowing the existence of a 'personal and superintending God'.[83]

In 1853, in *Of the Plurality of Worlds*, Whewell, by arguing against the probability of intelligent life elsewhere than on earth, contradicted what he had preached in his sermons in 1827 and what he had written in the Bridgewater Treatise.[84] As always Whewell had a long-time interest in his topic. In 1823 he had read Fontenelle's *Plurality of Worlds*.[85] In 1849, in a paper on philosophy that he read to the Cambridge Philosophical Society, Whewell provided a hint of what he was to argue in his *Plurality*.[86] Specifically, he was stimulated to write the book by his association with Thomas Chalmers and the reading of Chalmers's writing which had suggested the opposite thesis.[87]

In the *Plurality*, Whewell marshalled scientific evidence from numerous disciplines and cleverly manipulated it to show the uniqueness of the earth and the improbability of intelligence elsewhere, but it is obvious that he could have come down on either side of the question, and he knew that his arguments would not only go against the prevailing scientific beliefs of the time, but also against the religious faith of many. Indeed, some believed that Whewell was

[80] Whewell (1838*a*), Cambridge University Library, Cam. b. 838. 3.

[81] de Beer (ed.) (1960), ii. pt. 2, 87. [82] Ibid., ii. pt. 4, 166.

[83] Whewell (1845*a*), vii–x, 47, *passim*; [Chambers] (1858), 202–3, 213–15, 261, 264–9; Brooke, (1977*a*), 132–45; Yeo (1984).

[84] Whewell (1839*a*), 280–2.

[85] Whewell, Diary, Whewell Papers, R18 9[11], 22, entry for 23 Aug. 1823.

[86] Whewell (1848*b*), 615–16.

[87] Whewell to James Stephen, 13 Sept. 22 and 25 Oct. and 9 Nov. 1853, and to Forbes, 19 Feb. 1854, Todhunter (1876), ii. 380–1, 391–2, 394–5, 400–1.

using his tremendous intellect and knowledge to toy dangerously with the beliefs of others. Perhaps Whewell thought that the image of life on earth alone was more miraculous and made less likely the evolutionary theory that he saw gaining ground.[88] In the *Plurality*, Whewell wrote that God 'launched and governed man's course in a certain path which makes his lot and state different from all other creatures'.[89] In a dialogue-structured *Supplement* to the *Plurality*, through his mouthpiece 'z', Whewell argued that man did not evolve from some inferior form of life.[90] Also, because of the growing acceptance of the homologies among animals and because of the discovery of numerous organs with no evident purpose, Whewell had been forced to add to the concepts of purpose and final causes in biology, the concept of a unity of plan. To adjust to the latter, in which he could see no purpose or final cause, Whewell adopted from Owen the idea of archetypes. In the *Plurality*, Whewell argued that other bodies without life would not indicate superfluous (and thereby irrational) creations of God. Instead, Whewell compared uninhabited celestial bodies to organs for which man could not identify a useful function.[91] It is also possible, if we take the *Plurality* in the context of Whewell's full philosophy, that Whewell was edging towards an anthropic position. Finally, perhaps Whewell simply wrote the book in the belief to which he held (as Maxwell did later), that looking at a question from a different viewpoint was advantageous, his motivation for 'swimming his fiddle faddle corks'. And though few accepted Whewell's viewpoint, Whewell achieved the end at which he always aimed: a spate of reviews of his book filled the periodicals.[92]

In addition to moral philosophy and natural theology, Whewell took up poetry. As with architecture, poetry for Whewell began in his childhood. His mother contributed poetry to the local newspaper, and some feared that his 'Boadicea' manifested that he would follow in her footsteps. As an undergraduate, he read Samuel Taylor Coleridge and William Wordsworth, finding their philosophy muddled and their morality questionable at best. Only at the urging of his more romantic friends, and after having met the poets and established friendships with them, did he come to appreciate elements of their work.[93]

In his prose, Whewell often used beautiful metaphors with effect. Some of his writing, especially when he followed his heart into natural theology, elicits emotional response worthy of the poetic form. Indeed, part of his Bridgewater treatise was transformed into blank verse. Whewell often jotted down bits of poetry, especially in play or when personal tragedy tainted his life, and he

[88] Becher (1975); Brooke (1977).

[89] Whewell (1853), 74–5.

[90] Ibid. 39.

[91] Becher (1971), 36–40; Yeo (1979).

[92] For an extensive review of the plurality of world literature following Whewell's publications and for a most convincing demonstration that Whewell perceived a fundamental conflict between the existence of a plurality of worlds inhabited by intelligent life and the fundamental tenets of Christianity, see Crowe (1986), 265–355.

[93] Whewell to Rose, Whewell Papers, R2 99[101].

published a few pieces.[94] However, Whewell's major undertaking was to try to establish hexameters, which were popular in Germany, among an English population unfamiliar with the form.

In 1840 Whewell published *Isle of Sirens*, an adaptation into hexameters of some passages from Thomas Carlyle's *Chartism*. In 1847 Whewell edited *English Hexameter Translations from Schiller, Goethe, Homer, Callinus, and Meleager*. It contained Whewell's major work in English hexameters, a translation of Goethe's *Herman and Dorothea* which he had translated and printed in the summer of 1838 for circulation among friends.[95] It also included his renditions of Schiller's *Columbus*, *Odysseus*, *Dance*, and *Sexes*, and Goethe's *Metamorphosis of Plants*. In addition to his own works, Whewell gathered translations for the book from others, among whom were Hare and Herschel. In his review of Whewell's *History* in 1837, Herschel had reprinted, and complimented, Whewell's translation of eight lines from Goethe into English hexameters. In 1842 Herschel had also translated and printed for private circulation Schiller's *Walk*, which Whewell reprinted in *English Hexameter Translations*. Herschel and Whewell discussed English hexameters often and commented upon each other's work. Although they differed in particulars, they worked together to accustom the English ear to the form, and Herschel dedicated his translation of the *Iliad* to his old friend who had read and criticized the manuscript between 1862 and 1865.[96]

Whewell also published other translations, pieces of poetry, reviews of others' efforts in hexameters, and analyses of the structure of the form in various journals.[97] Here, as in his Bridgewater Treatise and his *Platonic Dialogues*, he aimed at the general reading public and not at the old boys of the public schools, accustomed as they were to Latin hexameters. Whewell insisted that, unlike the classical style, English hexameters rested upon accents as much as upon quantity and had to have trochees in places where the classical forms had spondees; after 1850, learning from Herschel's translations of Schiller's *Walk*, Whewell insisted on placing trochees rather than spondees at the end of sentences. Whewell also stressed the ease with which hexameters could be composed and the unaffected and meaningful manner in which they could express 'the simplicity and truth of reality' and 'the most trivial details in ordinary life'—in America as manifested in Longfellow's *Evangeline*, or in the Germany of *Herman and Dorothea*, or in the Book of Genesis, or in the patriarchal homilies of the Old Testament. But he met with opposition. In

[94] Todhunter (1876), i. 68–9; R. L. Ellis to William Walton [n.d.], Whewell Papers, Add. MS c. 67[85].

[95] Whewell (1840), Cambridge University Library, Cam. b. 840. 1; Whewell to L. A. J. Quetelet, 26 May 1839, Todhunter (1876), ii. 279–80.

[96] Herschel (1841), 238; Whewell to Herschel, 4 Feb. 1846 and 23 Nov. 1846, Royal Society of London, Herschel Papers, HS 18. 203, and Whewell Papers, O15 47[136]; Todhunter (1876), i. 298–9.

[97] Whewell (1847*c*), Cambridge University Library, Cam. c. 847. 121; id. (1842); id. (1847*c*); id. (1846*b*); id. (1847*b*); id. (1848*c*); id. (1849*b*); id. (1850*b*); id. (1853*a*); id. (1862*b*); id. (1862*c*).

1852 Whewell protested that 'the absurd injustice of our grammarian critics have somewhat disgusted me with the public to which I should have to appeal', and he complained that 'the common unperverted ear is well satisfied with the English Hexameter . . . and the persons who are opposed to it have had their habits formed by classical studies'. It 'is not the *Vox Populi*', he observed, 'but the *Vox Grammatici*, which cries down . . . [English hexameters]'.[98] The classicists won the day; hexameters did not become popular in England.

Whewell's attempt to gain a foothold for hexameters might be viewed as a revival of Poet Laureate Robert Southey's unsuccessful use of them earlier in the century. Whewell's work might also be seen as an extension of the German romanticism of his associates, Hare, Rose, and Connop Thirlwall, a Trinity Fellow whom Whewell hired as his classical lecturer in 1831.[99] Certainly Whewell's love of German poetry fitted well with theirs, but there was little romanticism in Whewell's labours. What stands out is Whewell the scientist. He was mainly concerned with technical aspects of the subject, much in the tradition of his researches in philology which extended from his student days to the late 1860s when he discussed Lyell's *Antiquity of Man* with him, specifically Whewell's ideas concerning the development of language which Lyell thought might have anticipated some of his own determinations.[100] Of course, in his translations, Whewell attempted to maintain the spirit as well as the letter of the poetry, but his concern with hexameters was as much that of the philologist as that of the poet.

While Whewell was moving into moral philosophy, he was also meeting with new personal circumstances. The Lancastrian lad, who as a student had had to save the material of unwearable shirts to patch those on his back, by the mid-1830s was financially secure, and could have lived out his days as a Trinity don unencumbered by his tutorial duties or any others. And he now had the prestige and the personal connections for a career in the Church. However, he considered a Church appointment only half-heartedly, and opted instead for marriage and the life of a scholar independent from his college.[101]

In 1841 Whewell married Cordelia Marshall of the industrial Marshalls of Leeds, who, reportedly, brought with her a dowry of fifty thousand pounds.[102] Whewell did not, however, leave his college. While on his honeymoon, he received from Peacock news that came 'like a thunderclap', news that he was to be offered the Mastership of Trinity. Whewell did not hesitate in accepting the Crown appointment since the Mastership was, as it was for Peacock, the

[98] Todhunter (1876), i. 284; Whewell (1848*c*), 295, 298; id. (1849*b*), 529, 531–2; id. (1850*b*), 37; Whewell to Edward Bulwer-Lytton, 19 Feb. 1865, D/EK C17; Whewell to J. I. Cochrane, 9 Sept. 1852 and 7 July 1853, Todhunter (1876), ii. 373–6.

[99] Cannon (1978), 45–50; Preyer (1981).

[100] Whewell to Lyell, 28 Feb., 14 May, and, 5 June 1863, Todhunter (1876) ii, 429–32.

[101] Whewell to his aunt, 27 Apr. 1814, and to Hare, 15, 26 Dec. 1840, 26 Feb. 1841, Whewell Papers, Add. MS c. 191[9], and Add. MSS a. 215[53, 54, 56]; Whewell to Jones, 21 Oct. and 10 Dec. 1833, Todhunter (1876), ii. 170–3. In Morrell and Thackray (1981), 25–6, Whewell is listed as holding a 'significant ecclesiastical preferment' without defining or specifying it.

[102] [Newspaper clipping], Cambridge University Library, University Papers, 1840–3, no. 406.

'summit of his ambition'.[103] As Master of Trinity, Whewell gained a powerful voice within his college and within the university. He used his new position as he said he would: to promote further the reforms for which he had laboured since becoming a tutor.[104]

Most historians of Cambridge present a dichotomy between 'progressives', who favoured a greater role for the university, and 'reactionaries', who, clinging to the sanctity of the colleges, fought any extension of the university. Within this division, Whewell was sometimes progressive. However, he did not, as has been claimed, put through the Senate a Grace for dissenters.[105] Whewell ardently believed that the alliance between the Church of England and the university was the very rock upon which English civilization rested, and he would allow nothing that weakened that foundation. We must look elsewhere to see how Whewell was pro-university.

Whewell's insistence that mathematics and classical languages remain introductory mainstays of a liberal education at Cambridge is well known. Also widely recognized is his untiring crusade to reform the mathematics curriculum at both the ordinary and the honours degree levels. What is less widely appreciated is that he promoted the employment of the university professors for the culmination of a liberal education in what he considered 'progressive' studies for students already well grounded in the two traditional disciplines.[106]

In 1828 the Professor of Mineralogy launched a campaign to provide the science professors with updated facilities. In 1830 that campaign seemed to be reaching fruition in the aforementioned plans to build a university complex. Then Whewell and Peacock fell out over the selection of an architect. Peacock backed Cockerell while Whewell championed Rickman, the results being a pamphlet war, division within the university, and the consequent stimulation, as only universities can be stimulated, into inaction. Delay proved to be lethal, for by 1835 the university finances had deteriorated from a state of affluence to one of bankruptcy.[107]

Undeterred, in the late 1840s, Whewell was at the forefront of an unsuccessful attempt to establish a standing university building fund. Additionally, as Vice-Chancellor in 1842, at the urging of Peacock, he formed a syndicate to study how the university professors might be better utilized. After much political manœuvring, Whewell managed the creation of the Natural Sciences Tripos and the Moral Sciences Tripos (first offered in 1851), and the mandate

[103] Whewell to George Peacock, 19 Oct. 1841, Todhunter (1876), ii. 301; Stair-Douglas (1881), 229; Whewell to his sister, 16 Oct. 1841, Rimmer (1960), 223.

[104] Whewell to James Henry Monk, 30 Oct. 1841, Whewell Papers, O15 47^{257}.

[105] Garland (1980), 77. As substantiation, Garland cites a letter from Whewell to Airy in 1842 (Whewell Papers, O15 47^{5}): 'I have passed through the Senate a grace for dissenting, but our dissent will affect only the station, not the main line.' However, this was not a metaphor. Airy was a consultant for the railroad, and the sentence before the quoted one reads: 'We have at last got the plans of the line of the intended Railroad as it passes this place.'

[106] Becher (1986), *passim*.

[107] Ibid. 74–7; [Peacock] (1831); id. (1831*a*); [Whewell] (1831*f*), Cambridge University Library, Cam. c. 500. $2^{1, 2, 5}$.

that all ordinary degree candidates attend the lectures of, and pass an examination given by, at least one of thirteen university professors. Having achieved this framework, Whewell, on an *ad hoc* basis, rallied the professors to ensure that the spirit as well as the letter of the reforms would be met.[108]

With the professors guaranteed a role in education, in 1853 and 1854 Whewell helped to resurrect the vision of 1830 for the science complex, and a syndicate selected as its architect a man who had consulted Whewell about German Gothic windows in the 1840s and had designed for Whewell a remodelling of the front of Trinity, one Anthony Salvin, the most successful of the later generation of Gothic architects. The plans to build again collapsed, this time for lack of funding. None the less, in 1861 and 1862, the persevering Whewell was on syndicates and the University Council that revived Salvin's plans and sought finance for the project.[109] Whewell was not as active in the 1860s as he had been in 1830. Perhaps, as one critic of Salvin's plans stated, Whewell was more interested in the construction of Whewell's Court, a hostel which the Master was then financing and building across the street from Trinity from plans by Salvin. Or perhaps Whewell was less interested because the economics of the times dictated that function determine form: the design was 'to be in the Workhouse style because it is intended to be a "work-house"'. There was little concern with aesthetics, Grecian, Gothic, or otherwise, except that the 'ugly' science buildings be built away from Free School Lane in hope that at some future date more pleasing buildings would be erected on the street and would hide the science complex.[110] It is even possible that Whewell had learned his lesson in university politics in 1830, that silence is golden. Most likely, however, Whewell had simply tired of the whole affair. It is not easy to maintain enthusiasm for a project for more than thirty years. None the less, between 1863 and 1865, the science complex was built. Before he died, Whewell could have repeated what he had written in 1832: 'We get on slowly with such things here.'[111]

In the 1850s Whewell had to face two parliamentary commissions, the first a commission of inquiry appointed in 1850, the second a statutory commission

[108] H. Philpott to Whewell, 20 Mar. 1849, Whewell Papers, Add. MS a. 63[99]; [Whewell], Flysheet [1851], Cambridge University Library, University Papers, 1843–7, no pagination; Pryme (1870), 332; James Cummings to Whewell, 1 June 1849, 27 Nov. 1850, 12 Feb. and 2 Apr. 1851, Whewell Papers, Add. MSS d. 63[91], a. 202[85], a. 59[3, 9]; John Henslow to Whewell, 16 Nov. and 12 Feb. 1848, ibid., Add. MSS a. 206[69, 70]; Robert Willis to Whewell, 31 Mar. 1851, ibid., Add. MS a. 59[5]; William Clark to Whewell, 30 Mar. 1857, ibid., Add. MS a. 59[4]; James Stephen to Whewell, 3 Apr. 1857, ibid., Add. MS a. 212[188].

[109] [Whewell] '[For Members of the Studies Syndicate]', 23 May 1853, Whewell Papers, Adv. c. 57[1]; Cambridge University Library, University Registrar, CUR 55. 1, 22 ff; Anthony Salvin to Whewell, 5 Oct. 1844 and 12 July 1854, Whewell Papers, Add. MSS c. 91[91, 93]; 'Minutes of the Museums and Lecture-Rooms Syndicate, 1853–1881', Cambridge University Library, Min. vi. 65, 38, 47, 51–3, 74–5, 84–5, 93; Dixon and Mathesuis (1978), 109.

[110] Cambridge University Library, University Registrar, CUR, 51–67; Salvin to Whewell, 4 June 1860, Whewell Papers, Add. MS c. 90[94]; Cambridge University Library, J. W. Clark Collection, 'Miscellaneous', Add. MS 5091, 186, 191.

[111] Whewell to his sister, 13 Mar. 1832, Whewell Papers, Add. MS c. 191[120].

appointed in 1855. Whewell found little quarrel with the commissioners' recommendations concerning university professors and curricular matters. And why should he have? Most of their prescriptions paralleled his own, and even though he did not favour a tax on colleges for the benefit of the professors, he advocated and got his college to agree to a voluntary contribution for financing university activities, only to find that other colleges would not go along with the proposal.[112] Whewell even acquiesced, albeit belatedly, in some political reforms which the commissioners commended. Only when the commissioners demanded more drastic legislation did Whewell become unyielding. From this point forward, Whewell objected to the parliamentary interference and to most of the suggestions aimed at democratizing the university political structure. But what he reacted to most violently was the revision of the Trinity statutes.[113]

Whewell, like almost all his peers, differentiated sharply between the university and the colleges. The university was public, but the colleges were *theirs*, property privately founded, inherited from previous Fellows, and the legacy of succeeding Fellows. Besides, for Whewell, university lectures could not be structured to inculcate in students the basics of a liberal education and moral character whereas the colleges were ideally suited for these functions.[114] With his view of the college, it is not surprising that when the statutory commission advocated that Trinity (and St John's) Fellowships be limited to a tenure of ten years and be thrown open to the university, and when the younger Fellows of Trinity manifested that they would support the commission, Whewell resorted to rabble-rousing, the rabble being the members of all of the governing bodies called together to discuss the commissioners' considerations concerning the statutes of the two colleges. Whewell gave a speech introducing a resolution to oppose the commissioners' recommendations for open Fellowships in which he viciously attacked the commissioners' proposals, and their motives and ethics. That Whewell was in accord with most members of the ruling bodies became clear when they supported his speech with cheers and his resolution overwhelmingly. Even Sedgwick, Vice-Master of Trinity but also a commissioner, seconded a motion against terminable Fellowships and gave a rousing seconding speech in which he made the remark that a Fellowship was a 'Freehold for life'. Only one of the members of the governing bodies present opposed the motion Sedgwick seconded, and none of them voted against a third resolution rejecting any 'measures which would tend to impair the existing connection between the colleges and the Church of England'.[115] Whig

[112] *Trinity College Conclusions Book* (1811–86), 255; 'The Commissioners and the Colleges', p. 97; Whewell, 'For the Fellows. Nov. 7, 1857', Whewell Papers, Add. MS a. 62^{2}.

[113] James Cartmell to Whewell, 25 May 1851, H. A. Woodham to Whewell, 27 Feb. 1852, and Whewell to Sidgwick, 25 Apr. and 21 May 1855, Whewell Papers, Add. MSS a. 53^{29}, a. 55^{30} and O15 $48^{78, 84}$; 'Communications from Royal Commissioners on University of Cambridge' (April 1855), Whewell Papers, Add. MS a. 54^{5}; Whewell (1854*b*); id. (1855*a*); id. (1855*b*); Whewell Papers, Add. MSS a. 62^{12} and a. $54^{6, 7}$.

[114] Whewell to his sister, 3 Nov. 1833, Add. MS c. 191^{133}; Whewell (1837), 83–140, *passim*; id. (1845), 104–15, *passim*.

[115] 'The Commissioners and the Colleges', 92–7.

historians from Whewell's day to ours have labelled the Master reactionary, but even in the areas where he fought against the parliamentary commissions, he was representative of the attitudes of his counterparts.

As Master, Whewell was autocratic, the not-so-benevolent, paternalistic ruler. Especially after the commissions, he felt himself surrounded by younger Fellows with ideas different from his. These were the men who, after 1854, could take honours in classics without first achieving honours in mathematics as had his generation, and men who had not attended professorial lectures as had his generation. These were men with different attitudes towards natural theology and strange ideas about secularization of both the colleges and the university. Of these men, one-third (22), none of whom had proceeded MA prior to 1850, had sent a letter urging drastic revision of the college statutes to the commissioners. These were also men like the Greek scholar, Joseph Lightfoot, who, in 1851, was more radical in his wishes to democratize the university than the commissioners. These were men like W. H. Thompson, the Regius Professor of Greek, who had not only been more of a democratic radical than the commissioners, but had angered Whewell in 1851 by fighting him to a standstill on a syndicate appointed to revise the statutes of the university. These were men like the classicists Francis Hawkins and John Llewellyn Davies who challenged Whewell within the college on the question of open Fellowships, disputed him (and Sedgwick) at the gathering of members of college governing bodies, and wrote pamphlets in opposition to those of their Master.[116] These were the men who, 'to seem fine in one another's eyes', demanded the introduction of silver forks, silver castors, and a hall butler to keep the servants in line. Whewell was not happy, and he feared for the future of the institution to which he had devoted his life. 'It is a very sad evening of my College life', he lamented, 'to have the College pulled in pieces and ruined by a set of schoolboys.'[117]

Whewell's last years were also marred by personal grief. Cordelia's death in 1855 after a lingering illness emotionally drained her husband. In 1858 Whewell married Lady Affleck, the widow of Sir Gilbert Affleck, and the sister of Robert Ellis. Ellis died in 1859 after a long, painful, debilitating and bed-confining illness, Lady Affleck died on 1 April 1865, and a little better than a year later, perhaps mercifully, Whewell himself died after suffering a head injury in falling from a horse. Lightfoot preached his funeral oration, and Thompson succeeded him as Master.

Before he died Whewell had recognized that developments in natural history had severely damaged the link between religion and science, and he could only hope that a future generation would reforge it.[118] Within months of Whewell's

[116] 'The Commissioners and the Colleges', 94–6; Becher (1986), 59, 80; Davies (1857), Cambridge University Library, Add. MS 745, f99; Winstanley (1977 [1940]), 49–50, 239–40, 345–7, 362–3.

[117] Whewell to his sister, 30 Oct. 1857, Stair-Douglas (1881), 497; Whewell to Forbes, 10 July 1863, Todhunter (1876), ii. 421–2.

[118] Whewell to Forbes, 4 Jan. 1864, Todhunter (1876), ii. 435–6.

death, the Board of Moral Sciences Studies deleted history of philosophy as a separate division for the Tripos, leaving it up to the examiners whether or not to introduce questions on the subject in other sections. It also changed the section on history and political philosophy to one on moral and political philosophy.[119] Whewell would not have been happy with that. Nor would he have rejoiced to have seen the introduction of the specialized ordinary degree system in 1866: witness that he had resigned in protest from the syndicate that had promulgated this reform.[120] His beloved liberal education was going the same way as natural theology.

Whewell was too much ingrained with the ideals of his time and too little anticipative of the future to become a centre of historical writing dominated by the Whig perspective. For the same reasons he was able to be a polymath. Other of his contemporaries could have done the same, for instance Herschel. But none of them pursued to the same degree broad academic interests (in the best sense of that phrase) on a 'Freehold for Life'. And Whewell was especially well-equipped for the task. He was not suited for postholing research, and he never displayed the creative knack of an outstanding scientist, not even in his studies of the tides. In this sense, science was his foible. Not given to creating new ideas, he was brilliant in analysing, synthesizing, and building upon the ideas of others. He could not make the pieces, but he could put them together in jigsaw fashion better than perhaps anybody else. And Whewell always *wanted* to build a broad system of knowledge with a 'theological bearing', not an abstract system, but one supported by and in turn supporting human knowledge. He determined the means for doing this, and executed it to the best of his ability for a lifetime. In his own eyes then, 'omniscience' was his forte.

From the historian's perspective, Whewell composed a symphony of learning, but it remained unfinished and attuned to his generation and not the subsequent one. Thomas Huxley, the epitome of the agnostics who came to the fore in the later nineteenth century, concluded that Whewell was 'always acute, if not always profound'.[121] From Huxley's viewpoint, of course, Whewell could not have been profound. Huxley's world also slipped into the sea of the past, but before it did it submerged Whewell's. Because of Whewell's make-up and also that of the era in which he thrived, we probably shall not meet the likes of him again, nor would we recognize him if we did, for like Huxley, we are not his *aequales*.

[119] 'Minutes of the Board of Moral Science Studies', Cambridge University Library, Min. v. 10, entry for 6 Nov. 1866.

[120] Whewell to the Vice-Chancellor, 17 Feb. 1863, Cambridge University Library, Add. MS 6580/35; id. (1865), Cambridge University Library, Cam. c. 865. 25.

[121] Huxley (1970 [1893–4]), v. 101.

2. *A Philosopher's Coming of Age: A Study in Erotetic Intellectual History*

by MENACHEM FISCH

1. Erotetic Contextualism

WHEWELL'S intellectual undertakings were enormous. He wrote widely on almost every topic on the early Victorian agenda. He was indeed a polymath, but, as noted in the introduction to this volume, the term polymath fails to do full justice to the nature of his universal scholarship. Whewell's works both literary and scientific, historical and philosophical, moral and theological were written with an explicit view not merely to embrace the world of learning but to synthesize it and render it a unified intellectual whole. As is evident from various of the contributions to this collective effort to assess Whewell's achievement, there exists a measure of disagreement as to the extent to which he succeeded in his ambitious interdisciplinary project. Contributors also differ in their location of the underlying unifying components of Whewell's multi-faceted enterprise. Yet there is no doubt that Whewell himself regarded his *History* and *Philosophy of the Inductive Sciences* as both the hallmark of his career and the unifying 'hard core' of his entire system of thought.

Historically speaking Whewell's *Philosophy* was unique in two major respects. First, in its having been the first full-blown critical and systematic alternative explicitly offered to counter Bacon's influential account of the scientific method (to be tellingly entitled *Novum Organon Renovatum*). And second, in the manner in which it was argued for against the backdrop of his three-tome *History of the Inductive Sciences, from the Earliest to the Present Time*—itself an impressive first.

A preliminary version of this essay was read at the joint meeting of the British Society for the History of Science and the British Society for the History of Philosophy, Churchill College, Cambridge, 11–13 April 1986. I am grateful to the following scholars for their comments and criticisms both in discussion and correspondence: Joseph Agassi, Gerd Buchdahl, Geoffrey Cantor, Ronald Curtis, Harvey Becher, Ivor Grattan-Guinness, Mary Hesse, Simon Schaffer, and Norton Wise. I also thank the Master and fellows of Trinity College Cambridge for their permission to quote from the Whewell Papers.

But Whewell's philosophical project was not only unique in its objectives and in the manner in which they were argued for. Various central and specific claims that it made seem, at first blush at least, peculiarly out of place. Few, I believe, would contest Robert Butts's observation that 'in keeping with his training, his interests, and the most influential trends of his age, Whewell should have been a philosophical empiricist',[1] just as few would contest the claim that Whewell ended up being anything but the philosophical empiricist he perhaps should have become. The epistemology of his 'Fundamental Antithesis of Philosophy' (1844), with its marked idealistic leaning, deviated sharply from the empiricism that prevailed among his philosophical peers.[2] His theory of scientific truth, with its emphasis on the ability of science both to arrive at and to recognize necessary truth, seemed to many to have been quite at odds with the more orthodox empirical methods of theory-appraisal that he preached.[3] And so different was his account of induction from that of Bacon, John Herschel, and later John Stuart Mill, that even his official biographer Isaac Todhunter seems at times to have missed the point,[4] whilst Charles Saunders Peirce (whose recently published early Harvard lectures show him to have been deeply impressed by Whewell)[5] saw fit to secure a special term—abduction—to denote such Whewellian reasoning. Finally, Whewell's historiography of science, which drew on important elements of his philosophical system,[6] yielded a number of surprising and highly uncommon historical judgements, such as a marked belittling of Newton's role in the inductive history of the laws of motion,[7] and a most positive assessment of Kepler's methods of enquiry.

Hence, in view of both the evident uniqueness and seeming perversity (to use Butts's somewhat degrading term)[8] of Whewell's philosophical project one would have expected historians of philosophy to have made him a prime target. Few thinkers innovated on so grand a scale, and in a manner so prima facie out of place as Whewell, whilst, at the same time playing, as Whewell did, so central a role in moulding the science, the learning, and the culture of their age. But such has not been the case. Whewell's unique and admittedly peculiar theory of science has, until very recently, all but failed to attract the attention of historians of philosophy.

As in so much of the history of philosophy, the closest one gets in the existing literature to a historical treatment of Whewell's philosophy of science are attempts to locate his philosophical influences and sources. Widespread have been views on precisely how far Whewell had gone along with, deviated from,

[1] Butts (1965), 161.

[2] On early Victorian Baconianism see Elkana (1984) and Yeo (1985).

[3] Cf. L. Laudan (1971), 383. For criticisms of Laudan see Rescher (1973*a*), 90, and (1973), 61; D. Wilson (1974), 85 n. 35); Fisch (1985*a*).

[4] Todhunter (1876), i. 107 ff.

[5] See especially Peirce (1865) and (1869).

[6] Though much less than is usually assumed. See below, section 6.1.

[7] In the *History*, Newton's formulation of the laws of motion belongs in the sequel to the inductive epoch of Galileo.

[8] Butts (1965), 161.

and/or misinterpreted the philosophy of Immanuel Kant. Some have reached even further back for possible philosophical inspirations. But discerning in Whewell's finished works, no matter how successfully, the philosophical footprints of Kant or Leibniz, of Reid or, as in a recent Ph.D., of St Thomas Aquinas (certainly the most interesting of all), helps little to explain what Whewell's works were written *for*. As though pointing out where the components of an instrument originally came from would be enough to explain what it was meant to do. Or, to take a less crude example, as though J. L. Lowes's monumental *The Road to Xanadu* can be said to explain *why* Coleridge wrote 'Kubla Khan' or 'The Ancient Mariner'. The historical assessment of a work of philosophy involves more than merely locating it within this or that philosophical tradition.

The formation of a philosophical system is a historical event, like any other, and as such it may enter history in a variety of ways. Whewell's philosophical deliberations and their eventual outcome, the *Philosophy*, were, as is evident from other contributions to this collection, no exception. The *Philosophy* undoubtedly reflected the culmination of Whewell's deliberations on a variety of issues, not all purely philosophical, and once completed certainly came to serve him in ways other than those pertaining directly to the philosophy of science. Yet notwithstanding, and certainly not detracting from its theological, educational, moral, and political significances, the *Philosophy* purported to say something about the nature of scientific enquiry, and something novel at that, and it took form in the course of keen philosophical reflection on such matters. Among other things, it was certainly intended as a work of philosophy proper, and the aim of my contributions to this collection is to assess it as such.

What is it then to assess a work of philosophy (*qua* philosophy) historically? The approach I shall adopt has been urged one way or another by such thinkers as R. G. Collingwood, Karl Popper, Hans-Georg Gadamer, and in somewhat different idiom, Quentin Skinner. According to this view the very notion of rationality is associated primarily with the confrontation of difficulties. Human endeavour, intellectual, practical, or political, acquires its meaning and worth, upon a Collingwoodian showing, only in relation to the problems it was or is intended to solve. Works of philosophy are hence to be viewed first and foremost as attempted solutions to philosophical problems.[9] To assess such a work historically requires the historian first to reconstruct the philosophical difficulties it was meant to overcome, and then to evaluate as far as possible the extent to which it succeeded in doing so, with reference to the set of alternative solutions which, to the best of his knowledge, were available to the thinker in question. Hence, at the heart of a historical reading of a philosophical system lies what one might term its *erotetic reconstruction*—i.e. a reconstruction of its emergence by tracing the process of philosophical questions and answers by

[9] For detailed expositions of erotetic contextualism and of the theory of rationality underlying it, see my forthcoming *William Whewell: Philosopher of Science*, section 1.2, and 'Towards a Rational Theory of Progress' (forthcoming).

which it initially took form. The logic of historical evaluation is thus what Collingwood termed 'the logic of question and answer', or to use the modern term, *erotetic logic*.

Viewed thus, the determination of an agent's sources (that is, the detection within his finished texts of echoes of other systems of thought) is at best a determination of materials reached for in his attempts to overcome his difficulties (if indeed he was genuinely grappling with problems and not merely engaged in expounding upon someone else's views), and can hence shed little light on the nature of the problems themselves.

The aim of this essay is to suggest in broad strokes such an erotetic reconstruction of Whewell's philosophical quest which culminated in the *Philosophy* of 1840. We shall naturally not be too concerned with the ways in which portions of the *Philosophy* were foreshadowed in his earlier publications, or echoed in his treatment of other topics thereafter. Rather we shall seek to analyse the philosophical problems Whewell encountered during each stage of his philosophical development, with a view to assessing in relation to them the solutions he eventually adopted. Intellectual innovation is not caused, in the simple sense of the term, but prompted by difficulties. Hence the process of awakening to problems and attempting to resolve them is the aetiology, so to speak, of intellectual progress. In this sense, to use a Whewellian term, this chapter purports to offer a 'palaetiology' of the *Philosophy of the Inductive Sciences*, in order to enable a more thorough philosophical assessment of the work in my second contribution to this collection.

2. A Working Hypothesis: Whewell as Whewellian

But where does one begin? Where does one look for the problems, philosophical and others, that Whewell encountered during each stage of his philosophical quest? Whewell himself did not make this quest at all easy. For all the praise he showered on Kepler for candidly making public his discarded hypotheses and their refutations, he himself did nothing of the sort. Despite the great emphasis he came to lay on history and historical research, he made things extremely difficult for his biographers. Like so many Victorians, he never discussed the problems he faced in writing. He perhaps pursued them more persistently than others, and refused to gloss over their solutions, but I know of no publication, or even draft essay, in which the difficulties themselves were aired. He kept nothing resembling a philosophical diary, and wrote no autobiography, intellectual or other. And to top it all, the manner in which he eventually elected to present his final conclusions in the *Philosophy*, is itself very misleading for the reconstruction of the manner in which they might have taken form.

In a word, the *Philosophy* addressed three main topics. First, and central to Whewell's entire system of thought, it presented a *theory of science* which endeavoured to explain what exactly an inductive science is, and consequently what such science can hope to achieve. Second, and following from his theory of science, the *Philosophy* offered a *prospective methodology*, or, as Whewell himself had it, a 'heuristic' which purported to explain how such science comes, and can be made to come, into existence. Lastly, both theory and prospective methodology of inductive science were supplemented by a theory of *retrospective theory-appraisal* designed to provide ways of singling out empirically true hypotheses once hit upon (comprising his much discussed criteria of factual truth, namely the ongoing tendency of a theory towards convergence and simplification, and what he dubbed 'consilience of inductions', that is, its repeated exhibition of explanatory surprises).

Writ large, Whewell's theory of science was presented in the *Philosophy* as an argument *from* epistemology *to* an account of science-at-its-best, which ran briefly as follows: All knowledge, he urged, comprises, *qua* knowledge, an objective element given by the world, informed and structured by conceptions furnished by the mind. Even the most common non-scientific perception, he argued, must go beyond mere sensation. For to perceive requires a singling out of sensation, a retention of it by the mind, and some awareness of the relations it bears to other sensations past and present; all of which cannot be said to be given by sensation itself. Knowledge hence embraces a fundamental antithesis of opposites. (It is, however, a fundamental antithesis of *philosophy*, one that can only be reflected on theoretically, for in practice no item of knowledge, he claimed, can be fully resolved into its primordial ingredients of fact and concept, since even knowledge of the purely factual, by this very view, will necessarily be *qua* knowledge itself antithetical.)

From the epistemology of the 'Fundamental Antithesis of Philosophy' Whewell goes on to argue for a theory of induction; that is, a theory of the acquisition of such antithetical knowledge. If each and every item of real knowledge consisted of a body of fact, unconnected of itself, 'colligated' and given unity by a concept provided by the mind, then the process of induction by which new such knowledge takes form, has to be one in which new such antithetical couplings are formed; namely, a process in which the mind suggests new concepts by which to colligate known but seemingly unconnected facts.

A perfect and mature science, Whewell argued further, is formed in the course of repeated colligations of ever higher generality. Such science in its finished form, he maintained, consists in a conceptual scheme by which an entire corpus of fact is colligated perfectly, and which, at the same time constitutes a perfect Euclidian deductive system of ideas.

Finally, such a view of excellent science, he argued, yields an essentially twofold theory of scientific truth. For it follows that the propositions of a reasonably mature body of scientific knowledge strive to attain two quite different and prima facie non-dependent forms of perfection: To be, on the one hand, well-formed theorems of a well-formed deductive conceptual scheme, whilst, at the same time, to perfectly colligate (i.e. correspond to) the facts.[10]

Such, in very broad and systematicized strokes, was Whewell's theory of inductive science as presented in the *Philosophy*: an argument which runs from the epistemology of the Fundamental Antithesis to a theory of induction as conceptual innovation; from induction to a theory of the structure of excellent science; and finally, from excellent science to a twofold theory of scientific truth.

Now, in view of the manner in which the entire argument of the *Philosophy* was thus made by Whewell to hinge on the Fundamental Antithesis, one might be inclined to conclude that the philosophical deliberations that initially prompted the *Philosophy* should be sought in the realm of epistemology. Accordingly, the story of the formation of Whewell's views on science would then run roughly thus: reared and bred in the British empiricist tradition Whewell gradually became aware of certain epistemological problems he felt the Lockean empiricist school fell short of solving (for example, the foundations of mathematical knowledge, as viewed by the Analytical Society with whom he was closely associated). As a result, (influenced perhaps by Coleridge or William Rowan Hamilton) he turned to consider Kant who proclaimed to offer an epistemological alternative better tailored to the epistemological challenge posed by modern mathematical science. Not satisfied by Kant's solutions (perhaps for religious reasons), he was eventually led to construct a theory of his own incorporating the best of both worlds—the Fundamental Antithesis—which swiftly proved, in his hands, to provide the basis for both a successful theory of scientific knowledge, and a fruitful foundation for a full-blooded historiography and methodology of science.

On the face of it there would seem to be quite a lot going for such an account of Whewell's philosophical coming of age. Whewell no doubt had an empiricist upbringing, and, as Michael Ruse has shown,[11] portions of his own scientific work could well be viewed, if one is thus inclined, to have been conducted along typically empiricist lines—as, for example, his and George Biddell Airy's Dolcoath Mine experiments to determine the density of the earth; and large parts of his work on the tides. It is also true that early in his career Whewell engaged in some issues that prima facie could have confronted an avid empiricist with grave epistemological problems—such as his involvement with the Analytical Society's campaign for the introduction of Lagrange's version of the calculus. Lastly, one could perhaps point to the fact that the Fundamental Antithesis was only fully elaborated in 1844, and interpret, as H. T. Walsh has done,[12] the way Kant (hailed 'the metaphysical genius in Germany') is reproduced almost verbatim on space and time in the first 1840 edition of the

[10] For a critical exegesis of Whewell's theory of scientific truth see my (1985), (1987), and 'Antithetical Knowledge', this volume Ch. 11.

[11] Ruse (1976).

[12] Walsh (1962). For a pointed criticism of Walsh's thesis see Butts (1965*a*).

Philosophy, as indicating a retreat on Whewell's behalf from a more orthodox Kantism thereafter.

But all this is at best only circumstantial. Upon a slightly closer examination such a working hypothesis falls flat in its main claim. For there is nothing, in either Whewell's published or unpublished work written during the 1830s, to indicate that deliberating the source(s) of knowledge was in any way a major concern of his until very close to the publication of the *Philosophy*. What he chose to publish during the 1830s contains nothing resembling a discussion of epistemological issues, and among the numerous letters and unpublished notebook entries he wrote during that period, few if any touch on such matters. Quite the contrary. On several occasions, both published and unpublished, one can sense a deliberate glossing over of matters epistemological. As, for example, in the following passage from his 1834 essay 'On the Nature of the Truth of the Laws of Motion', in which the epistemological status of the assumption that 'every change has a cause' is considered:

> Different people may, according to their different views of such subjects, call it a law of our nature that we should think thus, or a part of the constitution of the human mind, or a result of our power of seeing the true relating of things. Such a variety of opinion . . . would not affect the fundamental and universal conviction which the axiom expresses; and would therefore not interfere with our future reasonings.[13]

Needless to say, although his analysis of the laws of motion would be reproduced six years later in the *Philosophy* almost word for word, the above-quoted passage would be replaced by an entire chapter in which the epistemological status of 'the axioms of cause' is examined from the points of view of David Hume, Thomas Brown, Thomas Reid, James Beattie, Dugald Stewart, and Kant, with Whewell firmly taking sides with the latter. In 1834, however, Whewell seems to have maintained, similarly to Herschel and in line with the Scottish Common Sense philosophers, that the discussion of scientific method and the nature of scientific truth transcends epistemological considerations, and can be pursued regardless of one's epistemological denomination. Even much nearer the publication of the *Philosophy*, both in his essay on inductive reasoning appended to the *Mechanical Euclid* and in the historiographical introduction to the *History* (both published in 1837), though no longer feigning epistemological liberalism as in 1834, Whewell still hardly alluded to epistemology. Again his position seems to be that the process of induction can be rigorously investigated regardless of epistemological considerations.[14] Indeed, the fact that the three-volume *History* was completed before the *Philosophy* would itself indicate that the question of the sources of knowledge only later became a major concern.

I conclude, therefore, that the structure of the *Philosophy* does not recapitulate the process of its formation. But if that be the case how did Whewell's views evolve?

[13] Whewell (1847), ii. 574–5. [14] See below, Sect. 7.

The working hypothesis I shall adopt will be that the four theses comprising Whewell's theory of science were arrived at in almost the reverse order that they would eventually be argued for in the *Philosophy*.

The first component of his theory to form, I shall argue, was not the Fundamental Antithesis, but rather his theory of excellent *finished* science, which emerged in the course of his extensive textbook writing during the 1820s and 1830s. Then, equipped with a reasonably clear notion of what a good piece of science should look like, Whewell, I shall argue, embarked on two separate lines of enquiry: one to determine the nature of the truth of such science—the results of which were published in the 1834 essay on the laws of motion—and the other to determine the process by which such science emerged, a project that would take him until close to 1840 to fully complete. The first major step towards a full account of the emergence of inductive science was the formulation in 1834 of his theory of induction. This theory served as the basis for a detailed elaboration of his methodology during the late 1830s, leading, close to 1840, to the formulation of his epistemology.

Whewell is thus viewed, one could say, as conducting his philosophical quest in accord with the recommendations of his future system, namely 'inductively'; working his way up what would eventually become a downward deductive system. And doing so, as the *Philosophy* required, in a series of philosophical conceptual innovations, starting with his account of the structure and the nature of the truth of mathematical physics, and reaching completion in his all-embracing notion of antithetical knowledge. Whewell's philosophical coming of age, I thus propose, was brought about in two main stages: the first, in which the nature of the scientific end-product was examined—that is, the structure of excellent science and the nature of scientific truth—and the second, in which the way such scientific knowledge takes form was contemplated: an examination, one could say, first of the *being* and then of the *becoming* of modern science at its very best.

3. Towards a Theory of Excellent Science

3.1 *The Analytical Society*

Whewell first seriously encountered science during his undergraduate days at Trinity. He arrived in Cambridge in the autumn of 1812—the year following the foundation of the Analytical Society. Whewell was younger than the founding fathers of the Society (notably John Herschel, George Peacock, Charles Babbage, John Maule, and the brothers Edwin and Edward ffrench Bromhead), and perhaps initially less singlemindedly devoted than they to the study of pure mathematics, but there is no doubt that the young, awed, and outclassed Lancastrian was deeply impressed both by the members of the

Society and by their 'laudable object'. And, as is evident from the following letter to Whewell from Edward ffrench Bromhead, the attraction was mutual:

> Babbage and Herschel are both in town, and we had a kind of committee of notation . . . We have also agreed upon a Digest of which you must take part. It consists of a collection of all known algebraic results arranged in the order of Deduction. You must take the partial differential analysis . . . . Is there any other part you would fancy, except general functions, differential calculus, differences and common algebra, as these are distributed. Herschel is upon an Algebra.

'Pray give up the muses', Bromhead continues,

> the object of learning should be to keep men cool, stupid, and comfortable, and not to stimulate their passions into insubordination and vice by imagery and metre.[15]

Protesting against the disappointing standard of mathematics taught at Cambridge, the Analytical Society formed itself, as Babbage, parodying a theological broadsheet, put it: 'for propagating the gospel of '*d*'ism while consigning to perdition all supporters of the heresey of the dot'.[16] Babbage was referring of course to Leibniz's '*d*' notation, adopted by most eighteenth-century continental analysts, as opposed to the 'pricked letters' employed by Newton. But there was more to the analytical revolution in Cambridge mathematics than mere notational reform.

In all, three arguably non-dependent yet related issues were at stake. First indeed was the technical problem of introducing *d*s in place of dots without which most continental mathematics simply remained closed to the Cambridge-reared reader. Second, and certainly the Society's main objective, was to replace the geometrical approach of the *Principia* taught at Cambridge by the multi-variate and variational algebraic methods of analysis—both pure and mixed—developed on the continent during the latter part of the eighteenth century.[17] Lastly was the question of the mathematical foundation of the calculus itself, a question in dispute since the publication of Berkeley's *The Analyst* in 1734. Alongside their obviously laudable campaign for notational reform and the introduction of the powerful continental methods of analysis, the Analytical Society called with equal loudness, and seemingly equal clarity, for the adoption of Lagrange's algebraic and wholly formalistic version of the calculus.[18]

[15] Bromhead to Whewell, Whewell Papers, Trinity College Library, Cambridge University, Add. MS a. 201[105]. The letter is undated but Bromhead's allusion to Whewell's approaching fellowship examination would place it sometime during 1816–17.

[16] See Babbage (1864), 30, and Hyman (1982), 20 ff. For criticism of Babbage's and Hyman's dating see Schweber (1981), 56 ff.

[17] To what extent the dot notation was in itself responsible for the stagnation of 18th-century British calculus is a question in dispute. (De Morgan (1834) and Rouse Ball (1908), 391 f., claim that such indeed was the case, whereas Garland (1980) and Grattan-Guinness (1986), sect. 5, disagree—the latter dismissing such a view as 'largely rubbish'.)

[18] A position which by then by no means enjoyed consensus on the continent. See for example Grattan-Guinness (1979), 83, (1985), and (1986); Glas (1986); Daston (1986).

Rather than found the calculus on the (then still) loosely defined notions of continuity and limit, Lagrange defined the 'derived functions' of a given function $f(x,y, \ldots)$ simply as the coefficients of the Taylor expansion of the function. It was a move primarily designed to side-step problems of foundations, one which would allow for a rigorous refinement of analytical methods regardless of what, if anything, the calculus itself was intended to articulate—and it was a deliberately instrumentalistic move at that. (In fact, Boyer points out that Lagrange, like Berkeley, attributed the success of even his own version of the calculus to a compensation of errors rather than to sound foundation).[19] It so happened that Lagrange produced the most elaborate set of analytic apparatuses eighteenth-century continental mathematics had to offer. And it was for that, rather than for despair of the attempts to elucidate continuity and limit, that the Society initially opted for his *théorie des fonctions analytiques*.

However, a formalistic approach to pure mathematics is certainly not unfounded in principle. And notwithstanding the manner in which they came to adopt such a position in the first place, the three chief founders of the society—Herschel, Babbage, and Peacock—maintained their initial formalistic approach throughout their careers. George Peacock, in fact, remained faithful to Lagrange well after the publication of Cauchy's epsilon-delta treatment of limit, stating forcefully, as late as 1834, that to found the calculus, as Cauchy had done, on the notion of limit was 'altogether inconsistent with the spirit of symbolic algebra'.

Focusing primarily on the positive achievements of the Analytical Society, historians have so far failed to draw attention to the problems they created. In particular the Society's adoption of Lagrange's version of the calculus created a tension with respect to mathematical physics ('mixed mathematics' as it was then termed). Cast in latter-day terminology, such a move, as pointed out, implied taking an instrumental view of the *application* of the calculus—particularly of its application in physics. Yet the members of the Society—best represented here by John Herschel—maintained an orthodox Baconian realist view of physical knowledge; which for one thing was to be anything but an instrumentalist. The two positions—a Baconian view of natural knowledge and a Lagrangian view of the mathematical scheme applied to it—are by no means incompatible. Yet compounding the two does require a measure of philosophical finesse totally lacking in the Society's publications. Not surprisingly perhaps, the founding fathers of the Analytical Society all carefully avoided philosophical discussion of modern mathematicized physics thereafter. Peacock and Babbage steered well clear of discussing natural science altogether, while Herschel, who engaged in both physical research and the philosophy of the natural sciences, did an admirable job of side-stepping the problematic areas in which mathematics and natural knowledge met.

[19] Boyer (1949), 251.

Whewell's, however, was a different story. For he swiftly found himself confronted by those very problems Herschel chose to ignore. But unlike Herschel, Whewell rose to the challenge.

Though initially fully sympathetic to the Society's causes, from the start Whewell vehemently opposed their tactics. The reform of Cambridge mathematics, he urged, could not be accomplished by a head-on confrontation with the establishment in the realm of pure analysis, as Peacock (with Herschel's explicit blessing)[20] had attempted to do in the questions he set as Moderator on the 1817 Tripos. 'I do not think he took precisely the right way to introduce *the true faith*', he wrote to Herschel in 1817,

> he has stripped his analysis of its applications and turned it naked among them. Of course all the prudery of the University is up and shocked at the indecency of the spectacle. Now the way to prevent such a clamour would have been to have given good, intelligible, but difficult *physical problems*, things which people would see they couldn't do their own way, and which would excite their curiosity sufficiently to make them thank you for your way of doing them.[21] [my italics]

Still in full agreement with Herschel as to what 'the true faith' *was*—namely Lagrangian calculus—Whewell felt strongly that the revolution in pure mathematics had to start in mixed mathematics. And true to this utilitarian pedagogy, from 1817 he embarked on a series of mechanics textbooks designed to form the backbone of the university's mixed mathematics curriculum—a project that would swiftly confront him with the problematic compounded Baconian–Lagrangian position implied by the Society's programme.

3.2 *A Baconian–Lagrangian Account of Mathematical Physics*

What exactly did such a position involve? From a Lagrangian perspective the mathematicization of physical knowledge is in no way an act of *elucidation*. To represent a physical notion by means of a derivative is for the Lagrangian, to say no more than that the quantity in question is given by the value of a coefficient in a power-series. Assimilating the notion of acceleration, for example, to the analytical construct

$$\frac{d^2x}{dt^2},$$

did nothing to elucidate acceleration, and is hence prima facie in no way self-evident. The sole justification of such an assimilation, upon this view, lies wholly in the fact that certain analytical consequences of the formal notion of derivative are borne out, and none refuted, by experience.

[20] See letter from Herschel to Fearon Fallows, an examiner in 1817, urging the latter to help promote the Society's cause by standing by Peacock's choice of questions. Herschel concludes, 'I hope and trust you will excuse me for sending you all this exhortation and declamation. I love and reverence the University from the bottom of my soul and I cannot endure to sit quietly and see blemishes upon it.' (8 Dec. 1816, cited in Schweber (1981), 65.)

[21] Whewell to Herschel 6 Mar. 1817, Todhunter (1876), ii. 16.)

For the Lagrangian purist (that is, the fully fledged instrumentalist) the analytical systems science constructs serve to set empirical findings within an efficient and economical, integrated and well-formed, formal structure which has no pretensions to pose as a description of reality. Virtual velocities, general potentials, etc. together with the entire analytical superstructure in which they take their place, are not intended to mirror portions of reality, not even hypothetically. Similarly, the lines of analytical reasoning within such analytical systems are not constructed with a view to represent lines of actual physical causation. The mathematicization of the physical, for the Lagrangian, serves to save the phenomena, not to explain them. (I use 'explain' here in Pierre Duhem's rather than in Nancy Cartwright's sense of the word).[22] Science progresses, according to the Lagrangian purist, by seeking on the one hand to expand our corpus of experiential findings, whilst constantly striving to devise new elaborate systems by which to structure it more efficiently—systems that boast no more than completeness, rigour, and logical consistency.

For the Baconian realist, however, things are different. Unlike the instrumentalist purist, he maintains that physical knowledge exceeds bare empirical data. The inductive method, he holds, enables him to transcend raw experience. Science, if conducted properly, he believes, can scale the 'ladder of axioms' and gain insight into the true causes of phenomena. More importantly, it can do so unheeded by higher mathematics. The Baconian has no need to save the phenomena, since they are 'saved', as it were, by the actual physical causes he believes he can detect. Hence for him, to cast our knowledge of nature—axioms, causes, empirical fact and all—in terms of a totally formal system, as the Society's scheme implied, can serve two purposes only. First, in so far as a body of induced natural knowledge is found to fully translate in terms of a well-formed analytical system it can be said to enjoy the deductive rigour of the system. Second, and following from the first, in view of the formal dovetailing of the two—the analytical system and the body of induced scientific knowledge—the fruits of deductive reasoning within the formal system could be utilized, as in the instrumentalist's account, to probe the empirical and to yield new predictions. The point to emphasize is that in order for it to fulfil both objectives—both to vouch for the rigour of the body of induced scientific knowledge and to enable its broadening by prediction—the formal system the Baconian–Lagrangian elects to employ must admit to a one-to-one correspondence to the body of scientific knowledge it purports to structure.

Hence, one would expect a Baconian–Lagrangian's view of mathematicized natural science to differ from that of the Lagrangian purist on two accounts. First, since the entire enterprise of mathematicization turns for him on a one-to-one correspondence between the formal and the empirical, he will insist, contrary to the Lagrangian purist, that each and every step in the analysis fully

[22] For Cartwright, unlike Duhem, merely to 'set the phenomenon in a general theoretical framework' also counts as explaining it. See Cartwright (1983), 75, and cf. Duhem (1954), 7 f., and *passim*.

'cash out' in terms of the empirical. Years later, in *The History of the Inductive Sciences*, Whewell would express this sentiment in a telling comparison between the physics of Lagrange and that of Gauss:

> Lagrange, near the end of his life, expressed his sorrow that the methods of approximation employed in Physical Astronomy rested on arbitrary processes and not on any insight into the results of mechanical action. From the recent biography of Gauss, the greatest physical mathematician of modern times, we learn that he congratulated himself on having escaped this error. He remarked that many of the most celebrated mathematicians . . . had trusted too much the symbolic calculations of their problems, and would not have been able to give an account of the meaning of each successive step in their investigation. He said that he himself, on the other hand, could assert that at each step he took, he always had the aim and purpose of his operations before his eyes without ever turning aside from the way. The same, he remarked, might be said of Newton.[23]

Subsequently, the Baconian–Lagrangian and the Lagrangian purist will account differently for the manner in which such science emerges. The latter will view the formation of analytical science as an essentially phenomena-saving operation in the course of which a totally formal system is gradually erected and modified so as to structure an otherwise disordered body of fact. The Baconian–Lagrangian, on the other hand, will view the process as the happy culmination of two independent efforts: an inductive investigation of the empirical on the one hand, and the cultivation of a totally formal system on the other, found eventually to 'save', if you wish, the entire body of induced scientific knowledge *in toto*. Herschel, in his *Preliminary Discourse on the Study of Natural Philosophy*, describes the process (in the only passage remotely addressed to the issue) as two separate lines of investigation, one speculative and the other empirical, pursued independently until there occurs 'a sudden transfer of ideas from a remote station in one to an equally remote station in the other . . . to start the first idea of applying science'.[24]

What the two accounts of mathematicized science will have in common is their view of the calculus, and hence of the entire formal system employed in a mathematicized natural science as totally devoid of *elucidatory* power with respect to the empirical. For the Lagrangian purist the mathematical constructs employed serve merely as an elaborate pigeon-hole system to store empirical data,

[23] Additions to the third edition of the *History* (1857), ii. 439.

[24] Herschel (1830), 71–2. In the *Discourse* Herschel firmly distinguishes 'knowledge of nature' from what he calls 'abstract science'. The latter forever 'independent of a system of nature—of a creation—of everything except memory thought and reason' (p. 18); the former, on the other hand, wholly empirical: 'Experience once recognized as the fountain of all knowledge of nature, *it follows* that in the study of nature and its laws, we ought at once to make up our minds to dismiss as idle prejudice any preconceived notion of what might or what ought to be the order of nature . . . and . . . to stand and fall by the results of a direct appeal to facts.' (pp. 79 f.) 'It is to our immortal countryman Bacon', he says elsewhere, 'that we owe the broad announcement . . . and the development of the idea, that the whole of natural philosophy consists entirely of a series of inductive generalizations, commencing with the most circumstantially stated particulars, and carried up to universal laws, or axioms, which comprehend in their statements every subordinate degree of generality' (p. 104).

whereas for the Baconian–Lagrangian realist they constitute formal (and in themselves vacuous) mirrorings of scientific concepts, which are elucidated in turn by the facts that begot them *prior to*, and *regardless of* their mathematicization.

Yet though implied by the positions held by those associated with the Analytical Society, a compounded Baconian–Lagrangian philosophical account of mathematical science as outlined above was spelt out by no one. For all practical purposes, historically speaking it is a piece of philosophical fiction. Babbage and Peacock on the one hand initially endorsed Lagrange's version of the calculus for the way in which it liberated analysis from its former geometrical and physical tetherings, which were responsible for holding it back, and thereafter chose largely not to concern themselves with the problem of reapplying thus liberated mathematics back to the physical. Herschel, on the other hand, who had something to say both of pure analysis and of Baconian natural science, elected to keep them philosophically well apart. When it came to articulating a Baconian–Lagrangian philosophy of science all its potential advocates seem to have fallen conspicuously silent. All, that is, but Whewell.

Whewell's views on the nature of science and the manner it is achieved, I shall argue, took form in three phases, in which the three main tenets of a Baconian–Lagrangian position outlined above were closely and critically scrutinized.

Such a position, we found, (*a*) required the formal system employed in a science to fully mirror its empirical counterpart; (*b*) rendered the formal system employed devoid of elucidatory power; and (*c*) implied that the formal and empirical components of a science evolve separately and eventually combine by chance.

During the first phase of his career (roughly 1817–24), still fully committed both to the Society's view of the calculus and to Bacon's account of the scientific method, he set out upon a series of textbook expositions of mechanics in which (*a*), that is the requirement that the formal fully mirror the empirical, was made a central theme (in explicit contrast to French physics).

The second phase in Whewell's philosophical development was marked by a growing dissatisfaction and eventual dispensation with Lagrange's version of the calculus, a change of heart largely brought about by his increasing reluctance to endorse (*b*), namely ridding mathematics of its traditional elucidatory role, a point he became more and more aware of as he turned to address his textbooks to a more elementary readership. By the late 1820s Whewell had endorsed a D'Alembertian limit-based version of the calculus in place of that originally favoured by the Society. This would form the basis for his more general view of mathematical systems, pure and applied, as formal but not vacuous articulations of what he would call 'Fundamental Ideas'.

Finally, with what he felt was an adequate description of mathematical physics under his belt, he turned, from 1830 or so, to consider seriously how such science evolved. Reluctant to endorse the view implied by the Society's programme (and preached enthusiastically by the young William Rowan

Hamilton with whom he had by then come into lively contact), namely that such science came into existence by the chance crossings of two independent lines of investigation, Whewell embarked in earnest on the articulation and historical testing of an essentially unitarian theory of the growth of scientific knowledge. This would prove the toughest nut to crack, and would lead eventually to his abandoning the firmest of his earlier commitments: first the methodology and then the empiricist epistemology of Francis Bacon's philosophy of science.

3.3 *Whewell's Mechanics Textbooks: Reconsidering Lagrange*

Whewell's first two textbooks, *An Elementary Treatise on Mechanics* (1819), and *A Treatise on Dynamics* (1823), were designed to cater for third-year students studying for Cambridge honours. Only later would more general concerns for the wider aims of a Cambridge liberal education become the main issue for Whewell. At this stage his target was the Tripos, and his main objective to initiate the élite, that is the future dons, into the 'true faith' as effectively as possible.

The *Elementary Treatise* was originally conceived as a collection of problems in mechanics designed to promote the use of the new methods of analysis by way of demonstration. However, by the time the book was completed in 1819, Whewell had become intrigued by mechanics as a physical science in its own right.

He wrote to Herschel in 1819,

> I have the first volume of 'A System of Mechanics', containing statics and a little dynamics, ready for press. It consists in a great measure of a classification of problems. To serve as an elementary book introductory to the higher applications of mechanics is one objective, and another is to establish the science on simple and satisfactory principles, which I do not think has been done.[25]

This early shift of interest from viewing mechanics merely as a sounding-board for mathematical methods, to viewing (and more importantly), to teaching it as the model physical science, marks an important turning point in Whewell's career. From then on, unlike any of his fellow rebels, Whewell's interests came to focus on the most problematic area of the analytical revolution at Cambridge. Namely, the point in which formal analysis meets Baconian science. And from the very start Whewell seems to have realized that to maintain a Lagrangian view of the calculus alongside a Baconian realist view of natural science, one must require that the formal system employed should fully 'cash out' empirically. In the same letter to Herschel cited above, whilst fully aligning himself with the 'ultra analysts' and the need to 'further [their] laudable object', Whewell questioned the manner in which 'in French books on Mechanics' the physical differences between static pressure and dynamic force are glossed over by

[25] Whewell to Herschel, 1 Nov. 1818, Todhunter (1876), ii. 29 f.

applying to both the formula $F = ma$. French 'attachment to the forms and processes of pure analysis', he complained in the *Elementary Treatise*, had given them a disrelish for the more physical'.[26] Too slick a formalization seems, in the French texts, to have obliterated a real physical difference. Indeed D'Alembert's conflation of statics and dynamics would remain for Whewell the hallmark of the poverty of what I've termed the Lagrangian purist philosophy of science. And the blade was to cut both ways. Not only must it not be allowed that a real physical nuance be glossed over for the sake of analytical conciseness, but, Whewell urged, we must also disallow the introduction of empirically redundant analytical constructs however much neater they render the formalism. Thus Whewell's persistent objection to the use of virtual velocities, which, he urged time and again, 'are not produced by forces, nor indeed produced at all, but only supposed. It is therefore certain that such merely notional and hypothetical "virtual velocities" have no common connection with physical and *actual* velocities produced by force'. 'To invest these *geometrical phantoms* with any attributes arising from physical laws', he stormed, 'is a proceeding altogether *arbitrary* and *illogical*.'[27]

True to his initial (and perhaps partially tacit) commitment to a Baconian–Lagrangian view of science, Whewell swiftly came to regard mechanics as empirical *ab initio*. Each and every step in the formalism had to correspond to something real. It was not enough that the formalism *led* to empirical findings, it was required to refer to the real throughout. Hence his comparison in the *History* between Lagrange and Gauss cited above.

If during this first phase of his career, Whewell can be viewed as prudently working out the consequences of retaining a Baconian–Lagrangian account of up-to-date mechanics,[28] the following years were marked by a decisive retreat from both initial commitments in the face of the problems they posed—first

[26] Whewell (1819) vi.

[27] Whewell (1832*a*) p. ix (my italics). See also id. (1835) 24 f.; id. (1837*a*), 74–5; id. (1840), i. 192 f. Though clearly opposed to D'Alembert's conflation of statics and dynamics in the *Elementary Treatise* of 1819, there is still an entire chapter (ch. 9) on virtual velocities in the first edition, which disappears in the second edition of 1824. Similarly, in the *Treatise on Dynamics* of 1823 virtual velocities are used to prove the conservation of *vis viva* (app. H), and variational techniques are used to prove the principle of least action. Hence, only in 1824 did Whewell fully come to realize 'the fallacy' of virtual velocities. I am grateful to Dr Niccolo Guicciardini for drawing my attention to this point in private communication.

[28] Herschel seems to have largely misunderstood Whewell's insistence on utilizing analysis in mechanics only as far as it admitted to a full physical translation, and took the *Elementary Treatise* to be an attempt, not to articulate the Society's Baconian–Lagrangian position, but to meet the old regime half way: 'as far as I have been able to read it I like it very much, especially the very philosophical way in which you state your first principles and come broadly down upon experience for the solution of every difficulty of the obviation of every metaphysical subtlety . . . but I think your concessions to the cramming system have been rather too large and that the work would have been productive of more extensive good, have fought its way better . . . had you conformed a little more to the taste of the age and a little less to that of University. It has now a good deal of the aim of a stepping stone to some more finished and systematic *elementary* work, as if you intended it only to serve a temporary purpose, in the transition state of Cambridge reading . . . ' Herschel to Whewell, 19–20 Aug. 1818, Whewell Papers, Add. MS 207[4].

from Lagrange's version of the calculus, and later from Bacon's philosophy of induction.

Whewell's reassessment, during the late 1820s, first of the nature of the calculus and consequently of that of mathematics in general, was originally prompted by didactical rather than philosophical considerations. After completing the two *Treatises* (and modifying the first of them in the 1824 edition) Whewell turned to address a more elementary audience. First, he sought to further the reform of the Senate House textbooks so as to include students working towards regular degrees—the 'poll men' as they were called. Second, and perhaps more importantly, with the introduction of the Classics Tripos during the mid 1820s, a new class of future dons emerged, less mathematically inclined than the wranglers, yet none the less still required to take the maths Tripos.

In these more elementary texts[29] Whewell could take less for granted by way of mathematical representation. If in the first two *Treatises* he could start by *defining* velocity and force as derivatives, in the less advanced expositions he had to explain what he was doing. Here he ran into difficulties.

His 1832 *Introduction to Dynamics* is a good example. Whewell begins by defining velocity as 'the measure of the degree in which motion is quick or slow' (p. 15). Consequently, 'variable velocity' (instantaneous, that is) is taken as 'the ultimate value or limit of the quotient of the space by the time'. Formally, but not yet mathematically, that is expressed as:

$$(1)\ v = \text{limit of } \frac{\text{increment of } s}{\text{increment of } t}$$ [30]

which, Whewell remarks (since in the elementary textbooks the calculus is almost entirely avoided), 'by the principles of the Differential Calculus if s and t be expressed one in terms of the other or by means of any other analytical connexion, we have:

$$(2)\ \text{limit of } \frac{\text{increment of } s}{\text{increment of } t} = \frac{ds}{dt}$$

and hence:

$$(3)\ v = \frac{ds}{dt}$$ [31]

Subsequently, the 'accelerative force' is defined in likewise fashion as:

$$(4)\ a = \text{limit of } \frac{\text{increment of } v}{\text{increment of } t}$$

[29] Among Whewell's more elementary textbooks I would count *Elementary Treatise* from its 4th edn. onwards, *Introduction to Dynamics* (1832*d*), the *First Principles of Mechanics* (1832*c*), *Mechanical Euclid* (1837*e*), and *Doctrine of Limits* (1838).

[30] p. 16.

[31] Ibid.

and by the same token he obtains:

$$(5)\ a = \frac{dv}{dt}$$

For Whewell the realist equations (1) and (4) and the right hand sides of (2), (3), and (5) express actual physical realities. Prior to their mathematicization by means of derivatives, real physical velocity and acceleration were for him limit values of real changes of displacement and velocity in time. If, however one endorses Lagrange's version of the calculus, ds/dt and dv/dt are no more than the coefficients of the second terms in the Taylor expansions of $s(t)$ and $v(t)$ respectively. From which follows that for the Baconian–Lagrangian equating v and a with ds/dt and dv/dt in (3) and (5) can no longer be regarded as *definitions* of velocity and acceleration—which are defined by (1) and (4)—but only as mathematical representations of these definitions. Hence one could say (3) and (5) are the points where for the Baconian–Lagrangian, the 'sudden transfer' (to use Herschel's phrase) between the two sciences occurs.

But how does one explain, especially at such a basic level, *why* such a move is made? Derivatives thus construed, for all their algebraic rigour, do nothing to further elucidate the physical notions to which they are applied, just as their application, however fruitful in the long run, casts no reciprocal light on the nature of the derivative. From a Baconian-Lagrangian perspective (3) and (5) are justified solely by their future consequences but not by anything that goes into their construction. It is extremely difficult to teach something totally lacking prima facie justification. Yet (2), (3), and (5) constitute the very starting point for the entire enterprise of modern mathematical mechanics. Thus in the course of grappling with the problems of teaching such science at such an elementary level, Whewell, I conjecture, fully awoke to the sheer prima facie arbitrariness of the Herschelian 'transfer' (an arbitrariness paradoxically rendered all the more acute by the suggestive quotient form of the Leibnizian notation favoured by the Society.)

As a result Whewell abandoned Lagrangian calculus and promptly joined the opposition—namely, those who like D'Alembert and L'Huilier, and in a sense Cauchy, strove to found the calculus on a rigorous explication of the notion of continuity and limit. A move Harvey Becher has interpreted as reactionary, designed in his view to promote the restoration of 18th-century mathematics and the limitation of 19th-century analysis.[33] I think Becher is mistaken, and that his mistake largely distorts our understanding of Whewell. During the 1820s Whewell's textbooks underwent two major changes. One a move away from Lagrange's version of the calculus; a move, incidentally, very much in line with contemporary continental trends. And the other, a dramatic simplification of his exposition for the benefit of his new and far less advanced readership. Becher, in his seminal essay on Whewell and Cambridge mathematics, takes the two processes in oversimplified conjunction, wrongly

[32] p. 18. [33] Becher (1981), 25.

conflating Whewell's rejection of Lagrange with the emphasis he came to lay in the more elementary textbooks on intuitive geometrical problem-solving techniques reminiscent of the *Principia*. Becher however fails to note that regardless of the 18th-century mathematics employed in his elementary textbooks, Whewell spoke loud and clear of the decisive superiority of algebraic analysis for the mathematicization of the physics of his day. In 1832 he wrote 'The analytical mode of treating Dynamics is the only method which can now answer the requisites of natural philosophy'.[34] In the *History* his position remained exactly the same: 'As analysis was cultivated, it gained a *predominancy* over geometry; being found to be a *far more powerful instrument* for obtaining results; and possessing *a beauty and an evidence, which though different from those of geometry, had great attractions for minds to which they became familiar.*'[35]

Yet, though he retained a progressive and modern algebraic view of mathematical analysis, Whewell ceased to regard the calculus as a vacuous system (and consequently distanced himself from the developments in abstract algebra prompted by the work of Babbage and Peacock). 'Though it may be possible to establish, by means of algebraic operations without any new principle, a branch of analytical calculation which we may call the differential calculus,' he wrote in 1835, 'we cannot *justifiably* use it as the calculus is used by [mixed] mathematicians till we have introduced into it the notion of a limit.'[36] And the same, he maintained, was true of all mathematical systems. All branches of mathematics, he now claimed, were to be viewed as elaborate Euclidian systems of propositions that rigorously unpacked various of the features of some major notion—eventually to be dubbed: a Fundamental Idea.[37] Geometry articulated the Fundamental Idea of space; algebra that of number (later, following Kant, number would be replaced by the idea of pure time); and the calculus that of limit.

Consequently, Whewell—not unlike Herschel—came to regard well-formed *natural* sciences as combinations of two bodies of knowledge: a body of induced empirical knowledge on the one hand, *cast in terms* of such a well-formed 'pure' articulation of a Fundamental Idea on the other. 'Science', he wrote in a notebook entry of 1832 or 1833, 'is formed by acts of the mind, in addition to which there may or may not be employed, as materials, special facts given by experience. When science is free from any admixture of mere experimental knowledge, it is called *pure* science.'[38] Whewell's theory of the structure of fully fledged natural science was hence the first component of the philosophical system of the *Philosophy* to take form. It did so, I submit, by 1830 or so, possibly even a little earlier. In any event, it was stated clearly in his review of Herschel's *Discourse* published in 1831. A natural science, he claimed in the

[34] Whewell (1832*c*), xii.

[35] Id. (1837*a*), ii. 72 f. (my italics).

[36] Id. (1835), 18.

[37] To the best of my knowledge the term is used for the first time in print in his *Thoughts on the Study of Mathematics as Part of a Liberal Education* (1835*a*), 16–17.

[38] From an unpublished study entitled 'The Philosophy of the Pure Sciences', Whewell Papers R18 17[8], p. 37. The very last entry in the bulky essay (95 pp.) is dated 29 Dec. 1833.

review, was a body of knowledge

> which consists indeed only of phenomena as its materials, but in which phenomena are cemented by . . . successive modes of connected representation, and form a series of laws each supporting other laws more general; so that we obtain, combined with all the physical reality of the external world, all the permanence and symmetry of mathematical relations. That this exhibition of mere phenomena, and of pure ideal conceptions, as in harmony with each other, does constitute the supreme success of the progress of scientific inquiry, we believe will be most clear to the students of these sciences which have been carried to the furthest limit of generality and simplification; as for instance, mechanics and physical astronomy.[39]

4. Whewell's Problem of Induction 1830–1834

From 1830 onwards, Whewell turned in earnest to explore the philosophical ramifications of his theory of excellent science. His efforts were directed in the main to two problems: that of determining the nature of the truth of the propositions of a mature science thus construed, and that of articulating the process by which they took form. The first of the two—the problem of scientific truth—was resolved by late 1833, and was published in the spring of 1834 in his impressive 'The Nature of the Truth of the Laws of Motion'. His solution was original, and followed directly from his newly acquired account of mathematical physics. The propositions of a well-formed science, he argued, play a double role. On the one hand, they partake in a 'pure' conceptual scheme, and as such may be found *true* of the Fundamental Idea displayed by the system. On the other hand, and in no obvious way conditional on the first, the very same propositions 'colligate' phenomena, and as such may be formed true *of the facts*. Thus a true scientific proposition is at one and the same time, according to Whewell, both 'necessarily' true of a Fundamental Idea and contingently true of empirical reality. Whewell's theory of truth, as we shall see, has its problems, but they are not the problems commonly ascribed to it. For one thing, his notion of necessity is not, as Butts and others have interpreted it, Leibnizian. Whewellian necessities are *not* true in all possible worlds. Rather, they are inevitably to be adopted *if* the Fundamental Idea of which they are found to be true is recruited as the basis for the conceptual interpretative scheme of the science in question. But all this will be dealt with in detail in Chapter 11.

The crux of Whewell's difficulties with the account of the formation of modern physics implied by Herschel and preached enthusiastically by William Rowan Hamilton, with whom he debated the matter during the early 1830s, was his reluctance to endorse the claim that such science formed miraculously by the chance union of two initially unconnected efforts. Whewell's entire

[39] Whewell (1831*a*), 380.

induction project, one may say, was an attempt to explain in detail how an antithetically structured mathematical science, such as mechanics, was formed in the course of an essentially *unitary* effort to investigate the phenomena. Not only did he require, as a Baconian–Lagrangian view of science demanded, that in its finished form such science would exhibit an isomorphism between its ideal and empirical components, but that such would be the case with respect to each and every step in its development. An adequate theory of induction had to show how both the conceptual and empirical components of the *single physical law* emerged in the course of its discovery. In 1831 he had no such theory. Neither did Herschel, who chose to devote the *Discourse* to the formation of but the empirical component of scientific knowledge. And although in reviewing the book (whether for lack, as of yet, of a better alternative, or because of a reluctance to launder dirty linen publicly) Whewell refrained altogether from criticizing its shortcomings, one can still sense where he ran into problems. In the opening sections of the review Whewell declares that in view of the *Discourse* being 'indeed an admirable comment upon the *Novum Organum Scientarum*', it would be useful, before continuing to explain to the reader 'in the first place, the peculiar nature of modern science, and of that Inductive or Baconian method to which its marvels are attributed'.[40] His description in the review of the nature of modern science, as we've seen, was masterly, but of induction he had yet nothing to say. He continues as though all was well, but, in fact, all he had to say about the process of induction is not what it consists in, but only what it is supposed to *yield*. 'I shall employ the term "induction" ' he writes, 'to designate . . . the process of considering a class of phenomena . . . as represented by a general law, or a single conception of the mind.' However, he goes on to explain not the nature of the inductive process, but the nature of what it produces, namely,

> Such a single law . . . capable of including so many phenomena, in virtue of the natural connexions of our thoughts; and [which] will derive its validity from its being a connexion representation of the phenomena, and nothing more.[41] In this manner Induction agrees with mere Observation in accumulating facts, and with pure Reason in stating general propositions; but she does *more* than Observation, inasmuch as she not only collects facts, but catches some connexion or relation among them; and *less* than pure Reason . . . because she only declares that there *are* connected properites, without asserting that they *must* exist of necessity and in all cases.[42]

But of the inductive method itself—not a word. Of course, in its being so wholly positive, the review *implied* that Herschel's restatement of Bacon's original methodology was sufficient to explain the formation of mathematical scientific knowledge, which it certainly was not. Whewell must have had second thoughts about it, as his letter six months later to Richard Jones clearly indicates: 'I cannot say that I much like the review of Herschel now that I look

[40] Ibid. 378.
[41] Only by 1834 would *both* aspects of the nature of scientific truth be formulated.
[42] Ibid. 379.

at it in cold blood, and I have a strong persuasion that all the philosophical part will repel most and puzzle the rest, *except it be seized upon by somebody who resolves to prove it wrong.*[43]

The *Discourse* was certainly material in prompting Whewell's philosophical quest, but it did so primarily not for what it claimed, so much as for what it lacked. Not only did it fail to offer a satisfactory account of the formation of scientific knowledge as portrayed in Whewell's review, but in dealing solely with the empirical component of such knowledge it fell short of addressing what for Whewell was the main problem afloat. Herschel, hence, would never be regarded by Whewell as a philosophical *opponent.* Which explains why, prior to the full formation of his views, Whewell refrained from confiding in Herschel on philosophical matters (as he did, for example, in Jones).

In William Rowan Hamilton, on the other hand, Whewell found his match. The two men first met in 1832 during the young Irish prodigy's first tour of England. Hamilton was obviously impressed, writing back of Whewell that 'I thought with delight that I perceived a philosophical spirit more deep and true than I had dared hope for.'[44] Whewell too was equally taken by Hamilton and the two kept up a lively correspondence throughout the crucial years in which both men's major works were in the making. The story of their mutual impact is fascinating and certainly merits separate publication. Elsewhere, I've attempted to tell Whewell's side of the story,[45] but here space allows only a brief mention.

Unlike Herschel, in fact unlike anybody else in the English-speaking world, Hamilton spoke explicitly of modern mathematical physics as comprising two essentially different sciences, whose combination was, in the last analysis, an unexplained miracle. Also, as a keen follower of Coleridge, Hamilton was free of traditional Baconian commitment. He was outspokenly Kantian (though, due to Coleridge, initially influenced more by Kant's aesthetics and theory of the sublime than by the epistemology of the first *Critique*),[46] and hence shared Whewell's (by then well-formed) non-Lagrangian view of mathematics.

With regard to the antithetical structure of finished mathematical sciences, Whewell found in Hamilton a true ally, but a keen opponent to his unitarian and empiricalist view of the manner in which such science took form. Take for example Hamilton's letter to Whewell of 25 May 1833. He wrote

[43] Whewell to Jones, 15 July 1831, Todhunter (1876), ii. 123.

[44] Hamilton to Aubrey De Vere, 7 May 1832, cited in Graves (1975), i. 554.

[45] See my forthcoming *William Whewell: Philosopher of Science*, sect. 2, p. 331, and Fisch (1986), 113 ff.

[46] In his earlier writings—his introductory lectures on astronomy in particular—science is likened by Hamilton to art, and its truth viewed by him as pertaining to man's sense of beauty. In a letter of 27 June 1834 to the mathematician H. F. C. Logan, Hamilton admits to having not yet completed the reading of the 'Pure Reason' (Graves (1975), ii. 87). Three weeks later in a letter to Viscount Adare he reports that he has still not finished the *Critique* but has read enough to say, 'I long to compare Kant's views of physical science with my own.' Ibid. ii. 97. Incidentally, the idea of algebra as the science of pure time was embarked on by Hamilton only from the spring of 1834: see ibid. ii. 138 ff.

The opinion which I think I tried to express when I had the pleasure of talking the thing over with you was in substance this: that there are, or may be imagined, *two dynamical sciences*: one subjective *a priori*, metaphysical, deducible from our meditation on our ideas of Power, Space, Time; the other objective, *a posteriori*, physical, discoverable by observation and generalization of facts or phenomena . . .

For example [he continues] it is, I think, a subjective, *a priori*, metaphysical theorem, that when we think of a point as moving in a curve, we must think of it as changing direction and must consider this change as an effect, and must attribute the effect to a cause; which may firstly be called a deflecting force or power. And it is, I think, another theorem, objective, *a posteriori*, physical, discovered by observation and generalization of facts that when a body is found to move in a curve, we may expect to find some other body, resisting or attracting, or somehow physically influencing the former.

Hamilton however left no room for the possibility that the two, and in his view quite different, components of Newton's second law evolved, as Whewell would have it, in the course of one unified (not to say empirical) effort.

The metaphysical theorem I account indeed higher in dignity; but do not consider it as including the other, or as adequate ground to use for the expectation of any one appearance, though in the Divine Mind, indeed there may be some mysterious union between the causes or first springs of our thoughts and our sensations; of the ideas which seem necessary and eternal, and of the phenomena which seem casual and changing.[47]

Not only did Hamilton claim that the two sciences evolved separately, but, he wrote playfully two years earlier, they might well have developed in altogether different directions: 'The world which Newton *constructed* [*sic*] was like the outward world, but had it not been so, he might still have chosen to contemplate it.'[48]

Whewell willingly accepted Hamilton's 'tale of two sciences' as a retrospective description of the finished product of the scientific process. More so, he endorsed, as already mentioned, the twofold theory of scientific truth that it implied. But unlike Hamilton, Whewell was still a devout Baconian, and as such remained committed to Herschel's Baconian account of the manner in which the physical and experiential of the two sciences evolved. More importantly, as a Baconian, he refused to view the investigation of the physical process involving anything more than a single and directed effort to negotiate the facts. In other words his problem, at that stage, was to explain, contrary to Hamilton's account, how, *in the course of Baconian induction*, did the conceptual of the two sciences evolve side by side with the physical, given that once reaching completion it would resemble a fully fledged, pure Euclidian system of ideas?

In a notebook entry entitled: 'Types of the Progress of Science', dated 22 July 1831[49] (and hence, note, written after his review of Herschel), Whewell schematically outlined, in table form, his view of the inductive method:

[47] Ibid. ii. 48 f.

[48] 'Extracts from an Introductory Lecture on Astronomy Delivered in 1831', ibid. i. 503.

[49] Whewell Papers R18 17^{15}, p. 39(*b*).

Sciences of Observation	*of Experiments*
1. Common observation and *Collection of instances* (Natural objects and regular occurrences).	1. Common observation and *Collection of instances* (occasional occurrences).
2. *Decomposition of Phenomena* by perception of analogies.	2. *Decomposition of Phenomena* by perception of simple connexions.
3. Classification and *nomenclature*.	3. Insulation of facts and *Terminology*.
4. *Systematic observation and* technical description of facts.	4. *Systematic experiment* and measure-(ment) of insulated facts.
5. Induction 1. *Propositions concerning classes*.	5. Induction 1. *Laws of phenomena*.
6. Induction 2. *Causes of Laws*.	6. Induction 2. *Causes of Laws*.

The scheme was orthodox Bacon. Even the basic differentiation between sciences of observation and those of experiment owes its origins to Bacon,[50] as does the disproportion between the tasks of decomposition, analysis, and classification of phenomena (stages 1–4), and those of inductive ascent itself (stages 5–6).[51] The main point conveyed by the scheme is that scientific truths are disclosed by prudent inspection of the facts. The meaning as it were, is in the text. To induce a law boils down to *noticing* it in the course of inspecting a well-dissected and carefully tabled corpus of fact. If instances are properly collected, decomposed, named, and measured, to discover a law is essentially to stumble upon it.

In other words, in the course of discovery stages 5 and 6 involve the introduction of nothing really new.

That being the case, Whewell was faced with the problem of explaining how the outcomes of Baconian induction, namely the induced laws of phenomena and causes, acquired their dual and antithetical characteristics. How and when, he now had to show, was the conceptual component of a law introduced in the course of its discovery?

In keeping with the Baconian schema of 1831, Whewell's early attempts to locate the emergence of the conceptual component of scientific knowledge naturally concentrated on the stages of induction preparatory to discovery, namely stages 1–4. The direction his views were taking at the time was indicated in the review of the *Discourse* in a passage uncharacteristically critical of both Bacon and Herschel,

> there is one step in [the] formation [of science] of which we must say a word . . . it is one very necessary and important process in the inductive ascent, and has yet been entirely omitted by Bacon in his survey . . .

[50] See Bacon's *Novum Organum* (Spedding, Ellis, and Heath edn.), iv. 81. See also Jardine (1974), 136 f.

[51] Of the 36 sages of Salomon's House only three 'Interpreters of nature' were assigned the task of 'raising the former discoveries by experiments into greater observations, axioms and aphorisms' (*New Atlantis*, *Works*, iii. 165). See also Hesse (1964) and Jardine (1974), 147.

The step of which we speak is the marking and fixing of our generalizations by imposing exact *terms*—a process which may be included in the word terminology . . .[52]

It was in the coining of a precise nomenclature that Whewell first attempted to locate the introduction of the conceptual component of scientific knowledge.

These attempts reached their most complete form in a series of draft essays written in 1832 or 1833 under the collective title: 'The Philosophy of the Progressive Sciences'.[53] All knowledge, even factual knowledge, he observed in a section entitled 'Of the Steps of which Knowledge Consists', requires to be cast in *language*:

our knowledge begins with observation of the external world, and is *retained* by the aid of language. Observation necessarily supposes the powers of perception and *the language* by which its results are *retained and conveyed* cannot exist without nouns and verbs. Hence the power of perception and the faculty of conceiving things *so that nouns and verbs are applicable to them*, are essential to our capacity of knowledge.[54]

Knowledge, note, begins with observation of the external world and is thereafter retained by words—the words being the mind's exclusive contribution. The mathematics employed in physics is hence, upon this view, but a highly refined and elaborate form of language. None the less, for all its anticipation of his later views, Whewell's position remained at this stage still firmly tethered to a Baconian frame of mind. The *content* of scientific knowledge, he still maintained, hinges wholly on the input of the senses, whereas the role of mind, in contributing its linguistic setting, confines itself to the capturing and the codification—in a word, to the *representation*—of that which is first found in the facts. Nomenclature, hence, adds nothing *new* in the process of induction, but precisely and efficiently encodes the old. Conceptualization is thus a rear-guard operation, which serves to make induced knowledge available at a glance. Thus for Whewell at this stage, the language of a science, its conceptual scheme, in its essentially referential role, corresponds to what Michel Foucault has termed (pre-Kantian) 'classical language',[55] and the process of its formation is rendered wholly *parasitic* on the Baconian inductive process by which its empirical referendum emerges.

As soon as a law of nature *has been obtained*, it is desirable to express it in some precise and brief phraseology. This once done, a word or a phrase enables us to *adduce* the law, and thus to render available, at a moment's warning, all the toil and ingenuity which have been *embodied* in its discovery. Thus the terms we have used, of *conduction*, *radiation*, *latent head*, *caloric*, are words *impregnated*, as it were, *with a meaning derived from* the labours of many observers and calculators.[56]

[52] Whewell (1831*a*) 390 f.
[53] Whewell Papers, R18 17[8].
[54] p. 14.
[55] Foucault (1973), 311 and *passim*.
[56] Whewell (1831*a*), 390. Only the words: 'conduction, radiation, latent heat, caloric' italicized in the original.

This view however was swiftly abandoned.[57] What, I believe, prompted Whewell to reconsider his position was a bulky study he embarked on in December 1833 (the twenty-third to be precise), entitled: 'Of Practical Skill and of Speculative Knowledge, of the Differences between the Pure and the Physical Sciences'.[58] In it he attempted to draw the line between the pure and the physical sciences. The general approach consisted in emphasizing the manner in which the pure sciences were formed by reasoning from the top down, as opposed to the physical sciences which supposedly formed, à la Bacon, by reasoning inductively from the bottom up. By 1833 he was comfortably anti-Lagrangian and in the essay laid down in quite general philosophical terms the basic claims that would characterize his philosophy of mathematics thereon. 'The truths of pure mathematics', he stated,

> depend on the modes of presentation under which we necessarily apprehend all objects and occurrences, namely space and time. These modes of presentation give rise to certain fundamental characters of figure and number, which characters may be expressed by means of definitions and axioms; and from these fundamental principles and their combinations are deduced trains of other truths by logical processes, that is by the necessary rules of the connexion of truths, excluding all other sources of knowledge than those asserted.[59]

Hence, as far as the *formation* of mathematical systems goes: 'We shall have definitions and axioms to begin with; and of these a series of propositions with demonstrations depending entirely upon the definitions and axioms and upon preceding propositions.'[60] However, he continued, physical science progresses quite differently, namely, by reasoning, not downwards from self-evident axioms, but upwards from facts. Much confusion prevails, Whewell went on to say, and 'men . . . have been led into error when they have attempted . . . to extend to other provinces of knowledge the methods and forms which had been successful in mathematical science' where they had 'begun by assuming certain principles as self-evident, and . . . gone on to reason from them. This method of proceeding', he concluded, 'has however never led to any certain or progressive knowledge.'[61]

On the surface, all this sounds very traditional. True, there is a Kantian ring to his description of space and time as 'modes of presentation' etc. which might

[57] It is interesting to note in this context that for their larger part Whewell's famed contributions to scientific nomenclature were made prior to 1834. That is to say, during the period in which he ascribed to terminology so central a role in the formation of science. I refer mainly to his (1826), (1827*a*), and (1827*b*) on crystallographical notation, and his (1828*a*) on mineralogical nomenclature, and, most important perhaps in this connection, his well-known correspondence with Lyell and Faraday regarding specific terminology in their respective fields. For Lyell, see Whewell to Lyell, 31 Jan, 19 and 22 Feb. 1831, Todhunter (1876), ii. 109–14, and Lyell to Whewell, 3 and 4 Dec. 1831, Whewell Papers, Add. MS a. 208^{114-15}. For Faraday, see Whewell to Faraday, 25 Apr. and 5 May 1834, Todhunter (1876) ii. 178–83.

[58] Contained in an unpublished notebook, Whewell Papers R18 17^{6}. In the title of ch. 3 of the work (p. 13) the word 'skill' is crossed out, and the title reads, 'Of Practical and Speculative Knowledge'.

[59] Ibid. 81. [60] Ibid. 83. [61] Ibid. 105.

have been Hamilton's doing.[62] (Significantly, Hamilton's very Kantian work on algebra as the science of pure time was published for the first time in 1834.) But, as we have seen, in Whewell's case, the motivation for viewing mathematical systems as meaningful articulations of such Fundamental Ideas, owed its origin to his dissent from Lagrangian calculus rather than to the influence of Kantian metaphysics. In any event there was nothing Kantian at this stage about his account of physical science.

However, what to my mind is significant about this particular work is the fact that Whewell refrained from publishing it. It would take till 1835 for him to publish anything on the nature of mathematics, and even then he would do so only indirectly apropos of a discussion of mathematics as part of a liberal education. His 'Remarks on Mathematical Reasoning and on the Logic of Induction' was published in 1837, but again rather timidly as an appendix to a very elementary textbook, the *Mechanical Euclid*. Only in the *Philosophy* would his views be made a central theme. Why?

The reason for Whewell's reluctance to publish the work in 1833, was, I propose, that rather than *clarify* the differences between pure and physical science as it purported to, his analysis of the former rendered the task of analysing the latter—as he had by now come to view it—all the more perplexing. For as the study showed, for a pure conceptual scheme to form meaningfully it was to unfold downwards from intuitively grasped axioms and definitions. What then of antithetical physical science? In its finished form, the conceptual scheme of a physical science was supposed to resemble a well-formed pure science. Which meant, on the showing of the 1833 study, that the two components of such a science would have had to take form in the course of reasoning in two diametrically opposed directions, geared, as his investigations of scientific truth conducted at that time now indicated, towards attaining quite different standards of certitude. Reluctant to divorce the two processes, as Hamilton had suggested (or to assimilate pure to empirical knowledge as J. S. Mill would suggest), Whewell had attempted to portray the emergence of the pure component of a physical science as a step-by-step construction of its peculiar terminology accompanying the process of inductive ascent. However, in the light of the 1833 study of pure systems, this added up to saying that the pure component of a physical science formed itself *backwards*, and hence in so far as it ended up being a well-formed system, it did so *by chance*! This of course Whewell could not accept.

By 1834 Whewell's philosophical pre-commitments jarred with each other to the extent of rendering his philosophical problem—that of explaining the emergence of mathematical physics—virtually insoluble. By the spring of

[62] See for example the extract from the diary of the American journalist George Ticknor quoted by Graves, describing how in the course of the 1835 Dublin meeting of the British Association for the Advancement of Science 'Whewell made himself merry at a discussion about Kant's philosophy, in which Hamilton showed his metaphysical acumen against a German at table . . . ' Graves (1975), ii. 157.

1834 his philosophical position comprised the following four incompatible theses:

(*a*) Physical science, in its finished form, combines two well-formed and isomorphous structures: a pure Euclidian conceptual scheme and an induced body of empirical knowledge.
(*b*) Empirical knowledge is acquired in the course of Baconian inductive ascent.
(*c*) Well-formed pure sciences evolve in the course of reasoning down deductively from axioms found true of Fundamental Ideas.
(*d*) Physical science is the outcome of a unitary and directed investigation of the world without.

There was no way Whewell could retain all four theses as they stood. Earlier he had attempted to modify (*c*) with respect to the pure component of a physical science, suggesting that it somehow managed to form in the course of an upward construction of a scientific terminology, but to no avail. Hamilton had dispensed with (*d*) stating bluntly that 'the interest of the *has been*, the *is* and the *will be*, differs from that of the *may* and the *might*; and both these interests are combined in physical science in its perfection'.[63] Orthodox empiricism dispensed with (*a*). But Whewell could dispense with neither. And thus committed to remain an antitheticalist with respect to the end-product of scientific investigation, and a unitarian empiricalist with respect to the process by which it is achieved, Whewell was left no alternative but to lay aside the firmest of his initial philosophical commitments, namely thesis (*b*): Bacon's account of the scientific method.

And so, paradoxically, Whewell's antitheticalist view of finished science, and his empiricalist attitude to the process by which it formed, which both owed their origin to his early attempts to reconcile the Lagrangian and Baconian aspects of the Society's original programme, became eventually his main motive for abandoning both his initial commitments.

5. Whewell's Solution Phase I: Induction as Superinduction

When exactly Whewell's novel theory of induction was hit on is difficult to say. The stark a-philosophical setting and wholly retrospective presentation of 'The Nature of the Truth of the Laws of Motion' suggests that by the beginning of 1834 he had not yet reached a satisfactory account of induction. On the other hand, the *History* published in 1837, was written with his theory of induction (though not much more) very much in hand. As his letters to Jones indicate, the *History* was seriously embarked on during the summer of 1834. So far I have come across no datable notebook entries that pinpoint his

[63] From an introductory lecture on astronomy cited by Graves, ibid. i. 503. See also Hankins (1980), 384.

discovery.[64] But judging by the excited and enthusiastic tone of his letters on philosophy to Jones from August and October 1834,[65] I conjecture that it was sometime during the summer of 1834 that the discovery was made.

Whewell's theory of induction constituted the first and crucial step towards a full account of the emergence of mathematical physics. With respect to the formation of the conceptual component of scientific knowledge, it was the exact reversal of his earlier unsuccessful attempts. According to the Baconian schema of 1831 (supplemented by the draft essays on scientific language) the conceptual constituent of an induced law of nature was viewed as the upshot of a codification of knowledge formerly gleaned off the facts. His new and original theory of induction reversed the process thus rendering the importance of nomenclature as such secondary. According to his new theory, the central role that the invention of a new concept came to play, was not in capturing what had already been found in the facts, but in reading something *new* into them. The invention of a term or concept, Whewell came to realize, was not merely the invention of a catchy label for a discovery already made, it was discovery itself!

> in discovering the law of Mars' motion and in asserting that the planet moved in an ellipse, [Kepler] did this;—he bound together particular observations of separate places of Mars by . . . the *conception* of an *ellipse*, which was supplied by his own mind . . . Before this, the facts are seen as detached, separate, lawless; afterwards, they are seen as connected . . . as parts of one general fact, and thereby possessing innumerable new relations before unseen.[66]

'In collecting such propositions from the facts', he wrote earlier,

> there occurs a step corresponding to the term 'induction' . . . Some notion is *superinduced* upon the observed facts. In each inductive process, there is some general idea introduced, which is given not by the phenomena, but by the mind. The conclusion is not contained in the premises, but includes them by the introduction of a new generality.[67]

As Richard Yeo has pointed out,[68] Whewell's new theory of induction differed importantly from the Method of Hypothesis urged by several of his Baconian contemporaries. The type of hypothetico-deductive reasoning (to risk the anachronism of latter-day terminology) prescribed by Dugald Stewart, Thomas Brown, David Hartley, David Brewster, Auguste Comte, and John Herschel was a different thing entirely. For the Baconian who grudgingly allowed for hypothesizing, the hypothetical anticipation of a law was in essence a heuristic for the discovery of what could in principle be found *in* the facts. As

[64] I use 'discovery' in a Whewellian sense, since from then on, in keeping with his new sense of the term, the vast body of historical fact he compiled for his Bridgewater Treatise (1833*c*) and *First Principles of Mechanics* (1832*c*), now came to be seen in an entirely new light.

[65] Whewell to Jones 21 Aug. and 6 Oct. 1834, Todhunter (1876), ii. 185–8, 192–4.

[66] Whewell (1849), 28–30, later to be republished as ch. 22 of *The Philosophy of Discovery* (1860), Cf. pp. 247–8, 253 f.

[67] Whewell (1847), ii. 606–17.

[68] Yeo (1985), 268–70.

Comte would put it: 'One should, in order *to accelerate these discoveries*, construct directly and without scruples the most plausible hypothesis'. For the Baconian, hypothesizing was a short-cut, a premature guess at what careful empirical research could otherwise disclose. But for Whewell, the invention of a concept *creates* the law, not merely anticipates it. Whewell's induction, and central to it the indispensable role of mind, pertained to a new and different conception of knowledge from that of his Baconian predecessors and contemporaries. But to this he would awaken only gradually.

6. Phase II: The Authentication of the Ideal

6.1 *Whewell's Methodology of Pre-discovery*

To assess Whewell's novel move and the new problems to which it gave rise, it is not enough to read it in relation to the views of his opponents. If my analysis of Whewell's philosophical problem situation has so far been correct, his new theory of induction was designed first and foremost to compensate for the failure of his former Baconian attempts to explain the formation of the pure conceptual component of a science. To recall, their failure lay in the fact that for all the ingenuity they allowed in the construction of a scientific nomenclature, they rendered the formation of the conceptual scheme of a science (in the light of Whewell's study of pure science) wholly subordinate to empirical discovery, and hence as having developed backwards. His new theory of induction indeed liberated concept formation from pure empirical research, and thus constituted an important step towards the solution of his problem. None the less, much yet remained unexplained.

What Whewell had now to explain was what initially *authenticated* the search for an appropriate concept? Now that concept-forming had been divorced from the examination of facts, how was a Whewellian investigator, contemplating a disordered body of fact, to know in advance what kind of concept to look for? Blind trial and error was out of the question since the concept sought was also required to constitute a contrived and meaningful downward development of a pure deductive scheme of ideas. But in order for the conceptual scheme of science to be cultivated meaningfully and purposefully the axioms of the system had somehow to be known in advance. By reversing the processes of empirical discovery and concept forming Whewell had certainly made progress. The formation of the conceptual scheme of a science was no longer subordinate to empirical discovery. Yet it still formed backwards—that is, upwards towards axioms rather than downwards from axioms. In this respect his new theory was not yet an improvement upon the Baconian scheme of 1831.

The problem Whewell now faced was hence twofold; namely, to explain what initially authenticated the search for a concept (*a*) as a potential colligator of the facts in question, and (*b*) as a meaningful development in the downward unfolding of the conceptual scheme. As it stood Whewellian induction left both difficulties unresolved. Cast in the historiographical terminology of the *History*, Whewell's induction explained, one could say, what made for an inductive epoch; the question now was to explain what made for a meaningful prelude to an epoch.

Whewell's solution to the problem came in two stages; the first derived directly from his present philosophical position, the second took him way beyond it, and as close as he would get to Kant's theory of knowledge. However the *History* was embarked on and largely completed *before* any of these issues were faced. As a result, although conceived by Whewell to test his philosophical position 'that by means of it I shall be able to stop the mouths of all gainsayers of my philosophy',[69] the historiographical apparatus of the *History* in reality incorporated no more than his theory of induction; which his account of the various inductive epochs indeed demonstrated forcefully. But there is no trace of his later transcendental methodology of pre-discovery in the chapters that tell the tale of preludes to inductive epochs, and none of his criteria of theory appraisal in those addressed to the sequels.[70] From the point of view of the *Philosophy*, the *History* was in important respects premature. Unfortunately, later editions never put this right. The *History* was never rewritten in the light of the *Philosophy*, but only supplemented once and again by new up-to-date material. But to return to the problem of concept forming.

What directed the search for a concept at any stage of an inductive investigation, argued Whewell, was that portion of the conceptual scheme already in operation. For by the very nature of Whewellian induction, the facts under consideration at any stage are themeselves the outcome of former inductions. In this way the facts under consideration at any stage embody the *inductive history* of the science in question. Whewell hence conceived of a stage preceding induction, one preparatory to discovery, in which the facts in question were to be 'decomposed' repeatedly as far as possible into their conceptual and factual components. In this manner, what one might term the 'conceptual history' of the science was to be exposed as far as possible—at least that portion of it to which the facts under consideration pertain directly. Whewell maintained that this portion of the science's conceptual scheme, when rendered explicit, provides the investigator with both the setting and the idiom within which his discovery is to be made; that is, with the conceptual backdrop against which his conjectured concepts are to be 'explicated'. For a concept to be conceptually 'appropriate' to a body of fact, it must belong, he urged, within the system of ideas to which the facts already pertain.

[69] Whewell to Jones 6 Oct. 1834, Todhunter (1876), ii. 193.

[70] Consilience and simplification are not discussed as 'the logic of induction' as he would have it in the *Philosophy*, although the story told in the *History* undoubtedly demonstrates his later analysis. In other words, unlike induction proper, his theory of appraisal was not part of the historiographical apparatus of the *History*, but rather a lesson he would learn from it.

Although, as we said, we can give few precise directions for the cardinal process, the selection of the Idea, in speculating on phenomena, yet there is one Rule which may have its use: it is this:- *The Idea and the facts must be homogeneous*: the elementary Conceptions, into which the facts have been decomposed, must be of the same nature as the Idea by which we attempt to collect them into laws. Thus, if facts have been observed and measured by reference to space, they must be bound together by the idea of space; if we would obtain a knowledge of mechanical forces in the solar system, we must observe mechanical phenomena.[71]

Once the 'explicative' conceptual context is exposed by decomposition, discovery itself, Whewell argued, remained a matter of trial and error, though no longer blind.

6.2 *Transcendental Deduction*

The 'decomposition of facts' and the 'explication of conceptions' still fell short of solving the problem of pre-determined concept candidacy within a Whewellian framework. For if all one has to go by is that part of the conceptual scheme inherent in the facts—the part disclosable by repeated decomposition—suggestions of new colligating concepts can still not be considered meaningful steps in the formation of a pure system such as the conceptual scheme was supposed to form. That is because repeated decomposition of the facts will take one back down previous inductive ascent, and will therefore reveal deductive chains of propositions leading *up* to the present stage (in relation to which the sought concept will enable the construction of a proposition situated even higher *en route* to the axioms). In other words, if all that is known prior to conceptual innovation is that portion of the conceptual scheme already incorporated in the facts, the pure system of the science would still be forming backwards (that is, upwards). For the process of induction to resemble also the formation of a pure science, Whewell somehow had to show that in the course of negotiating the facts and contemplating the conceptual history of the science, certain of the *axioms* of the system also came into view. So that side by side with the upward inductive process of repeated colligation the formation of the concepts could also be shown to participate in a downward and deductive process of reasoning *from* axioms.

The move that in conjunction with the 'decomposition of facts' and the 'explication of conceptions' opened the way for a satisfactory methodology of pre-discovery, was Whewell's adoption of a distinctly Kantian type of transcendental deduction. Although he never described it as either Kantian or transcendental, the move, which I conjecture was made sometime between the summer of 1837 and that of 1839,[72] is unmistakable. Crucial axioms of the

[71] Whewell (1858), 193.

[72] So far I have found no explicit notebook entry to either confirm or refute my conjecture that Whewell's methodology of pre-discovery was arrived at only after the completion of the *History*. There exists a draft essay on the idea of cause which is almost identical to his discussion of cause in the *Philosophy* (Whewell Papers, R18 17[9], pp. 2–58) and even refers to Kant as the 'metaphysical

system of ideas, he urged, were in fact disclosed in the course of empirical investigation, without them actually having been learned from experience—that is, without them being, in Whewellian terms, the actual outcomes of colligations of the facts in hand. In the course of inductive research, he pointed out, investigators gradually become aware of the fact that general laws, such as 'every effect has a cause', '*ex nihilo nil fit*', and 'all matter is heavy', are *presupposed* in the course of their work. Such laws, he urged,

> are not the results of experience, nor can be. No experience can prove them; they are necessarily assumed as the interpretation of experience. They were not proved in the course of scientific investigations, but brought to light as such investigations showed their necessity. They are not the results, but the conditions of experimental sciences.[73]

The adoption of such transcendental reasoning thus enabled Whewell to solve to his satisfaction both the problem authenticating concept formation, and that of presenting induction itself as a meaningful step in the formation of the pure element of antithetical science, over and above its empirical colligatory function. Moreover, in stressing the fact that axioms of the system came into view in the course of empirical research, he was able to emphasize further the unitary, and more so the essentially empirical nature of scientific investigation.

7. The Fundamental Antithesis

Transcendental reasoning enabled Whewell to preserve the working ethos of an empirical researcher without having to concede to an empiricist epistemology or methodology. The inclusion of transcendental deduction as part of his methodology of pre-discovery thus finally rendered the process of induction both meaningful, deliberate, and purposeful with respect to *both* antithetical components of well-formed science. He had solved his problem. But as much as he might have liked to, he could not stop here, for by exhibiting so vividly

genius in Germany' (p. 10). Unfortunately the notebook is undated, and carries nothing which might hint at its date. My case rests therefore on the total lack of any such theory in his discussion of the preludes in the *History*, which are described vaguely, devoid of 'decomposition', 'explication', or transcendental reasoning, as periods 'during which the ideas and facts on which they were called into action;—were gradually evolved into clearness and connexion, permanency and certainty; till at last the discovery which marks the Epoch, seized and fixed forever the truth which had till then been obscurely and doubtfully discerned.' (1857), i. 10. Indeed, he wrote to his sister Ann in November 1837, 'My old complaint of being overwhelmed with business especially at this time of the year, is at present . . . more severe than ever. For besides my usual employments . . . I am printing a book which *I have not yet written*.' 18 Nov. 1837, cited in Stair-Douglas (1881), 189, my italics.)

[73] Whewell (1860), 349. No amount of empirical evidence can prove that all matter is heavy, none the less, it can be inferred from the fact that 'practically speaking, quantity of matter is always measured by weight, both by mechanicians and chemists: and as we have proved that this procedure is utterly insecure in all cases, on the hypothesis of weightless matter, the practice rests upon a conviction that the hypothesis is false. And yet the production is universal. Every experimenter measures quantity of matter by the balance'. (Whewell (1847), ii.629.

the a priori nature of its conclusions, transcendental deduction left Whewell no choice but to climb down reluctantly from his perch on the methodologist's fence, and face up to the epistemological consequences of his position.[74]

Whewell undoubtedly did so grudgingly, and in a letter to Herschel announcing the completion of the *Philosophy* he sounds almost apologetic:

> I had no rest till I had thought it all out, and am right glad to have it off my hands . . . My argument is all in a single sentence. You *must* adopt such a view of the nature of scientific truth as makes universal and necessary propositions possible; for it appears that there are such, not only in arithmetic and geometry, but in mechanics, physics and other things. I know no solution to this difficulty except by assuming *a priori* grounds; but I am most willing to look at any proposed solution.[75]

But as before, regardless of prior commitments (though by no means taking them lightly), Whewell rose to the problematic challenge of the position he had come to adopt.

Prior to the introduction of transcendental reasoning, Whewell need not have assumed more than that by nature scientific knowledge was structured by concepts, and could have avoided altogether, as did so many Baconians, the epistemological question of the fundamental sources of knowledge. As late as 1837, as for example in the introduction to the *History*, Whewell can still be read as maintaining that the mind's contribution to the acquisition of knowledge consists only in the organization of purely factual knowledge, which can, in turn, be talked of independently of mind—as though his theory of induction applied only to the acquisition of 'high' scientific knowledge. Thus he writes in the introduction of 1837: 'The impressions of sense unconnected by some rational and speculative principle, can only end in a practical acquaintance with individual objects.'[76] And if science is understood as a process by which such 'practical', mind-independent 'acquaintances' are only organized by mind, a case for an essentially empiricist epistemology could still be made. Not that Whewell assented to such a position in 1837, but he still seems to have been toying with the (largely Scottish Common Sense) Baconian myth of an epistemologically free theory of scientific method. With the endorsement of transcendental reasoning the myth was exploded. Even if purely sensual knowledge were possible, transcendental deduction demonstrated the fact that certain laws concerning 'individual objects' were as a matter of fact presupposed *in*, but not learned *from*, experience. The mind therefore not only organized such 'acquaintances', its contribution, in the event, was what made them possible. Whewell was hence left no choice but to transform his theory of the antithetical structure of scientific knowledge into a theory of the *sources* of all knowledge. Whewell was in deeper waters than he had bargained for. For one thing, he could no longer avoid discussing the epistemological status of the

[74] For a discussion of the manner in which epistemology was side-stepped by late 18th- and early 19th-century Baconians, see my (1986), 204 ff.

[75] Whewell to Herschel, 22 April 1841, Todhunter (1876), ii. 298.

[76] Whewell (1857), i. 5.

axioms of a science, as he did in the 1834 analysis of the laws of motion. 'The part I am now revising', he wrote to Jones in July 1839,

> alarms me sorely by its abstract and almost mathematical aspect; and, what is worse, I can neither do without this part nor make it more popular. It is about the metaphysical principles of mechanics and astronomy, and a number of the more weighty questions turn upon it . . . in truth it puts me out of heart about the reception of the book.[77]

Whewell's epistemology, 'The Fundamental Antithesis of Philosophy', was fully formulated by 1840, although it received its name only in 1844. It was a theory of the fundamental sources, and hence of the structure of knowledge.

> In every act of my knowledge, there must be concerned the things whereof I know, and thoughts of me who knows: I must both passively receive or have received impressions and I must actively combine them and reason on them. No apprehension of things is purely ideal: no experience of external things is purely sensational. If they be conceived as *things* the mind must have been awoke to the conviction of things by sensation: if they be *conceived* as things, the expressions of the senses must have been bound together by conceptions. If we *think* of any *thing*, we must recognize the existence both of thoughts and of things. *The fundamental antithesis of philosophy is an antithesis of inseparable elements.*[78]

The Fundamental Antithesis thus constituted the final step in Whewell's philosophical quest; and, to use Whewell's own terminology, one could say it furnishes the Fundamental Idea governing his mature philosophical system. With the formulation of the Antithesis Whewell takes his place with the great epistemologists. It parts early with traditional empiricism, but it also goes an important step beyond Kant, for although the role he assigned to mind in the acquisition of knowledge is similar to that preached by Kant, Whewell's stock of Fundamental Ideas is essentially open. Mind is viewed not as manipulating a prefixed and bounded repertoire of forms of intuition and categories, but as inventively creating interpretative conceptual schemes as the case arises.

Yet Whewell's theory of knowledge was not born of epistemological controversy. In a sense its birth was made possible by his long ignoring of such controversy. It was born, I have argued, of an ongoing effort, spanning the best of twenty-five years, to come to terms, first didactically and eventually philosophically, with the modern physical sciences he admired so much. Whewellian antithetical knowledge has its roots not in a wish to reconcile Locke and Kant, but in the problematic, powerful, yet rash grafting of continental analysis upon traditional British natural philosophy as promoted by the Analytical Society. And both Whewell's importance and possible impact owe as much to the problems he posed and the manner in which he faced up to them, as to the actual solutions he eventually came to adopt.

It is fitting perhaps to conclude this sketch of Whewell's philosophical coming of age with the words with which the young Charles Saunders Peirce

[77] Whewell to Jones, 14 July 1839, Todhunter (1876), ii. 218.
[78] Whewell (1847), ii. 651.

chose to end his own 1869 Whewell lecture, for they apply equally well to Whewell as they do to the manner in which we have told his story. 'Facts involve IDEAS', Whewell had claimed, to which Peirce added:

> Hence *observation* as distinct from mere *gazing* consists in perception in the light of a question. (Enlarge on this—Whewell has not remarked this. Why(?))[79]

[79] Peirce (1869), 345.

3. *Between Rationalism and Romanticism: Whewell's Historiography of the Inductive Sciences*

by GEOFFREY N. CANTOR

1. The Roles for History

WRITING to Richard Jones in the summer of 1834, Whewell outlined his planned series of three books, the first of which would deal with the '*History* of Inductive Science . . . historiographized in a new and philosophical manner'.[1] Three years later this first stage of Whewell's project reached fruition with the publication of three volumes entitled *History of the Inductive Sciences* (1837*a*). This work ranks among the most impressive nineteenth-century contributions to the genre, containing not just the history of one science, but of a range of sciences stretching from astronomy to mineralogy and covering the period from the ancients to Whewell's own day. However, the significance of the *History* for the history of science does not lie either in its scope or in Whewell's interpretation of specific events, but in the 'new and philosophical' historiography he brought to the subject.

In order to assess the significance and novelty of Whewell's *History* we must examine the development of his views about the nature of history and the reasons why in the mid-1830s he turned his attention to the history of science. The problems he confronted at that period, and his deployment of history in the resolution of those problems, are the subject of this essay.

There is little evidence that Whewell was interested in the history of science prior to the mid-1830s and his few earlier excursions into the subject indicate that he only developed his more mature historiography at that time. One of his earliest references to history appeared in his 1831 review of Herschel's *A*

I should like to thank Menachem Fisch, Jonathan Hodge, Don McNally, and Simon Schaffer for their most helpful comments on an earlier draft of this paper. I am also greatly indebted to the Master and Fellows of Trinity College, Cambridge, for permission to cite from the Whewell Papers.

[1] Todhunter (1876), i. 90. The second book was to be a philosophy of the inductive sciences, while the third book, which was never written, would have investigated the extension of the inductive method to such areas as 'philology, art, politics and morals'.

Preliminary Discourse on the Study of Natural Philosophy, where in a footnote he briefly praised histories of science on the sole ground that they are useful to the practising scientist since they contain summaries of recent developments in a particular field.[2]

A year later the history of science played a more prominent role in Whewell's *First Principles of Mechanics* which was written to provide students with the basic theory of mechanics set out in a 'logical manner'. In this work Whewell was reacting against the established didactic tradition in which mechanics was portrayed as a deductive system, founded on supposedly self-evident premisses, totally independent of empirical evidence. Although deduction was obviously crucial for problem solving, Whewell insisted that mechanics is an inductive science and that students should appreciate its dependence on experiment. He considered that the best method to achieve this end was to provide a brief history of each of the major principles of mechanics so that students would be aware of the empirical basis of the subject and also of the path, strewn with errors, leading to the present highly ordered state of the science.[3] Probably for the first time, Whewell had to confront the difficult question of explaining the current state of a science in historical terms. This was to be one of the central problems he later engaged in writing the *History*; however, the resources he deployed to solve this problem in 1832 bore little relation to those employed five years later.

In the earlier work Whewell adopted as his model the conjectural historiography of Adam Smith's 'History of Astronomy', although shorn of the psychological factors that feature so prominently in Smith's account.[4] Instead, Whewell concentrated on compiling several speculative histories to show how the principles of mechanics, such as the first law of motion and the equilibrium of the lever, would have emerged from man's early experiences of the physical world. For example, the third law of motion grew out of early observations of nails being hammered. There were clearly some significant stages in this conceptual development, such as the rise of artisans, but much of this history was the history of mankind, divorced from names and dates. Aristotle, who established the theory of the lever on 'clear and satisfactory principles',[5] was the first writer named by Whewell, while Stevin and Galileo were the main characters among the moderns. In these conjectural histories Whewell followed Smith's view that man has a natural propensity to generalize from his empirical experience. It is also significant that Whewell rejected the view that the history of science consists of successful inductions; instead he pointed out that some

[2] Whewell (1831*a*), 375–6.

[3] Whewell (1832*c*), pp. iii–v.

[4] A. Smith (1872).

[5] Whewell (1832*c*), 20. A similar conjectural history is to be found in a contemporary manuscript entitled 'Of Practical and Speculative Knowledge' (Whewell Papers, Trinity College Library, Cambridge University, R18 17[6], fos. 33–41), where Whewell argues that man initially developed the useful arts and only later began to frame more abstract speculative views about the world.

earlier theories were found to be erroneous and had to be rejected and replaced by better theories. Excepting these general points, Whewell offered no insights into the processes of science.

Whewell's manuscripts of the period further confirm this view. In the early 1830s he filled several notebooks with reading notes and the first draft of a work on induction. As he stated in June 1830 his aim was to examine critically the inductive philosophy in order to clarify its nature and to show how it had been utilized by scientists. Although this project involved him in reading numerous primary sources, his aim was to contribute to the theory of induction, not to the history of science.[6] Excepting a few brief notes on various forms of history,[7] the first occasion on which Whewell acknowledged his intention of writing a historical work was in December 1833. In a notebook entitled 'The Philosophy of the Progressive Sciences' he stated as his aim a complete survey of 'the present state and past history of many of the most complete branches of human knowledge'.[8] However, the subsequent text is largely ahistorical and it indicates that Whewell did not at that time possess a clear conception of historical development.

Whewell's *History* contrasts strikingly with both the *First Principles of Mechanics* and his notebooks on induction; the contrasts being indicative of the development of Whewell's conception of history over a short period of time. These changes will be outlined briefly and then subjected in turn to more extensive discussion since they relate directly to problems with which Whewell was grappling in the mid-1830s and to which history appeared to supply answers.

The four problem areas are as follows. Firstly, Whewell was developing his theory of induction and had in the mid-1830s achieved an early but not fully worked-out version. One of his main aims in turning to the history of science was to test this theory of induction against the historical record. Secondly, he was deeply concerned with ethics and sought through history to combat the rationalist, positivist, and utilitarian doctrines espoused by many of his contemporaries by showing that the development of science depends not only on rationality but also on the power of the imagination, and hence on a romantic view of man. Thirdly, as in the *First Principles of Mechanics*, he used the *History* to account for the progress and present state of science, but achieved this end with a much more sophisticated historiography. Finally, Whewell's account of historical progress also related to his involvement in contemporary scientific problems, in particular his defence of the wave theory of light.

[6] Whewell Papers, R18 17^{12}, fos. v–xix.

[7] Ibid., R18 17^{15}, fos. 55–6. There is a reference here to Saint-Simon's view that history progresses through organic and critical periods, but notes dated 12 Feb. 1832 indicate that Whewell was more interested in Saint-Simon's epistemology than his historiography.

[8] Ibid., R18 17^{8}, fos. 1–13, 136.

2. History, Historiography, and the Theory of Induction

One cannot proceed far with either the *History* or the *Philosophy of the Inductive Sciences*, published in 1840, without recognizing how closely the two enterprises were connected. They were written over approximately the same period, and Whewell even claimed that when he found his researches 'becoming metaphysical and transcendental' he closed his historical notebook and recommenced writing in the philosophical one.[9] Thus it is hardly surprising that his contributions to the history of science were deeply marked by his philosophical interests which both underpinned and sharpened his historical interpretations. Similarly, in the opening pages of his *Philosophy*, Whewell argued that the best way to understand how real knowledge is produced is to examine 'how truths, now universally recognized, have really been discovered'—in other words, by studying the history of science.[10] These obvious connections between the two works encourage the view that the *History* should be read from the perspective of Whewell's philosophical commitments. Although the *History* is undeniably underpinned by Whewell's philosophical apparatus, this reading has its limitations. In this section I shall indicate some points of connection between the two ventures, but also show that such a simplistic characterization of the *History* is untenable since it not only overlooks much of the sophistication of Whewell's historical writing, but also because there are many points in the book where he departs from his stated philosophical and historiographical principles.

The main philosophical insight that Whewell employed in his *History* involved the relationship between facts and theories, sensations and ideas. In the introduction he claimed that facts without theories cannot take us beyond knowledge of individuals, while theories without facts lead to 'empty abstraction and barren ingenuity'. By contrast, 'Real speculative knowledge demands the combination of the two ingredients—right reason, and facts to reason upon.' At several places in the *History* Whewell emphasized that science did not progress when either facts or theories prevailed to the exclusion of the other class. For example, the failure of early botanists to develop their subject into a proper science was, he claimed, due either to their excessive empiricism or to their tendency to employ premature speculations. By contrast, rapid, epochal advances were marked by the apposite fact–theory relationship. Thus Copernicus's success must have been due to 'a clear and steady possession of abstract Ideas, and an aptitude in comprehending real Facts under these general conceptions'.[11]

The requirement that both facts and theories were equally necessary for science represents an important development in Whewell's philosophy of induction that dates from the middle of 1834; indeed McNally has identified Whewell's first announcement of this doctrine in a letter to Jones of 5 August

[9] Todhunter (1876), ii. 193. [10] Whewell (1840), i. 4.
[11] Whewell (1857), i. 5–6; ibid. iii. 223; ibid. i. 282.

1834.[12] The synthesis between this 'fundamental antithesis'—as he later called the relation between fact and theory—provided a ncessary condition for advance in all subjects with empirical content, not only science. As Whewell explained to Jones in 1835, 'the beauty of my Induction is, that it is like the Devonshire man's pie, into which he puts every thing which he catches'.[13] All the inductive sciences formed part of the catch. Moreover, into the pie Whewell later threw the history of epistemology which he interpreted even more self-consciously than the history of science in terms of the opposition between sensationalists and idealists, with success falling to those who combined these two antithetical approaches. Among those who achieved this synthesis were Richard Price, Thomas Reid, Dugald Stewart, Whateley, Ampère, Kant, and of course Whewell himself.[14]

As Fisch shows in his contribution to this volume, Whewell's philosophy of science was not fully developed in the mid-1830s and he had not at that time formulated his transcendental view of pre-discovery. What he had achieved by then was principally the version of the fundamental antithesis discussed above. It is important to notice that just two weeks after Whewell first announced this doctrine to Jones he penned another letter in which he outlined his plan to write a historical work which would 'shew *clearly* that in order to arrive at knowledge or science we must have, besides impressions of sense, certain mental bonds of connexion, ideal relations, combinatory modes of conception, sciential conditions or whatever else you can help me call them: they are what I called *Ideas* in my former letter'.[15] Thus one of Whewell's reasons for suddenly turning to history at this crucial stage in the development of his philosophy of science was to test whether the actual development of science confirmed this version of his theory of induction.

As the same letter makes clear, the new theory of induction was intimately related to one of Whewell's major historiographical innovations which he was later to employ in the *History*. In that letter Whewell extended his theory of induction into a sketch for a three-stage theory of scientific development which, he hoped to show, applied to all sciences. According to this theory a science proceeds through three distinct, but possibly recursive, stages: the *prelude*, *inductive epoch*, and *sequel*.[16] Most sciences begin with premature speculation and the arbitrary collection of facts, and although some of this activity proves useless in the long term, there is a slow, although often

[12] McNally (1982), 115. The crucial document is Whewell to Richard Jones, 5 Aug. 1834: Whewell Papers, Add. MS c. 51[174].

[13] Todhunter (1876), ii. 210–11.

[14] Whewell (1856), 213–24.

[15] Todhunter (1876), ii. 186. The letter cited is dated 21 Aug. 1834.

[16] Although the word 'epoch' frequently occurs in earlier historical writing, Whewell appears to have been innovative in connecting these three terms and giving them meanings specific to his philosophy of induction. According to *OED* the word epoch was in wide usage by 1837 and had even been used to refer to important scientific developments during a specific period. It appears that Whewell was the first to use 'sequel' to mean an age or period following or influenced by a preceding period.

quickening, progress when facts and ideas are 'gradually evolved into clearness and connexion, permanency and certainty'. Whewell termed this period the *prelude* since it often precedes the discovery of an important inductive truth. This discovery forms the *inductive epoch*, when rapid progress is achieved though a grand inductive generalization imposing order on the field and connecting disparate facts and ideas. Usually this induction is the work of one or two men whose names become associated with their achievements. There follows a third period, the *sequel*, 'during which the discovery has acquired a more perfect certainty and a more complete development among the leaders of the advance'. During this period of consolidation the new generalization is extended to further problems, becomes diffused throughout the scientific community and its articulation is often attended by controversy.[17]

Although Whewell expressed his intention to apply this tripartite scheme to all the sciences, the published *History* shows that he was unable to apply the scheme outside the few paradigmatic cases, such as astronomy which he termed his 'pattern science'.[18] In only six out of seventeen sciences was he able to identify *preludes*, *inductive epochs*, and *sequels*—see the final column of Table 1—although he sometimes used these terms more loosely when discussing other sciences. While the fundamental antithesis provided him with the main organizing principle in writing the *History*, he remained sensitive to the differences exhibited by the histories of specific sciences. Indeed, he frequently emphasized the different patterns of development manifested by the sciences.[19] Moreover, he sometimes departed from his tripartite scheme by introducing issues that fell outside his theoretical superstructure. For example, in the history of physical astronomy Whewell devoted a chapter to stressing the importance of both instrumental innovations and such institutional aids as observatories, scientific societies, patronage, and expeditions in the *sequel* to Newton's discovery.[20]

Whewell's taxonomy of the sciences encountered similar difficulties. Like the contemporary British Association, Whewell held the view that science could be divided naturally into several distinct subject areas, according to the different topics and methods practised. Thus chemistry was essentially different from electricity, and electricity was different from mechanics. In order to write the history of each science he had first to devise criteria for distinguishing one from another. The taxonomy that Whewell adopted is set out in Table 1, which shows the different types of science (genus) and their defining characteristics. In the final column the specific sciences are listed under each genus. Thus, for example, formal astronomy (which is concerned with the mathematically expressed laws, not causes, of planetary motion) and geometrical optics (which deals with the geometry of light propagation) are both formal sciences since they are founded on the ideas of space, time, and number.

The history of science did not, however, fit unproblematically into this

[17] Whewell (1857), i. 9–10.
[18] Todhunter (1876), ii. 199.
[19] e.g. Whewell (1857), ii. 129–70.
[20] Ibid. 207–29.

TABLE 3.1 *Whewell's Taxonomy of the Inductive Sciences*

Genera of the sciences	Characteristics[a]	Species	Epochs
1. Formal	Space, Time[b] (and Number)	Formal Astronomy	Hipparchus Copernicus Kepler
		Geometrical Optics	
2. Mechanical			
Primary	Matter and Force *or* Cause[c]	Mechanics Physical Astronomy	Galileo Newton
Secondary	Matter and Force + Medium[d]	Physical Optics Acoustics Thermotics and Atmology	Young/Fresnel
3. Analytical	Composition/Decomposition[e] (Substance)	Chemistry	Lavoisier Davy/Faraday
4. Mechanico-Chemical	2 and 3[f] (Polarity)	Electricity Magnetism Galvanism	
5. Classificatory	Classifying, Naming[g] (Natural Affinities)	Systematic Botany Systematic Zoology	Caesalpinus
6. Analytico-Classificatory	3 and 5[h] (Likeness)	Mineralogy	
7. Organical	Life: Function and Organization[i] (Vital Powers, Final Cause)	Physiology Comparative Anatomy	
8. Palaetiological	Historical Causation[j]	Geology History[k]	

[a] In his *Philosophy*, ii. 277–82, Whewell sought to characterize each type of science in terms of its 'Fundamental Ideas'. The list he produced (p. 281) differs in many respects from the way he characterized the sciences in his *History*. Some of the major changes introduced in the *Philosophy* are indicated in this table by parentheses. However, not all the changes are recorded here.

[b] Whewell (1857), i. 272; ii. 5. [c] Ibid. i. 272; ii. 5. [d] Ibid. ii. 233. [e] Ibid. ii. 90, 96, 153. [f] Ibid. 5. [g] Ibid. 164. [h] Ibid. 164–5. [i] Ibid. 315. [j] Ibid. 397.

[k] Unlike the other subjects listed in the third column, Whewell did not discuss History in his *History*.

taxonomy. To cite one of several such examples, towards the end of his discussion of thermotics Whewell argued that while some parts of that science appeared readily reducible to mechanics, other parts were

> connected with properties and laws of a nature altogether different—those of chemistry [which, for Whewell, was an analytical science involving the composition and decomposition of substances]; properties and laws depending upon a new system of notions and relations, among which clear and substantial general principles are far more difficult to lay hold of, and with which the future progress of human knowledge appears to be far more concerned.[21]

This example illustrates the difficulty Whewell encountered in trying to divide the mechanical from the analytical sciences. The distinction between them was all the more confused by the subjects of electricity, magnetism, and galvanism which Whewell classified as mechanico-chemical sciences since, like thermotics, they displayed some of the characteristics of each genus.

Although there appear to be many close connections between Whewell's account of the history of science and his early theory of induction, it is debatable whether the *History* should be read as confirming his theoretical views. It is certainly true that he frequently accounted for major epochal developments in terms of the correct interrelationship between fact and theory. However, as we have seen, his theory of epochs proved applicable in only a minority of sciences.[22] Also his taxonomy of the sciences was severely undermined by numerous problematic instances. Finally, there is much material included in the *History* that does not relate to Whewell's project of framing a new philosophy of the inductive sciences. Taken together these points suggest that the book should not be read simply as an articulation of Whewell's philosophy of scientific development. We must therefore look elsewhere for some of the other historiographical ingredients of the *History*.

3. Whewell as Moral Philosopher

History was also an ingredient of 'the Devonshire man's pie', and Whewell clearly recognized that, in contrast to so many earlier histories of science which merely catalogued discoveries, he had to interpret facts with the aid of general theories about the nature of scientific change. Thus the historian of science synthesizes historical facts with historiographical—and thus philosophical—theory. On the other hand, he avoided the excesses of some speculative historians by obtaining a firm grasp and extensive knowledge of primary sources.[23]

[21] Whewell (1857), ii. 437–8.

[22] Notwithstanding Whewell's assertions in the preface to the third edition: ibid. i. vii.

[23] Whewell's knowledge of primary sources is uneven, for while he was exceedingly well read in the physical sciences he was relatively unfamiliar with the primary sources in other areas and often drew excessively on secondary works, such as Cuvier's *Historical View of Ichthyology*: ibid. iii. 297.

Whewell's attempt to synthesize the facts and theories of history appears to be a further example of his general theory of induction. While this is certainly true, it is important to recognize that Whewell was not alone in deploying this strategy but that a number of contemporary historians were likewise emphasizing the need to integrate historical facts and theories. Duncan Forbes has drawn attention to a group of 'Liberal Anglican' historians which he has located in respect to two opposing historiographical traditions drawing support from different views of man. The rationalist, sensationalist tradition emphasizes the growth of empirical knowledge and records the vast accumulation of historical facts. By contrast the more speculative tradition, which in Britain was fuelled by Coleridge and Wordsworth, tends to ride roughshod over facts with their grand historical schemata. For Liberal Anglicans history was not a panoply of facts to be collected by the historian; instead, since documentary sources are interpretations of events overlaid with the author's perspective, the historian has to examine critically all evidence and weigh its truth. Likewise he has to employ his imaginative powers to impose order and unity on the evidence, but must not allow his pet theories to lead him into mere fiction.[24] In seeking to combine facts and ideas, sensation and imagination, the Liberal Anglicans mediated between the rationalist and romantic historical traditions, but were in many respects closer to the romantics.

Not only was Whewell's fundamental antithesis a reflection of Liberal Anglican historiography, but like Forbes's Liberal Anglicans Whewell insisted that historians must achieve a high scholarly standard and possess command of primary sources. As his notebooks and publications show, Whewell was exceedingly well read in the sciences, especially the physical sciences. He could also stand his ground against Comte, Mill, and other writers who, he complained, incompetently pressed historical examples into the service of their pet philosophical systems. For example, in reviewing Comte's positive philosophy in 1866 Whewell undermined Comte's three-stage theory of development by showing that his history was falsified by numerous factual errors.[25]

Whewell was composing his *History* at a time when new levels of sophistication were being achieved by historians working in a number of different problem areas. For example, Thomas Arnold, Thomas Macaulay, and Thomas Carlyle all contributed important essays on history and offered significant examples of the genre. Of direct relevance to Whewell was the new historical scholarship emanating from Germany, particularly from the pen of Barthold Niebuhr, and the group of Liberal Anglican historians who were active from the 1820s and who derived their inspiration principally from Niebuhr and Vico. Among this group was not only Arnold, but also two Trinity dons, Julius Charles Hare and Connop Thirlwall, who were close to Whewell and belonged to the so-called

[24] Forbes (1952), 135.

[25] Whewell (1865–6), especially 358. Cf. David Brewster's review of Whewell's *History* in Brewster (1837), 110–51.

'Cambridge Network'.[26] The new standards of historical scholarship would have been discussed frequently at High Table by the members of this group who turned to Germany for intellectual stimulus. From his published letters it is clear that Whewell was in frequent contact with Hare and Thirlwall at the time when he was writing the *History*.[27]

Forbes locates the Liberal Anglican approach to history in the social and political uncertainties of the early nineteenth century when, following the end of the Napoleonic wars, the future of the nation seemed particularly bleak. Liberal Anglicans such as Hare, Thirlwall, and Arnold responded to this situation by seeking to reconstruct the world on more humane principles. Their opposition was not merely to degrading social conditions but more directly to the false theories of 'optimism, . . . utilitarianism and individualism' which underpinned Britain's moral condition. The present state of affairs was seen as the natural culmination of the eighteenth-century political, social, and religious ideologies propounded by rationalists, utilitarians, and fellow travellers. The crisis of the 1820s was therefore not only social but moral and intellectual.[28]

In responding to these contemporary trends the Liberal Anglicans turned to both the romantic tradition of Coleridge and Wordsworth, and its German sources, where a very different view of man was to be found; instead of a pale rational being, manipulated by an alien political and economic system, romantic man was free, imaginative, and able to relate to his neighbour and to God. One cannot proceed far with, say, *Guesses at Truth* by Julius and Augustus Hare without encountering the many manifestations of this view of man and the key writers in the romantic tradition. This orientation towards Germany can readily be seen in the group's concern with German literature, philosophy, philology, and, of course, history. The Hare brothers characterized German historians as 'endeavouring to estimate all prior systems according to their historical position in the progressive development of human thought, to shew what truths it was the especial province of each to bring out, and how each fulfilled its appointed work'.[29] History was not to glorify the present, but the time and place of each actor had to be reconstructed with integrity so as to expose his or her situation. For the Liberal Anglicans historical scholarship demanded detailed and extensive knowledge of sources and their critical and impartial appraisal.

Both Liberal Anglicans and their German mentors recognized that writing history is an interpretive act involving the interaction of facts and general

[26] S. Cannon (1978), 29–72. See also Preyer (1981); Robson (1967). Hare and Thirlwall translated the first two volumes of Niebuhr's *The History of Rome* (1818–32) and the former also rushed to Niebuhr's defence in 1829 with his *A Vindication of Niebuhr's History of Rome*. Moreover, Julius Hare collaborated with his brother Augustus in writing *Guesses at Truth by Two Brothers* (1827, revised 1838 and 1847–8) which was the vehicle for propagating the views of Coleridge and of German writers on many subjects, history included.

[27] See especially Todhunter (1876), ii. 160.

[28] Forbes (1952), 1–11.

[29] Hare and Hare (1873), 470. In this late edition the authors proceeded to point out that the same method had been adopted by Whewell in his historical essays.

ideas. The general ideas on which their own historical writings were based were derived largely from romanticism and Forbes draws particular attention to the rejection by early Liberal Anglicans of the idea of progress, which had been accepted simplistically and unproblematically by rationalist historians. In its place Liberal Anglicans viewed history as the natural, law-like unfolding of a nation's inner dynamic; an unfolding which did not presuppose progress since a nation finally decays and becomes extinct. Yet Forbes is concerned to emphasize that such a view of history was not naturalistic but providential, manifesting God's will. Providence was also seen at work in the moral and intellectual characteristics exhibited by each nation through its history. The Liberal Anglicans emphasized those imaginative individuals who both reflected a nation's values and shaped its history. Thus in place of utility and determinism they sought through history to show the dominance of mental and moral characteristics, and thus of the romantic view of man.

Although, as we shall see in the next section, Whewell departed from the early writings of the Liberal Anglicans on the issue of progress, he rejected utilitarianism and the stronger versions of rationalism, but his attitude towards romanticism underwent a significant change; while he had earlier dismissed Coleridge's metaphysics as 'muddy with their own turbulence' and rejected German philosophy,[30] he became increasingly sympathetic towards Coleridge's philosophy and by the end of 1833 he had adopted a Coleridgean notion of intuition.[31] Intuition was to play an important role in Whewell's account of the history of science while, more generally, his *History* and subsequent works bore many marks of romanticism.

One of the central tenets adopted by rationalist historians is that all men possess the same set of mental qualities and that any account of science must start from a discussion of the nature of mind. In his *Elements of Morality* (1845*b*) Whewell adopted a similar approach and portrayed man as possessing an innate desire to reason and to form ever more abstract notions; thus as mind develops and attains these abilities, it becomes able to engage in science.[32] However, it is important to notice that he did not extend this form of analysis to history, but drew a sharp distinction between the analysis of man's mental constitution *per se* and the historical study of the development of the human mind.[33] When writing history he rejected the former mode of analysis with its rationalist implications and instead adopted the view, similar to that held by the Liberal Anglicans, that the types of mental operations performed by individuals have to be understood in the context of their historical situations, which may be very different from our own. Intellectual and moral development is therefore an intrinsic and inseparable part of a nation's development, and the

[30] Stair-Douglas (1881), 29; Todhunter (1876), ii. 61; McNally (1982), 38–49.

[31] Ibid. 152–60. See also Sanders (1942); Elkana (1984*a*); this volume Ch. 8.

[32] Whewell (1848*e*), i. 13; id. (1857), iii. 13.

[33] Cf. Whewell's preface to Mackintosh (1862), pp. xiii–xiv.

'Moral Nature of Man is moulded into shape by the History of each Nation'.[34] This Liberal Anglican position was accepted by Whewell who conceived that science, especially progressive science, is not to be found among all men at all times, but is limited to certain cultures and to a few individuals, possessing highly developed mental qualities, within those cultures.

In Whewell's view science is strongly affected by culture. Thus he claimed that the speculative frame of mind, one of the essential conditions for science, had first appeared in ancient Greece, and yet the Greeks possessed another trait which rendered so much of their science void. While the Greeks loved reason and did not neglect facts, their enthusiasm for ideas led them to become trapped in linguistic philosophy; by attending too closely to abstract terms they courted metaphor and vague, indistinct notions. Moreover, not only did their ideas lack the clarity necessary for science, but by failing to connect these ideas to appropriate facts they were debarred from discovering the fundamental principles of science. Thus, for example, in framing his theory of motion Aristotle had become trapped in a sequence of linguistic errors arising from the apparent opposition between light and heavy, and in this exercise he had forsaken all interest in defining his terms clearly and in relating them to facts. Likewise Platonic philosophy, which contained some valuable scientific ideas, especially in the area of astronomy, was soon diverted into sterile scholasticism.[35]

Hardly any scientific progress occurred during the medieval period, owing to several cultural factors which diverted men from seeking the necessary co-operation between ideas and facts. One such factor was mysticism, which turned men's minds from the rational contemplation of nature to enthusiasm and the search for supernatural causes. On the other hand dogmatism over-emphasized the role of reason and led men to concentrate on language and abstract thought, while neglecting observation. Mysticism and rationalism, being opposing but equally fallacious philosophies, signified not only the intellectual failings but also the moral bankruptcy of the medievals. Those who knelt before authority were surrendering their courage and originality; similarly, the enthusiast allowed himself to slip from reality into a world of illusion. The prevalence of these false views ensured that indistinct ideas reigned, and little worthwhile science was produced. Other cultures, such as Arabic and Oriental cultures, likewise failed to provide suitable conditions for the advancement of science, and Whewell even confided to Jones that 'I shall be glad if I find the Arabian [Alhazen] discovered nothing, for I have a strong *European nationality*.'[36]

Whewell attributed some importance to the progress of the mechanical arts in the medieval and renaissance periods, but drew a sharp distinction between science and art, claiming that radically different mental operations were involved. The two activities nevertheless bore a *historical* relationship; while

[34] Whewell (1848*e*), i. 48; see also Preyer (1981), 50.

[35] Whewell (1857), i. 1–66. See also Whewell Papers, R18 17^{15}, fos. 24–35.

[36] Todhunter (1876), ii. 261. See also Whewell (1857), i. 174, 178, 259; ibid. iii. 227; Whewell Papers, R18 17^{6}, fos. 43–9.

acknowledging the importance of technological innovations, such as the telescope, in the development of scientific ideas, Whewell claimed that improvements in the arts occur prior to developments in science. However, since 'Art is the parent, not the progeny of science,' science brings little, if any, benefit to the arts.[37] History, then, undermines the utilitarian justification for science. Like the Liberal Anglicans Whewell was staunchly anti-utilitarian and he attacked utilitarianism not only in the *History* but also in such works as his edition of Mackintosh's *Dissertation on the Progress of Ethical Philosophy* and his *Lectures on the History of Moral Philosophy in England*.[38]

Although technological progress continued into the Renaissance, there then occurred a more important, if unexplained, change in man's mental outlook. In place of indistinct thought, linguistic disputations, and intellectual servitude, which had dominated the earlier period, men began to cultivate pure mathematics and astronomy and to describe natural phenomena in distinct terms. This clarity of mind was also responsible for breaking the yoke of authority. The age of Copernicus was at hand, and the beginning of new, exciting developments in all branches of science. These examples illustrate Whewell's view that specific mental characteristics are usually required before either the inductive philosophy or science can flourish in a particular culture. Together with these general trends, certain individuals display mental characteristics apposite for the practice of philosophy or science. Thus in a generally hostile intellectual environment Francis Bacon, the first philosopher of the inductive sciences, was able to break free from the chains of authority and to interrelate ideas and facts in the proper manner. Likewise Newton possessed the psychological characteristics necessary for practising the true method of discovery and thus for becoming a successful scientist.

It is instructive to examine Whewell's portrayal of Newton's mentality in more detail. Whewell argued that Newton's 'mathematical talent' *must* comprise the following characteristics: 'distinctness of intuition, tenacity and facility in tracing logical connexion, fertility of invention, and a strong tendency to generalization', and he offered historical evidence to support each of these claims.[39] Whewell had also detected in Newton the ability to become detached from other activities and to pay full attention to his train of ideas. It is important to notice that this list of attributes includes not only rational but also imaginative qualities; indeed, 'distinctness of intuition' appears first on the list. The creative scientist was not for Whewell armed solely with the tools of reason, for while the reasoning powers were necessary intellectual traits for a Newton or a Faraday, a creative imagination was also a necessity. However, he backed away from analysing creativity and declared that the 'hidden fountain of our unbidden thoughts is for us a mystery'.[40] Other successful scientists were similarly endowed: Hipparchus's 'diligence in collecting observations,

[37] Whewell (1857), i. 253; Todhunter (1876), ii. 209.

[38] Mackintosh (1862), p. xxix; Whewell (1852), pp. ix–x.

[39] Id. (1857), ii. 139. [40] Ibid. 140.

and clearness of idea in representing them'; Faraday's 'peculiar steadiness and clearness in his conceptions of space',[41] to cite but two of the many examples. For Whewell the successful scientist had to possess certain mental qualities, and especially the ability to frame abstract ideas clearly.

This point gains further significance when we recognize its basis in Whewell's moral philosophy: thus in his history of moral philosophy he rejected the view that human action is directed towards some external objective, such as happiness or utility; instead he followed in the tradition of Plato who had argued that action is regulated by principles internal to man, such as duty.[42] While internal principles were thus the springs of human action, these principles are activated by the will. It is therefore incumbent on the historian to draw attention to the highly developed intellectual and moral qualities possessed by scientists who achieve major discoveries. This emphasis is also to be found in Liberal Anglicans such as Hare who argued that 'Human character, as modifying and modified by circumstances, . . . must be the historian's ultimate object.'[43]

Through his *History* Whewell sought to show that great scientists are marked both intellectually and morally; thus not only was Newton amply endowed with the apposite intellectual qualities, but he was also, in Whewell's opinion, morally virtuous—'candid and humble, mild and good'.[44] Moral considerations also entered into Whewell's pamphlet *Newton and Flamsteed* (1836) where he defended Newton against an anonymous reviewer who had argued that Flamsteed had been ill-treated by Newton over the lunar observations.[45] A copy of the pamphlet was sent to S. P. Rigaud who expressed the hope that Whewell would pursue the matter further, 'for if Newton's character is lowered, the character of England is lowered and the cause of religion is injured'.[46]

History was therefore a tool in Whewell's battle against the utilitarians since, if closely studied, it showed the cultural conditions necessary for progress and the intellectual and moral characteristics of great scientists. These conditions combined not just facts and theories but reason and imagination, intellect and morality. As Whewell concluded in his *History*, the aim of the history of science is not merely to display 'a barren Knowledge of the external world, or [as utilitarians would claim] a few Arts ministering to our gratification', but it is 'to lead us from matter to mind, from the external to the internal world'.[47] History thus displays both the successful attainment of scientific knowledge and the higher moral purpose of science. The history of science is thus ultimately a form of moral discourse.

[41] Mackintosh (1862), i. 144; Whewell (1852), iii. 86.
[42] Whewell (1852*a*), p. ix.
[43] Quoted in Forbes (1952), 131.
[44] Whewell (1857), ii. 141–3.
[45] Id. (1836). Houghton (1966), i. 717, identifies the reviewer as Sir John Barrow.
[46] S. P. Rigaud to Whewell, 25 Jan. 1836; Whewell Papers, Add. MS a. 211[79].
[47] Whewell (1857), iii. 519–20.

4. The Problem of Progress

'The present generation finds itself heir of a vast patrimony of science', wrote Whewell in the introduction to the *History*, 'and it must needs concern us to know the steps by which these possessions were acquired.'[48] The question of how science had reached its present state underpinned both Whewell's excursion into the history of mechanics in 1832 and his larger project for writing the history of the inductive sciences. Present-day science was the outcome of a long-term process, but one which was neither arbitrary nor the result of accident. Like the Liberal Anglicans Whewell considered that history was governed by laws and he included history among his palaetiological sciences since the historian, like the geologist, aims 'to ascend from the present state of things to a more ancient condition, from which the present is derived by intelligible causes'.[49] Such an investigation drew on both facts and theories, such as the three-stage theory of epochs which Whewell utilized with only partial success in his *History*.

In writing his *History* Whewell confronted a major historiographical problem for which the rationalist and romantic traditions offered very different solutions. For Enlightenment, rationalist historians, such as Priestley and Condorcet, the growth of science exemplified the progress of the human mind. Present-day scientific knowledge was the result of a cumulative process which Priestley likened to ascending a mountain; from our present position we can look back on the upward-sloping path we have taken.[50] The history of science charted that path and it was therefore an account of man's progressive uncovering of nature's secrets. Such a history is concerned with the discovery of truth, and there is no reason to discuss erroneous theories other than to warn against the use of deductive systems and to urge induction as the proper scientific method. Moreover, these empiricist histories concentrate on instruments, discoveries, and inductive generalizations but exclude consideration of the scientist as a product of his time and as a moral agent.

At the other extreme romantic historiography is highly relativistic. For example, Niebuhr and the early Liberal Anglicans utilized the histories of the Greek, Roman, and other nations to show that progress, in its broadest sense, was illusory since each nation underwent a similar three-stage cyclical transformation leading to its decline. Since a nation possesses its distinctive moral and intellectual characteristics, it is unable to export or bequeath its culture to any other nation. The historian has thus to set each person within his national context and construct his experience in terms of the romantic conception of man. Thus history served to preserve the integrity of the past, without explicit reference to presentist concerns. Yet the Liberal Anglicans were sympathetic

[48] Ibid. i. 3.

[49] Ibid. iii. 397.

[50] Priestley (1775), i. p. vi. Whewell employed the same metaphor in the introduction to Whewell (1857), i. 3–4.

to this cyclical view of history precisely because they were seeking to undermine the optimistic pretensions of the utilitarians and instead articulate a view of history which legitimated the perceived decline of contemporary society dominated by uncongenial social, political, and moral values.

However, as Forbes argues, Thomas Arnold did not subscribe to this extreme relativism but sought to combine it with his view that God improves the lot of both individuals and nations. In the 1840s Arnold resolved this implicit conflict between a fatalist, cyclical history and progressive providentialism by combining the cyclical pattern with an overall cumulative tendency, whereby the positive traits acquired by one nation do not die but can be inherited by later nations. Thus, for example, Greek rationalism did not decay with the Greek nation but was transmitted to Western Europe during the Renaissance and became a feature of progressive civilization.[51]

Whewell was the heir to both the rationalist and the romantic conceptions of history, although on the question of progress he was much closer to the rationalist camp. It was manifest that science had progressed by impressive strides, achieving real, secure knowledge; for example, the theory of gravitation was an epochal induction, a real acquisition to man's knowledge of the physical universe. In writing the history of science the historian must therefore show how earlier discoveries contributed to our present store of knowledge. Yet for Whewell progress was not a purely secular concept; instead progress in our knowledge of the physical world was analogous to moral progress and strengthened our conviction in the latter. As he claimed in a notebook dating from December 1833, his review of the progress of human knowledge concerned man's 'moral and religious condition as much as that which refers to the material world. . . . All things work together for good.'[52] Progress was therefore a crucial part of Whewell's providentialism.

Not only were cumulative, progressive elements present in Whewell's historiography, but also cyclical ones. In particular, his theory of epochal development involves a cyclical view of history since each science passes through the same law-like pattern of development, sometimes several times over. Hipparchus, Copernicus, and Kepler were epochal figures who discovered grand inductive generalizations in formal astronomy, and in each case their discoveries were preceded and succeeded by activities which Whewell characterized respectively as the *prelude* and *sequel*. However, these three revolutions in formal astronomy were not discrete but interconnected, since Copernicus stood on the shoulders of Hipparchus, and Kepler, in his turn, built on Copernicus's work; the subject progressing by the accumulation of increasingly powerful inductions.

The relation between cyclical and cumulative elements proved particularly problematic in Whewell's account of theory change during an *inductive epoch*. He argued that while there appears to be a discontinuity between old and new

[51] Forbes (1952), 55–62.
[52] Whewell Papers, R18 17[8], fos. 12–13, 136.

theories, examination of the fine structure of theory change shows that science develops continuously, and that this development consists of

> a series of changes; a repeated progress from one principle to another, different and often apparently contradictory. Now, it is important to remember that this contradiction is apparent only. The principles which constituted the triumph of the preceding stages of the science, may appear to be subverted and ejected by later discoveries, but in fact they are, (as far as they were true,) taken up into the subsequent doctrines and included in them. . . . The history of each science, which may thus appear like a succession of revolutions, is, in reality, a series of developments.[53]

Thus, when viewed historically, rival hypotheses do not appear as incommensurable alternatives. However, there exists a tension between Whewell's account of macroscopic, revolutionary changes and small-scale, continuous ones; a tension that Whewell tried further to resolve in his 1844 article on the transformation of hypotheses where he argued that hypotheses are modified over time such that the one which is ultimately defeated loses its false components and becomes transformed into a vehicle of truth.[54] Whereas in his *History* Whewell clearly drew on both rationalist and romantic historiographies, in that article he articulated a more rationalist view of science.

5. History and Contemporary Science

Since Whewell was involved in several branches of science in the 1830s, it would be surprising if his excursions into the history of science were not related to his scientific interests. In order to illustrate one such connection I will discuss how the controversy over the wave and particle theories of light, which reached its peak in the 1830s, was portrayed in the *History*. In writings that predate his interest in the history of science Whewell expressed his commitment to the wave theory as the true theory of light and to the luminiferous ether as the medium for optical vibrations.[55] However, in the first half of his *History* Whewell effectively showed that the formal and primary mechanical sciences progressed according to his three-stage historiography. Then turning to physical optics he mobilized this historiography to show clearly that the wave theory was the epochal development in optical theory and therefore the natural successor to the particle theory that had held sway for over a century. Although he opened this section by claiming that the history of optics would 'illustrate the nature of science and the conditions of its progress',[56] his exegesis of recent developments also provided support for the wave theory of light. To appreciate how he achieved this we must briefly outline his account of the history of optics.[57]

[53] Whewell (1857), i. 10.

[54] Id. (1851*a*).

[55] Id. (1833*c*), 128–41. See also Whewell's 'View of the Modern Theory of Light', Whewell Papers, Add. MS R18 17^3, and Cantor (1981).

[56] Whewell (1857), ii. 269.

[57] Ibid., 312–73. See also Cantor (1983).

In the *prelude* Whewell identified two rival theories; one equated light with the emission of matter, the other explained optical phenomena in terms of undulations. Rather surprisingly he attributed the first of these theories to Descartes, who had initiated the programme to mechanize optics. The rival wave theory was developed by Hooke who obscurely formulated a principle of interference, and by Huygens, who incisively utilized the theory to explain such phenomena as reflection, refraction, and double refraction. Yet despite Huygens's undoubted success, Whewell refused to attribute to him an epochal advance since the theory 'was not *established* till a later era, when the fringes of shadows, rightly understood, made the waves visible, and when the hypothesis . . . was found to contain also an explanation of polarization'.[58] By comparison with Huygens, Newton was severely censured; he supported the bankrupt emission theory and his negative influence was felt throughout the eighteenth century, when only Euler dissented from the majority opinion and advocated the wave theory.

The epochal advance occurred with the extension of the wave theory by Thomas Young and Augustin Fresnel who were able to explain, in numerical terms, the various phenomena of periodic colours, as well as the complex phenomena of double refraction, polarization, and dipolarization. Unlike its rival, which had to be sustained continuously by a plethora of *ad hoc* hypotheses, the wave theory manifested the signs of a true theory:

> all tends to unity and simplicity . . . Truth may give rise to such a coincidence, falsehood cannot. But the phenomena become more numerous, more various, more strange—no matter: the Theory is equal to them all. It makes not a single new physical hypothesis; but out of its original stock of principles it educes the counterpart of all that observation shows. It accounts for, explains, simplifies, the most entangled cases; corrects known facts and laws; predicts and discloses unknown ones; becomes the guide of its former teacher, Observation; and, enlightened by mechanical conceptions, acquires an insight which pierces through shape and colour to force and cause.[59]

Here, then, were the signs of a great truth, and one which was revolutionizing physical optics at the time Whewell was writing his *History*. He concluded his account of physical optics by showing that during the *sequel* the wave theory had given rise to a number of highly successful lines of research.

Whewell's interpretation of the history of optics implicitly contrasted the lack of theoretical development in the eighteenth century under the thraldom of Newton's emission theory with the grand and rapid successes produced by the wave theory. By studying its history Whewell drew attention to the signs of truth manifested by the wave theory, whereas its rival typified error. The moral from history is clear. But there is another point relating to the controversy that Whewell made in his *History*. According to some of its proponents the emission theory was vastly superior to its rival because it alone offered a way of accounting for the chemical interaction between light and matter, such as the

[58] Whewell (1857), ii. 315. [59] Ibid. 340–1.

optical blackening of silver salts. To nullify this claim Whewell argued that such phenomena fall outside physical optics (which he classified as a mechanical science) and belong instead to the domain of chemistry. Thus, while ignoring some of the problems raised by his method of classification, Whewell employed his taxonomy of the sciences to rule optico-chemical phenomena outside the domain of optics and thus outside the controversy over theories of light.[60]

In the 1830s, when the scientific press was filled with papers arguing the merits of the rival theories of light, Whewell entered the fray with lessons drawn from history. Since the progress of science was, for Whewell, governed by historical laws, history seemed to offer a clear, naturalistic perspective on the controversy and one which totally favoured the wave theory. Thus he deployed history in the service of science.

6. Further Departures from a Rationalist Historiography

In this paper I have argued that Whewell's historiography was a response to several problems he encountered relating to the philosophy of induction, moral philosophy, scientific progress, and contemporary developments in science. I have also suggested that in the *History* Whewell often achieved a high standard in historical scholarship. Moreover, I have located Whewell in terms of two opposing historiographical traditions—one rationalist, the other romantic. In conclusion I shall draw attention to some further issues raised by this last point, but ones which also emerge strongly in Whewell's later writings.

Since most earlier historians of science subscribed to rationalist historiographies, one useful way of characterizing Whewell's *History* is in terms of his departure from this tradition. As already emphasized, Whewell insisted that the history of science was not an account of accumulated facts but of the relationship between facts and theories. But his conception of history also included a further class of concepts noticeably absent from rationalist accounts. For Whewell the history of science must not only address positive discoveries, it must also encompass those imaginative, non-empirical ideas that shape man's conceptualization of nature. Thus in the preface to the third (1857) edition of the *History* he emphasized the importance of such ideas as 'the nature and necessary attributes of Matter, of Force, of Atoms, of Mediums, of Kinds, [and] of Organization' in the development of science.[61] Such concepts were discussed in the *History*, but their significance emerges more strongly in Whewell's 1866 review of Comte, where he declared that the 'radical mistake' of positive philosophy was its rejection of metaphysics. Whewell cited historical examples to show that metaphysics is an essential part of science and not a useless appendage which disappears once the laws relating phenomena have been discovered. Science, and thus its history, must be concerned with 'all abstract conceptions, causes, theories, and the like'.[62] This contrast illustrates

[60] Ibid. 319–20, 478.
[61] Ibid. i. p. ix.
[62] Whewell (1865–6), 353, 358.

a further important difference not only between Comte and Whewell, but also between the historical agendas of Whewell and the rationalist, positivist, and utilitarian historians.

Comte, among others, viewed scientific progress as naturalistic, and therefore in *Philosophie Positive* he portrayed individual scientists as cyphers representing a particular stage in human development. If individuals are discussed at all in rationalist histories they are characterized as confronting nature and, if successful, as wresting its secrets. Thus in his *The History and Present State of Discoveries Relating to Vision, Light, and Colours* (1772) Joseph Priestley divided his subject by period and then by topic, including the names of scientists in whichever section dealt with the discoveries they had made. Although a whole 'period' was devoted to Newton, Priestley concentrated on his discoveries 'drawn from the clearest and most decisive experiments'. He certainly praised Newton, but only for his experimental skill and mathematical ability.[63]

Whewell portrayed Newton very differently in his account of the history of physical astronomy; he was the prime example of an inductive philosopher, combining rational and creative traits (as discussed above). He was also the heir to the vast accumulation of facts and theories (constituting the *prelude* to this history), and the active agent who reformed past knowledge into a new, powerful induction. Progress was the result of the human will; it was forged by individuals possessing intellectual and moral strength. Thus unlike rationalist historians Whewell centred his account of science on individual scientists. His debt to romantic historiography and his Liberal Anglican colleagues is clear from his emphasis on the role of national, intellectual, and moral components in the creation of science.

Finally Whewell and the rationalist historians differed on another key issue. Like other historians working in that tradition Priestley portrayed the scientist as an explorer visiting a new country and charting its geography, flora, and fauna.[64] The scientist, then, discovers what is 'out there', bringing no baggage with him on the journey. By contrast, Whewell conceived the scientist as mediating between nature and the acquired facts and theories and the scientist's own metaphysical ideas. Thus for Whewell the scientist has one foot firmly planted in his contemporary intellectual environment; the history of science therefore becomes part of intellectual history.

In this paper I have directed attention to the problems that Whewell confronted in the 1830s and his deployment of history in their resolution. His use of elements drawn from two often conflicting traditions in the history of science also shows how far he diverged from rationalist historiography and helps us appreciate why the *History* deserves the epithet 'new'. Moreover, this divergence explains why the *History* appears much closer to much modern writing than does its rationalist predecessors.

[63] Priestley (1772), 233–353.
[64] Priestley (1775), i. pp. i–xxiii.

4. *William Whewell: Omniscientist*

by MICHAEL RUSE

THERE was virtually no area in early nineteenth-century science in which William Whewell failed to make suggestions, comments, experiments, measure ments, linguistic improvements, and (always dear to Whewell's heart) critiques of the work of others. This indefatigable industry, breadth of interest, and inventiveness of mind, although highly admirable, unfortunately sets a problem for one, such as myself, who would presume to write on Whewell's science—the whole of Whewell's science—in a short span. But, such is the aim of this essay.

Fortunately, help in presenting and understanding the work of William Whewell the scientist can be sought in the writings of William Whewell the philosopher. There is a way in which we can get a meaningful overview, hence producing more than a mere catalogue of achievements. Following Whewell's own beliefs, his philosophy of science must be seen as the inspiration for his actual science. Conversely, his actual science must be seen as the guide to, and confirmation of his philosophy of science.

> I have been employed all the term hitherto upon a thumping paper on the Tides, which I intend to be a step of some consequence in the theory. I wish I could explain to you how useful my philosophy is in shewing me how to set about a matter like this, and how good a subject this one of the Tides is to exemplify it.[1]

This passage, written to one of his most intimate friends, will be my thread of Ariadne. Thus, I begin my discussion of Whewell's science with an account of Whewell's philosophical thinking about science. Then, and only then, can I move through the various areas of actual science in which Whewell worked, or (towards the end of my list) on which Whewell commented. These will prove to be physics (broadly construed), mineralogy and crystallography, political economy, geology, and biology.

Philosophy

Whewell drew from many different elements and sources to make his philosophy of science: from philosophy itself, from science, history, theology and religion,

[1] Todhunter (1876), ii. 172.

and even his status as a fellow at (what he undoubtedly believed to be) the leading college at the University of Cambridge. To lead our way in, beginning with what I shall refer to as Whewell's 'metaphysics' of science, let us take this last point, namely that Whewell was a life-long member of Trinity College, that institution which, nearly two hundred years before, had housed Isaac Newton. For Whewell always there was the model and example of Newton, most particularly of the author of the *Principia*. This, more than anything, Whewell saw as the apotheosis of human thought, never to be exceeded, always to be emulated.

Whewell's philosophy of science, therefore, was structured around the theory of the *Principia*. In particular, Whewell interpreted Newtonian mechanics as what today we would call a 'hypothetico-deductive' system, starting with a few high-powered premises or axioms, and then going formally (deductively) down to lower-level claims. Such a theory Whewell saw as being essentially general, although (as in the laws of Kepler) applicable to the particular. This means that Whewell saw a scientific theory, in ideal, as being composed of universal statements, or *laws*. He would not have sympathized with those of today's theorists who deny the importance of laws in science, although, as we shall learn, Whewell certainly did not want to claim that law-bound scientific knowledge constitutes the whole of our understanding.

All laws are equal, but some are more equal than others—at least, some parts of a theory are more equal than others. For Whewell, it is essential to distinguish the upper and the lower levels of a theory. Not only is there a formal difference, there is also a material difference. The upper levels of a theory tell you *why* things happen. They refer to the powers of nature, or *causes*. The lower levels of a theory tell you *what* things happen. They refer to the empirical effects or *phenomena*. The Newtonian thinking about forces is the paradigmatic example of a 'theory of causes'. The consequences drawn by Kepler and Galileo are likewise paradigmatic examples of 'laws of phenomena'. Whewell spoke also of these two aspects of a theory as the *physical* and the *formal* parts of science, respectively. It goes without saying (although Whewell was happy to say it many times) that the best science extends beyond the formal to the physical. Normally, as in mechanics, one expects the formal part of science to be articulated first; but Whewell knew too much history to think that this is necessarily or inevitably the case. Sometimes the ideas come first, and you have then to go out and check the facts.

Trinity was not only the home of Newton. It was also the college of Francis Bacon, that other great hero (to the Victorians) of scientific method. Whewell was not about to let this point go unnoted. Indeed, he started his major philosophical treatise, the *Philosophy of the Inductive Sciences*, in Baconian fashion, with a series of aphorisms. But, essentially, Whewell was totally unsympathetic to the popular and much-praised crude empiricism that went under the name of 'Baconian inductivism'—a kind of random gathering of the facts with the hope that something good will emerge. Here, Whewell broke the mould, for (*qua* philosophy) he had fallen under the influence of the great eighteenth-century German philosopher Immanuel Kant. Like Kant, Whewell denied absolutely that knowledge, specifically scientific knowledge, is simply a disinterested report of naked, uninterpreted experience. Rather, again like Kant, Whewell saw knowledge as being the result of the interplay between mind and the world. Thus, ringingly, the *Philosophy* began with an aphorism proclaiming this participant-observer stance: 'Man is the interpreter of Nature, Science is the right interpretation.'[2]

Yet Whewell went beyond Kant in at least two significant and interrelated ways. First, Whewell added a historical dimension to his epistemology of science.[3] We do not, as it were, just sit down and intuit out the truth. We have to work away at our science and gradually the truth becomes revealed. Second, as we succeed in our labours, we come to see the truth not simply as true, but as *necessarily* true. This is the status now of Newton's laws of motion. We have gone beyond contingency, to the totally binding—you cannot deny them without some level of inconceivability.

Of course, Kant believed in necessary truth, the 'synthetic a priori'; but Whewell's notion was stronger. This became more explicit in later years, as Whewell started to veer from Kant somewhat, towards Platonism. But it was always there, and (I believe) is explicable in the light of another element in Whewell's thinking, namely his deep commitment to Christianity. (Remember, as a fellow of Trinity College he had to be an Anglican clergyman.) For Whewell, the world could not be a contingent hodge-podge, just thrown together, or even something merely existing, in a Humean fashion. The world is as it is, necessarily, because that is the way it has been created by a good God—it is the way it is because it is held immanently by this good God. There is almost a Leibnizian strand in Whewell's thought here.

Doubling back, as we explicate Whewell's notion of necessity, the historical emphasis starts to come into focus. Why did Whewell incorporate a temporal dimension into his thought, especially into his thought about the invariable connections binding the world's phenomena? Obviously, on the one hand, we have Whewell the historian of science at work here. He just knew, from study after study, that you do not get at the truth easily. Light comes only after time and much labour. On the other hand, however, the Christian influence is equally significant. We are not put here on this earth simply to have fun. We have roles to play in the cosmic unfurling of God's creation. At the one level, we have our obvious moral tasks. At the other level, we have our intellectual duties. God made this magnificent world, and he expects us to work at discovering its guiding laws. As we do this, and inasmuch as we do this, we show our love of and respect for the Lord. History is the precondition and the

[2] Whewell (1840), i. p. xvii.

[3] Whether Whewell was influenced here by Hegel is a nice question. Whewell castigated Hegel severely for having the temerity to criticize Newton. Nevertheless, there might have been some influence.

result of this gradual process. Necessity has to be set against a developing background.

So much for what I have called the 'metaphysics' of Whewell's philosophy of science. Now, let us turn to the 'methodology' of the philosophy. How is science actually done and made? We must pick up again on the Kantianism. The great German categories of the understanding, shaping, influencing, and making possible of our thought. Whewell's equivalents are his 'Fundamental Ideas' which structure and guide experience: 'those inevitable general relations which are imposed upon our perceptions by acts of the mind, and which are different from anything which our senses directly offer us'.[4] We see the world, as it were, through the lens of these Ideas: as with lenses, making experience possible, but, as with lenses, transforming and constituting experience.

In fact, in real life, Whewell thought that the Ideas generally remain one step removed. In thinking, we work through 'modifications' of these Ideas; secondary notions, which Whewell termed 'Ideal Conceptions'. We have the Ideas of space and time and cause; but these themselves are altogether too abstract for daily use in the laboratory, as it were. Here we need Conceptions, like that of circle (from the Idea of Space) and mutual attraction (from the Idea of Cause). The process of scientific discovery, for Whewell, therefore came down to putting the Ideas, in their manifestations as conceptions, in touch with the real world of facts. The trick is to find the right conception for the right fact.

Digging deeper into this meeting of mind and matter, Whewell isolated three processes—isolated conceptually, that is, for (as intimated earlier) Whewell knew full well that real-life discovery is messy, and may not go in logical sequence. First, we have to work at the empirical material level, the *Decomposition of Facts*, where we must 'resolve the complex appearances which nature offers to us, and the mixed and manifold modes of looking at these appearances which rise in our thoughts, into limited, definite, and clearly-understood portions'.[5] Second, running alongside this process at the material level, we have the mental equivalent, the *Explication of Conceptions*. 'The conceptions must be, as it were, carefully *unfolded*, so as to bring into clear view the elements of truth with which they are marked from their ideal origin.'[6]

This notion of an explication is somewhat more complex than appears at first sight. There are at least two things happening here. First, and most obviously, the scientist is simply trying to get the conceptions clear. What is a circle? How does one define it mathematically? And so forth. Whewell was a sufficiently sophisticated thinker to realize that the simplest notions are often those most difficult to grasp. Second, less obviously but as importantly, there is the task of making sure that a conception is *appropriate* for the area of science to which it is being applied. Let me illustrate this point with the example of biology. The key distinguishing mark of the organic, pointed out Whewell, is that it

[4] Whewell (1840), i. 26. [5] Ibid. ii. 199. [6] Ibid. 172.

works—it is organized, with the parts serving ends. (In this view, Whewell was obviously and explicitly showing the influence of the natural theology of his day.) In this functioning, we can distinguish the world of living things from the world of mere objects.

Stones slide from a rock down the side of a hill and cause it to be smooth; the smoothness of the slope causes stones still to slide. Yet no one would call such a slide an organized system. The system is organized, when the effects which take place among the parts are *essential to our conception of the whole*; when the whole would not *be* a whole, nor the parts, parts, except these effects were produced; when the effects not only happen in fact, but are included in the idea of the object; when they are not only seen, but foreseen; not only expected, but intended: in short when, instead of being causes and effects, they are *ends* and *means*, as they are termed in the above definition.[7]

What does all this mean? Clearly, argued Whewell, we must be applying a specific idea to, and only to, the organic realm. (Despite his careful distinction, Whewell frequently, as here, runs together Ideas and Conceptions.) There is a mind-injected element which makes sense of biology.

Thus we necessarily include, in our Idea of Organization, the notion of an End, a Purpose, a Design; or, to use another phrase which has been peculiarly appropriated in this case, a *Final Cause*. This idea of a Final Cause is an essential condition in order to the pursuing our researches respecting organized bodies.

This idea of Final Cause is not *deduced* from the phenomena by reasoning, but is *assumed* as the only condition under which we can reason on such subjects at all.[8]

The important point is that the Idea of Final Cause is not merely essential, it is uniquely appropriate. Note, incidentally, that one interesting consequence of this position is that although, like today's neo-positivists, Whewell is enthusiastic about universal laws and axiom systems, unlike them, he could never go on to take the next step, arguing that ultimately all science will become one, with (perhaps) physics proving the foundation for all else. Whewell is an extreme 'anti-reductionist'. Each branch of science has its own Fundamental Ideas, and they cannot be subsumed beneath or explained away by the Ideas of other branches. The Idea of Final Cause is highly inappropriate in the physical sciences. (Whewell was at one with Bacon on this.) It is absolutely demanded in the biological sciences. (Whewell was at one with Paley on this.)

Decomposition of Facts; Explication of Conceptions; and so we come to the third part of the discovery process, the *Colligation of Facts*. Facts and conceptions are to be brought together, in such a manner 'as to give rise to those general propositions of which science consists'. Once again, it is Newtonian physics which sets the pace.

Thus the facts that the planets revolve about the sun in certain periodic times and at certain distances, are included and connected in Kepler's law by means of such

[7] Ibid. 78 [8] Ibid.

conceptions as the *squares of numbers*, the *cubes of distances*, and the *proportionality* of these quantities. Again the existence of this proportion in the motions of any two planets, forms a set of facts which may all be combined by means of the conception of a certain *central accelerating force*, as was proved by Newton. The whole of our physical knowledge consists in the establishment of such propositions; and in all such cases facts are bound together by the aid of suitable conceptions.[9]

This then was the heart of the Whewellian discovery process—what Whewell thought of as the key act of *Induction*. It should be noted, however (continuing somewhat anachronistically to compare Whewell with the present) that he would have agreed with the neo-positivists that attempts to provide a logic of discovery are bound to fail. He was highly critical of Mill's 'Methods', at least in so far as they are offered as recipes for discovering new truths. Whewell as scientist and as historian knew full well that genuine advance requires a spark of creative genius not given to all men. Putting facts and Ideas (or Conceptions) together requires real inventiveness.

This is not to overlook the fact that Whewell was aware that there are things one can do to help the creative process. It is not necessary to sit helpless until inspiration strikes. At the level of facts, for instance, there are techniques ('Methods of Observation') which can be employed more accurately to discern the way the world is. For example, one can (and should) repeat measurements, averaging out to get a more robust result. Then, at the level of Conceptions, there are analogous procedures ('Methods of Induction') which help to ferret out nature's laws. For instance, there are well-tried tricks for estimating curves. Likewise, there is the 'Method of Residues':

When we have, in a series of changes of a variable quantity, discovered *one* Law which the changes follow, detected its argument, and determined its magnitude so as to explain most clearly the course of observed facts, we may still find that the observed changes are not fully accounted for. When we compare the results of our Law with the observations, there may be a difference, or as we may term it, a *Residue*, still unexplained. But this Residue being thus detached from the rest, may be examined and scrutinized in the same manner as the whole observed quantity was treated at first: and we may in this way detect in *it* also a Law of Change.[10]

All of this hardly guarantees the finding of laws, but it does tell you when they are needed.

I have written of Whewell's agreeing with the neo-positivists on the question of a logic of discovery. (There isn't any!) Actually, however, this is a little too quick. The neo-positivists draw a strict line between the context of *discovery* and the context of *justification*. Never the twain shall meet. A scientist may be as crazy as you like—and frequently is—but how he or she finds the ideas is quite distinct from the rational endeavour to find supporting evidence. Whewell would have sided with today's critics of the positivists, thinking that no such sharp line can be drawn between such fundamental parts of the work of the

[9] Whewell (1840), ii. 202 [10] Ibid. 556.

scientist. Whewell implies this, inasmuch as he talks of the process of discovery as an 'inductive' process, a term the positivists reserve exclusively for the realm of justification (or would-be justification). And indeed, explicitly, Whewell's whole position ignores (defies, transcends—choose your verb) the discovery – justification dichotomy. A colligation presents you with something new (discovery). Yet, at the same time, if done properly, it contains within itself the ring of truth (justification). Given the mind-injected element in Whewell's philosophy, as soon as you have grasped a concept, proof necessarily hovers nearby.[11]

None of this is to deny that (whether or not the division be sharp) just as Whewell is sensitive to aspects of discovery, so we find he is also sensitive to aspects of justification. As we colligate, we test and check our work by *predicting* from our new-found general statements. As the predictions (at this point, of the kind which led to our general statements) succeed, we build again. As they fail, we tear down and start once more. Even the best scientists—especially the best scientists—try out many different hypotheses before they hit on one which is adequate. Kepler's work on the motion and path of Mars proves this.

Thus, the scientist builds his theory, starting with facts and colligating, and then using these colligations as facts themselves and beginning all over again. Eventually, he or she arrives at the already-discussed axioms, which have the ring of truth—the ring of necessity—and realizes he need go no further. Realizes he *can* go no further. He has reached the point from which the whole of his branch of science can be deduced. Mind and matter combine to give ultimate truth.

Yet, for all his idealism, it was at this point that Whewell introduced one of his most important contributions to our understanding of science—a contribution which quite rises above his own particular metaphysics. Truth, however necessary, is not just handed to you on a plate. Science—and God—make you work for it. You must look for the marks of absolute authenticity. In Whewell's opinion, there are two, although he was inclined to think that they may well be different sides of the same coin.

First, we have the *Consilience of Inductions*. The real mark of scientific excellence, the proof that you have delved into the way that the world works, is the coming together of different areas of science under the same all-reaching hypothesis (or small set of hypotheses):

> Accordingly the cases in which inductions from classes of facts altogether different have thus *jumped together*, belong only to the best established theories which the history of science contains. And as I shall have occasion to refer to this peculiar feature in their

[11] This is the crucial point where Whewell differs from any kind of positivist. They think that science reports in a disinterested fashion on the real world 'out there'. In this respect Whewell is closer to those who would break down the discovery/justification distinction like Hanson (1958) and Kuhn (1962).

evidence, I will take the liberty of describing it by a particular phrase; and will term it the *Consilience of Inductions.*[12]

Again, the paradigm is Newton. From ancient times, people had drawn a strict line between the eternal motions of the heavens and the corruptible world of decay down here on Earth. Newton showed definitively that the same laws apply throughout.

Whewell has spoken already of the importance of prediction. He invokes this activity again, when speaking of consilience, but with a subtle and vitally important difference. Previously, prediction had been of phenomena of the *same* kind (as led to the initial colligation). Now we have prediction of *different* kinds.

We have here spoken of the prediction of facts *of the same kind* as those from which our rule was collected. But the evidence in favour of our induction is of a much higher and more forcible character when it enables us to explain and determine cases of a *kind different* from those which were contemplated in the formation of our hypothesis. The instances in which this has occurred, indeed, impress us with a conviction that the truth of our hypothesis is certain. No accident could give rise to such an extraordinary coincidence. No false supposition could, after being adjusted to one class of phenomena, so exactly represent a different class, when the agreement was unforeseen and uncontemplated. That rules springing from remote and unconnected quarters should thus leap to the same point, can only arise from *that* being the point where truth resides.[13]

Second, as a mark that we have reached absolute truth, following on Consilience, we have the progressive *Simplification of the Theory*. Whewell felt that true laws and theories have a note of clarity, a mark of simplicity, not possessed by the inaccurate or fumbling. Indeed, when new facts come along, true theories (if anything) get simpler, rather than more complex:

The Cartesian system of vortices might be said to explain the primary phenomena of the revolutions of planets about the sun, and satellites about planets. But the elliptical form of the orbits required new suppositions. Bernoulli ascribed this curve to the shape of the planet, operating on the stream of the vortex in a manner similar to the rudder of a boat. But then the motions of the aphelia, and of the nodes—the perturbations—even the action of gravity to the earth—could not be accounted for without new and independent suppositions. Here was none of the simplicity of truth. The theory of gravitation on the other hand became more simple as the facts to be explained became more numerous.[14]

As noted, Whewell saw this process of simplification as probably really being consilience by another name. We bring diverse areas of science beneath the same hypothesis—we create simplicity out of complexity. And, when we have done this, we know we have arrived at the truth. Thus, Whewell's methodology brings him back to his metaphysics: 'in order to make the conclusion demonstrative, which in perfect examples of Induction it is, we ought to be able to declare that the results can be clearly explained and

[12] Whewell (1840), ii. 230. [13] Ibid. [14] Ibid. ii. 234–5.

rigorously declared *only* by the [claims] which we adopt. And in reality, the conviction of the sound inductive reasoner does reach to this point.' There is no more to be said. At this point, we have achieved the highest kind of understanding: an insight into the ultimate causal workings of the world. Deftly, Whewell took one of the more opaque of Newton's dictates about proper scientific methodology and interpreted it to his own ends. With Consilience and Simplification, we have reached a 'true cause'—a *vera causa*.

So much for Whewell's philosophy of science. To the modern mind (more accurately, to *my* mind) Whewell is a funny mixture of the old and the new, of the ludicrously dated and of the percipiently forward-looking. These extremes are highlighted if you compare him to his contemporary John Stuart Mill. The latter never quite sounds as stuck in the ideology of his day as does Whewell, and yet (to me) he never conveys the very feel of science that oozes from every one of Whewell's pages. Whewell knew real science as only one who has been a scientist knows it. I believe him to be an extremely good philosopher of science—to use a hackneyed phrase, much under-rated. But defence of Whewell (as such) is not my aim in this essay. I want to play off his philosophy against his science, as I play off his science against his philosophy, showing (as he himself thought) that they interacted to their mutual benefit. We have now the philosophy. Let us move on to the science.[15]

Physics

Whewell first made his name as a writer of textbooks on mathematics and physics, beginning in 1819 (the year after he became lecturer in mathematics at Trinity College) with a 350-page volume, *An Elementary Treatise on Mechanics*. This was a major shot in the campaign by Whewell and friends (especially George Peacock and John F. W. Herschel) to introduce continental mathematical techniques into Cambridge, replacing the obsolete system of fluxions. To this end, differential and integral calculus were much used, and demanded of the student.[16]

Apparently one of life's great constants is the practice of textbook writers and publishers of forever bringing forth new editions. No doubt this is one of those happy consiliences, where the interests of the student (in having up-to-date texts) coincide with the interests of the producers (in having ongoing royalties). Suffice it to say that Whewell succeeded well enough in his efforts

[15] Much of the science, with the major exception of the work on tides, was finished before Whewell actually set about writing his *Philosophy of the Inductive Sciences*. But letters and other comments show clearly that Whewell was thinking hard about philosophical issues from the beginning of his academic career.

[16] Todhunter (1876), i. 13–28 has a full discussion of Whewell's efforts as a writer of textbooks. Cannon (1978) discusses Cambridge science at the time of Whewell, noting especially the work of the reformers. See also the contributions of Becher and Fisch to this volume.

that his *Mechanics* went through some seven editions (until 1847), and was followed by several other (multi-edition) works of the same ilk. Of course, neither Whewell nor anyone else was under the illusion that textbook writing was a substitute for genuine research. However, this work undoubtedly helped fix Whewell's image in the public mind as an expert on science, particularly the end of the scientific spectrum dealing with the great triumphs of physics. In 1831, it was Whewell who was asked to contribute the volume on '*Astronomy and General Physics*' to the series of tomes on natural theology commissioned in the will of the late Earl of Bridgewater.[17]

Undoubtedly also, Whewell's textbook work helped fix his image in his own mind as an expert on science, particularly at the end of the scientific spectrum dealing with the great triumphs of physics. Certainly it is here that we find what was Whewell's most sustained scientific project. It was one which was to bring him professional credit, including a medal from the Royal Society, the opportunity to carry forward the work of his hero Newton, and the hope of doing that which his philosophy taught him was the mark of good (if not great) science.

Between 1833 and 1850, Whewell wrote some fifteen papers on the subject of 'tidology' (a name, like so many, of his own creation).[18] What are the tides, and more importantly (especially to one like Whewell who put so much emphasis on causes) why are the tides? Taking the second question first (here was a case where causes were suspected before phenomena were discerned) Newton had clearly shown that the tides depend on the varied gravitational effects of the heavenly bodies. Following this, two rival causal approaches had been sketched. First, we have Daniel Bernoulli's equilibrium hypothesis, where the causal question is simplified by assuming that the earth's waters are in equilibrium. Second, we have a hydrodynamical hypothesis, proposed by Laplace. The latter hypothesis, as Whewell knew it, also incorporated a major simplification, namely that the globe is covered with water of a uniform depth.

Nevertheless, since the waters of the earth are so obviously *not* in equilibrium, one might have expected that one like Whewell, who was to make a major effort to understand the tides, would have preferred Laplace's option. However, he did not, working rather against a background of presumed equilibrium.[19] To be honest, the reason why Whewell took this course was probably in major part due to personal inadequacy; he admitted that he lacked the mathematical skills demanded by the hydrodynamical approach.[20] But quite legitimately, Whewell stressed that his main aim lay in getting the facts right, and for such an early part of the development of a science, we might properly use (what we

[17] Whewell's volume, published in 1833, was the first and (generally judged) most popular of the Bridgewater Treatises.

[18] The tidal work was published in the *Philosophical Transactions*. It is reviewed in Todhunter (1876), i. 75–88 and in Ruse (1976).

[19] Whewell (1833*b*); id. (1834*b*).

[20] Todhunter (1876), ii. 269.

today would call) simplified 'models', to co-ordinate and give some meaning to what is (primarily) basic empirical research:

> by treating the tides as a problem of the oscillations rather than of the equilibrium of fluids, undoubtedly [Laplace] introduced the correct view of the real operation of the forces, but it does not appear that in this way he has obtained any consequences to which *Newton's* [and *Bernoulli's*] mode of considering the subject did not lead with equal certainty and greater simplicity . . . [21]

I should add that Whewell always recognized the importance of models (as a stage on the way to the truth), writing to his friend George Airy that the value of theories which are not strictly true 'is so far from being strange, that it is the universal course of scientific history'.[22]

Whewell's avowed interests and intentions, therefore, were essentially focused at the level of the empirical data, the gathering of the facts about the tides, and (presumably) their decomposition. When Whewell entered the field, there was much work to be done. For fairly obvious commercial reasons, families living at or near ports, knowing the details of the local tides, tended to keep such information to themselves, revealing it only when they received (what they took to be) adequate recompense. There had been little or no effort to put tidal readings into systematic order, or to co-ordinate and compare tidal effects across Britain, let alone across the globe.

No one was better suited that Whewell for changing all of that. With his customary energy he guided a squad of workers into getting systematic readings in London, and then in Liverpool, eventually widening his scope so people all around the world were engaged in the work.[23]

Moreover, he massaged the collected data, trying to reduce it to some kind of sense and (expectedly for one who put so much emphasis on the law-like nature of the world, as reflected in science) to discern low-level regularities.[24]

Whewell's theoretical efforts were rewarded with some success, for he was able to discern regularities, particularly those governing the *diurnal inequalities* of the tides, that is the inequalities of the tides on the morning and evening of the same day, at the same place. Particularly gratifying for one whose philosophy of confirmation put great emphasis on successful prediction was the fact that regularities first noted of the London tides were later found to hold also of the Liverpool tides.[25]

Not only did Whewell show his philosophical commitments at this point, inasmuch as he aimed to find low-level ('formal') laws of the tides (the first rungs of a deductive system which would, presumably, end with the Newtonian 'causal' laws of physics) but also, as he was happy to note in both his scientific and his philosophical writings, Whewell found his tide work fertile ground (or water!) for using and elaborating his so-called 'Methods of Induction'—methods

[21] Whewell (1833*b*), 147.
[22] Todhunter (1876), ii. 307.
[23] Whewell (1834*b*); id. (1836*a*); id. (1836*b*); id. (1837*c*); id. (1837*b*); id. (1848).
[24] Human calculators were hired to do tedious arithmetical work at this period.
[25] Whewell (1834*b*); id. (1836*b*).

supposedly used by the working scientist. Thus, for instance, through repeated uses of the method of residues Whewell found ever more accurate formulae governing the effects of lunar declination on the times of the tides.[26]

All of this was worthy work. However, viewed from afar, it really does seem fairly humdrum 'normal' science (to use Thomas Kuhn's (1962) term for non-revolutionary science).[27] And, if the truth be known, I think Whewell would have agreed with this assessment. Certainly, judged by his own philosophy, Whewell thought his work on tides to be a failure, or at least inadequately forward moving. We have seen that the key to his understanding of science lay in the notion of a Fundamental Idea, which in turn reveals itself in modified form through Ideal Conceptions. In scientific labours, along with the decomposition of facts, we get the explication of conceptions and the colligation of facts.

Obviously, therefore, were Whewell to have thought himself even moderately successful, he would have needed to make some progress in the direction of conception explication, and even fact colligation. But here, he confessed, he had quite failed. There is little doubt that, at first (perhaps even attracting him into his tidal work), Whewell thought he was on the track of an appropriate conception. He set about drawing up maps, showing 'cotidal' lines, that is to say, lines joining simultaneous points of high water. Undoubtedly, Whewell hoped that some sort of pattern would emerge, leading presumably to a conception and thus to a colligation.

At first, Whewell sensed that his maps did show something, namely that tides are governed by rules which hold of any channel of water.[28] Thus not only would he find that the rules apply to tides in obvious channels, like that between England and France, but on a larger scale apply to less obvious channels, like that between the Americas and Africa and Europe, namely the Atlantic Ocean. If such did hold, then Whewell was well on the way to laying claim to the title of 'Kepler of tidology'.

Unfortunately, as Whewell himself soon realized, no such universal patterns do really obtain. The tides are far more complex than Whewell had originally suspected. He was, therefore, left only with the satisfaction of proving himself wrong. No doubt this would have been gratifying to a Popperian; but was significantly less so to one with an ego the size of Whewell's. Fortunately, not all of his scientific eggs were in the basket of physics.[29]

Crystallography and Mineralogy

Thus far, we have seen little really to engage our attention. I have confessed my admiration for Whewell as a philosopher of science. To this I would add an

[26] Whewell (1834*b*). [27] Kuhn (1962). [28] Whewell (1833*b*).

[29] In fairness, although Whewell had his prejudices about science he was prepared to admit his own failures and inadequacies. Moreover, he was sensitive to what Popper calls 'falsifiability', even though he thought that ultimately in science we reach necessity.

equal admiration for Whewell as a historian of science. Perhaps this is all that can be expected of one man. As a commentator on science he is first class. As a scientist he is not. (Sorry wine makes good vinegar.) Whewell's tidological work is, almost predictably, pedestrian. Yet, perhaps as predictably, as soon as we get Whewell away from physics—from work within the paradigm, to work where he can use the paradigm—things start to look up significantly. I turn now, in this section and the next, to subjects where, although his engagement was brief, Whewell produced results which today are acclaimed (by experts in their fields) as brilliant anticipations of the way the respective sciences should go. As we shall learn, in significant respects, it was his philosophy which drove Whewell forward, although, paradoxically, in some ways it was his philosophy which held him back.

Whewell was professor of mineralogy at Cambridge from 1828 to 1832. This was hardly a post to which he aspired from a life-long devotion to the subject. Indeed, he calmly considered the prospects of other professorships. The status was the thing. However, in his usual fashion, once the professorship was a real possibility, Whewell set about learning the subject in a full and comprehensive manner.[30]

In this area, Whewell should be remembered (and celebrated) for two seminal achievements. First, there was a fundamental paper on crystallography, for which the claim has been made that it 'laid the foundations of mathematical crystallography'.[31] The paper's title explains all: 'A general method of calculating the angles made by any planes of crystals, and the laws according to which they are formed'.[32] Essentially, Whewell's achievement was to deduce a formula representing the angle between any two planes (specified by Cartesian co-ordinates of the form $Ax + By + Cz = m$), and then to show how the face of any crystal can be represented by such a plane.[33]

In particular, following from the suggestion by René-Just Haüy that the cleavage of crystals can be analysed by supposing them constituted of many regular small volumes, Whewell argued:

Let there be a rhomboid, *A*, divided into a number of small equal rhomboids by planes parallel to its faces. Let any one of the points of division of each of its three upper edges be taken as *P*, *Q*, *R*; and let a plane pass through these three points, *P*, *Q*, *R*. Let the small rhomboids which are above this plane be removed, so as to leave a uniform assemblage of cavities. Then, the remaining surface *P Q R*, being composed of the trihedral angles of small rhomboids, if we suppose the small rhomboids to become smaller than the least distinguishable magnitude, the surface *P Q R* will appear a plane. And if we suppose these rhomboids to represent the primary form of a crystalline body,

[30] That one should acquire a science professorship in a subject about which one knew absolutely nothing was certainly then no oddity. In 1818 Adam Sedgwick had campaigned for the geology professorship under the slogan that hitherto he had turned no stone, but if selected would leave no stone unturned. When successful, Sedgwick set about turning himself into a first-class geologist, a pattern which Whewell was following. See Ruse (1979).

[31] Deas (1959). [32] Whewell (1825). [33] Cf. id. (1827).

P Q R will be a secondary surface deduced from a certain arrangement of these primary elements.[34]

Whewell then showed that if the upper edges of the rhomboid *Ax*, *Ay*, *Az*, be our three axes of co-ordinates, and if the number of small rhomboids along *AP*, *AQ*, *AR* are h, k, l respectively, then the plane *P Q R* has equation

$$\frac{x}{h} + \frac{y}{k} + \frac{z}{l} = 1$$

which is of the required form. Finally, Whewell generalized his result to other types of crystal.

Historically, Whewell's essay proved most important, for his successor in the chair of mineralogy, W. H. Miller, took his result (labelling it the 'law of rational indices of crystal faces'), and making it an axiom, provided a deductive system which formed the basis of crystallographic study for the rest of the nineteenth century.[35] Sadly, however, taking the result as an axiom (as well as Miller's use of spherical trigonometry rather than Whewell's co-ordinate geometry), meant that all of Whewell's reasonings went unmentioned. Thus was eliminated the crucial use of Haüy's ploy, namely that of regarding larger crystals as sub-divisible into smaller ones. Without denying the intrinsic value and actual historical importance of Miller's work, today's crystallographers (using modern x-ray techniques) look back beyond Miller to Whewell; for, like the earlier man, today's workers depend crucially on the supposition that crystals can and must be divided into much smaller units, conceptually if not actually.

Then there was Whewell's second achievement. Working directly as a mineralogist, Whewell (1828) wrote a stimulating monograph which (as in crystallography) incorporates ideas in use today. The big problem in mineralogy (*the* big problem then, and still *a* big problem now) is that of classification. How shall we classify minerals, and is there one system which commends itself above all others? In particular is there one system which is natural rather than artificial? If there is, then perhaps we have hope of getting genuine laws rather than mere coincidences.

Here, with a theoretical problem like this, Whewell the scientist and Whewell the philosopher really merged into one.

> Do not imagine I am doing all for the material sciences with my mineralogy. If I do not fail altogether, it will be seen that this is one of the very best occasions to rectify and apply our general principles of reasoning; and my science shall, without ceasing to be good and true mineralogy, be also a most profitable example of that higher philosophy of yours which legislates for sciences. Remember also that we have got to do something for that same philosophy one day. I have got hold of some good ends of speculations which I think will wind out well.[36]

[34] Id. (1825), 88.
[35] W. Miller (1839).
[36] Todhunter (1876), ii. 61–2.

Whewell argued that if we can find two quite independent ways of classifying minerals (not just separate, but logically distinct), and that if the classifications produced by these ways coincide exactly, then we can be sure that we have more than mere chance. We have a reflection of genuine reality—we have a natural classification. Moreover, our laws come immediately, as we connect claims made on the one classificatory basis with claims made on the other basis.

In the mineralogical case, Whewell thought that just such a natural classification was a possibility. You can classify minerals physically or chemically; on the basis of their physical external features and on the basis of their chemical compositions. Hence, Whewell could write:

> Now whatever may be thought of the respective merits of these two modes of classification, this will probably be allowed;—that either of them would receive a remarkable confirmation if it conducted us to the other:—That if the classes selected in consequence of their external properties, were found to have each some peculiarity or common analogy of chemical constitution;—or if the divisions made by the chemists were found to consist of minerals with striking external resemblances;—we should acquire a sort of criterion that we were approaching to a system which was both chemically and mineralogically true; and its claims to be adopted would probably be readily acknowledged.[37]

Following on actual investigations, Whewell concluded that natural classification is possible at a certain general level (the level he called 'order'), but that more detailed natural classification proves impossible. So-called mineralogical 'species' cannot be natural because they permit the phenomenon of 'isomorphism'—some minerals like garnet, distinguished as one mineral physically, can have different chemical compositions, even within the same specimen.

Some ten years later, when Whewell was writing his *Philosophy*, the ideas on which his essay was founded were incorporated without change into his discussion of classification. The fundamental idea pertinent to that part of the scientific effort is *Natural Affinity*. Of this Whewell wrote: 'And the Maxim by which all systems professing to be natural must be tested in this:—*that the arrangement obtained from one set of characters coincides with the arrangement obtained from another set*.'[38] The illustrative example was, naturally enough, mineralogy.

Whewell's views on the naturalness of classification are those held by mineralogists today, as for instance, Berry and Mason (1959). Thus, as in crystallography, we can praise him for having made a lasting contribution. And we can praise his philosophical *alter ego* for having thus guided him in his scientific efforts. Indeed, as John Herschel (1841) noted, in reviewing Whewell's work, his philosophy of classification reflects the very core element Whewell saw in successful science—a consilience of inductions. In both cases, reality is being revealed by coincidences which are just too close to be merely that.

But, as we leave Whewell's work in this area, let us pause for a moment. His

[37] Whewell (1828*a*), p. iii. [38] Id. (1840), i. 521 (his italics).

ideas were as brilliantly suggestive here as they were humdrum in the field of tides. And yet, Whewell's work here has had to be discovered late in the twentieth century. Part of the reason for this has been explained (or hinted at) already. Whewell did not stay with mineralogy, soon moving on. And, by its very nature, Miller's work concealed the originality and fertility of Whewell's crystallographical speculations. Yet, at least part of Whewell's neglect was self-imposed, brought about, as I have said, by those very factors which spurred him forward; in particular those very *philosophical* factors. Ironically, Whewell himself endorsed Miller's concealment of that part of his reasoning where he used Haüy's suggestion of breaking down larger crystals into smaller sub-units.

Why, the answer is as follows. Although, as we have seen, Whewell was prepared to countenance the use of models and like tools to forward the march of science, for him they could be no more than temporary (in a conceptual sense at least). And this was all the status he was prepared to allow his own speculations. Their consequence was a phenomenal law ('of rational indices of crystal faces'), and hence as such was quite acceptable. But to Whewell the philosopher, the reasoning itself (supposing that crystals might be broken into small sub-units) was highly suspect. And the reason for this was simply that the resultant minute crystals yielded by division—what Whewell, naturally enough, labelled 'molecules'—could, in his opinion, be no more than fictions.[39]

This will strike the modern reader as incredible. Molecules are as well established today as chairs and tables. Modern x-ray crystallography has little doubt about the reality of its sub-units, virtually considering them observable rather than theoretical. But Whewell drew back. Partly this was because he was not convinced that atomic hypotheses adequately accounted for all of the phenomena. But I suspect chiefly it was because he feared that any sub-unit hypothesis would violate one of his fundamental ideas, namely that of *space*. To Whewell, an essential feature of this idea is that it is infinitely divisible. But sub-unit hypotheses suggest that the dividing might have to stop. And so Whewell was led to suspect and reject all such hypotheses, even though this meant a down-grading of his own seminal achievements.

Political Economy

We switch now, quite drastically, from the study of rocks to the study of rents; although, when we learn that the political economist Richard Jones was perhaps Whewell's closest friend, the switch becomes rather less surprising. Whewell seems never to have been tempted into empirical work in the human sciences. Although he spent his life mixing with and dealing with people, and indeed as a philosopher turned increasingly to moral issues his scientific tastes

[39] Whewell (1828*a*), i. 405–22.

ran more to areas where strict quantification and calculation were possible. However, at least at two points of his life, Whewell felt able to bring his mathematical powers to bear on human issues, specifically as they obtained in the area of political economy.

In particular, around 1830 (when he had finished his mineralogical work, but before he took up the tides), we find Whewell writing two papers on taxes and rent.[40] Then, in 1850, Whewell wrote a third paper on international trade.[41] In 1860 came what was probably the greatest satisfaction of all that Whewell was to derive from political economy, when, at the express wish of the Prince Consort, he was commanded to give a course of lectures on the subject to the Prince of Wales.[42]

Whewell's papers are highly mathematical. They symbolize and use as axioms various claims about economic factors, and then they trace the consequences. The earlier two papers, on rents and taxes, have different aims. One is positive, the other is negative. In the positive paper, Whewell formalized the views of Perronet Thompson, who had claimed that taxes on the produce of land fall primarily on landlords rather than on consumers. Whewell thought that Thompson had incorporated a number of simplifying assumptions into his arguments, but none of them concealed the essential truth of what happens. Thus, Whewell was happy to provide a formal model, showing that Thompson's conclusion does indeed follow from certain fundamentally correct assumptions about land, taxation, supply, and demand.

In the negative paper, Whewell went on the attack against David Ricardo, who had claimed the very opposite of Thompson, namely that it was the consumers who suffered, not the landlords. Whewell showed how one would reach Ricardo's conclusions, and then, using figures collected by his friend Jones, he argued that several of the axioms needed by Ricardo were, in fact, false. For instance, Whewell thought Jones had shown conclusively that a number of Ricardo's assumptions about rent simply were too limited to be trustworthy. For instance, Ricardo assumed that, if a farmer could make more money elsewhere, that farmer could and would remove his capital from the land. Relying on Jones, Whewell argued that such an assumption applied at most to one-hundredth of the cultivated world. In most places, for instance where there were serfs and similar, there was simply no question of reducing capital at will. In other words, Ricardo's assumptions were false, because they followed from false premises.

Whewell's third and final paper on political economy, directed against John Stuart Mill and his support for international trade, was likewise negative. By formalizing and showing the assumptions Mill needed to get his conclusions, Whewell again (as with Ricardo) felt able to point to incorrect appeals to the physical evidence—inadequate 'inductions'. Most particularly, Whewell faulted Mill for assuming implicitly that there was a mutualness of demand between

[40] Id. (1830); id. (1831*b*). [41] Id. (1850). [42] Id. (1862).

countries. When this (in Whewell's opinion) obviously incorrect assumption went, so also did the supposed benefits of free trade.

Whewell's excursions into political economy were brief, but the reasons why he got involved at all are very easy to see. The issues at stake impinged directly on the self-interests of one who was a member (and, by the time of the Mill paper, leader) of an institution which got most of its income from rents.[43]

Remember, the first papers were written at a time when the much-hated Corn Laws, jacking up the price of imported corn and thus keeping domestic agricultural rents high, were in force. The last paper was written (although just after the repeal of the Laws) when there was still much debate between the radicals (like Mill), supporters of industry and trade, who wanted food prices low and free access to foreign markets, and conservatives (like the then Master of Trinity) who wanted a good return from the rents of their lands. Thus, whatever else may be the case, Whewell's productions were certainly not disinterested studies for the sake of pure research.

One suspects that the extremely applied nature of Whewell's political economizing was at least part of the reason why he never pushed his researches any further, into a comprehensive theory. He worked at the subject only tangentially, without interest in it for its own sake. But one suspects there was more to the story than this, especially given that Whewell's work was so good, and so far ahead of its time (being very highly praised by today's historians of the subject.)[44]

The most obvious reason why Whewell did so little (and suffered almost total neglect) is precisely that which makes us appreciate him today. He was just too much in advance of his time for his fellows to appreciate him. Even now most social scientists have trouble with mathematics, and there is little reason to think things were much otherwise a hundred and fifty years ago. But, while there is probably truth in this suggestion, it hardly tells the whole story. If anyone was in a position, both in status and in force of character, to push ahead with his ideas, Whewell was.

The major reason, especially for Whewell's own lack of enthusiasm, must be sought elsewhere, and again I think we come back to Whewell's philosophy. At one level, what Whewell did in political economy stemmed from and, indeed, strongly confirmed his philosophical views. Given Whewell's veneration of Newton's labours in mechanics, and his (Whewell's) consequent extreme endorsements of hypothetico-deductive systems, Whewell's formal political-economic models are at once explained. Nor is this explanation negated by the fact that Thompson's model contained simplifications, and that Whewell used the other models to disprove (rather than prove) premises. We know that Whewell saw the development of science as relying on simple or even false

[43] Checkland (1951).

[44] 'Although Whewell's original presentation contained errors, it emerges, when refurbished, as a truly outstanding treatment of the labor theory of value in a multi-input context.' Cochrane (1970), 422.

ideas. The important point is not to take these ideas literally, but to use them in the quest for the truth.

This, Whewell certainly did in political economy. For instance, in attacking Ricardo, Whewell showed that if (as Ricardo claimed) the rise of rents in Britain was caused by the high price of corn, then 'the whole rent must necessarily become a larger fraction of the produce'.[45] This was not so, however. Although rents had risen, the fraction had not. And so, Whewell was led to search for the real rise in rents, which he located in the use of Auxiliary Capital—'that is, capital employed in machines, (ploughs, carts, and c. manure), draining, working cattle, and all other contrivances by which the agricultural labour of man is assisted'.[46] Further, Whewell showed that plausible assumptions about this kind of capital lead to correct conclusions about the actual and fractional rise in rents.

However, for all of this, Whewell had philosophical qualms. Could any study with human beings as its subject ultimately be truly scientific? Whewell was virtually schizophrenic over the answer to this question. At times, he confidently assured others—and himself—that human nature was no bar to true scientific insight. He wrote to one correspondent: 'I am quite satisfied both of the truth of Jones's general views, and also that they possess the great property of true doctrines, that they are capable of being proved at any step of their induction between the most general and the most particular'.[47] And to Jones: 'But what your wiseacres mean by the Metaph. of Polit. Econ. I cannot tell; for there are no peculiar principles of observation or deduction employed in that science—they may as well talk of the metaphysics of Chemistry.'[48]

Yet at other times Whewell was far less sure. Of the laws of political economy, he wrote that they 'will necessarily at first, and probably always be too limited and too dependent on moral and social elements, to become the basis of mathematical calculation'.[49] And, significantly, although the early political economy papers preceded the *Philosophy* by ten years, the subject found no place within the first edition's covers.

It is true that political economy squeezed into that part of the third edition of the *Philosophy* retitled *The Philosophy of Discovery* (1860). But here we find the same ambivalence, being told on successive pages that: 'It may therefore be doubted whether the methods which have been effectual in the discovery of physical theories will not require to be greatly modified, or replaced by processes altogether different, when we would make advances in ethical, political, or social knowledge.' Yet in the social and political arena, it is quite possible that 'the knowledge which we thus obtain may lead us to bodies of knowledge which we may call *Sciences*, and compare with the other Sciences as to their form and maxims'.[50]

[45] Whewell (1862), 66.
[46] Ibid. 67–8.
[47] Todhunter (1876), ii. 86.
[48] Ibid. 49.
[49] Whewell (1831*g*), 43–4.
[50] Id. (1860), 292–3.

Looking at his expressions of doubt, one suspects that Whewell, the moral philosopher, the natural theologian, the Christian clergyman, the teacher of the youth, had trouble harmonizing his beliefs in human freedom and moral responsibility with his philosophical beliefs about the binding necessity of laws. Perhaps they can be reconciled (certainly, many thinkers today believe this to be possible); but Whewell himself was not sure. And so he was inclined to keep political economy at arm's length.

Geology

As our survey of Whewell's scientific achievements draws along, we start to change direction somewhat. To this point, we have been looking at areas where Whewell was actually involved as a working scientist; areas where, as it were, he rolled up his sleeves and got his hands dirty. Now we turn to Whewell's broader interests; to areas where he himself did no actual experimental or theoretical work. Areas where, rather, Whewell was a commentator, critic, judge, and resource person.

However, it would be a mistake to think that these areas were only of peripheral interest to Whewell the scientist. Merely padding Whewell's scientific curriculum vitae, as it were. This was not so, at all. They were very much part of Whewell's scientific fabric. This point applies especially to the topic to be discussed in this section, namely geology. Whewell was no practising geologist, like Adam Sedgwick or Roderick Murchison, but he had a keen interest in the subject—hardly surprising, given that he and Sedgwick were fellows at Trinity together and that geology was *the* trendy subject of the day (a position occupied today by such subjects as sociology and psychology). Moreover, Whewell's knowledge and expertise was valued by the geological community. This was recognized when they made him president of the (London) Geological Society during two of its crucial years.[51]

Furthermore, Whewell's importance as a thinker about matters geological was recognized by that seminal figure, Charles Lyell, who made some effort to ensure that the first exposure that many would have to the ideas of his *Principles of Geology* would be through prominent reviews by Whewell.[52] Although one may perhaps doubt whether Lyell's achievements were quite as epoch-making as he himself was inclined to think (and as history has tended to agree), that they were very important goes without argument, if only for their crucial impact on Lyell's young supporter, Charles Darwin.[53] It was Whewell, more than anyone, who helped to define Lyell's project in the *Principles*, who showed what Lyell was trying to do, what (in the opinion of many of his contemporaries) he failed to do, and what he should rather have done. Thus, in

[51] In his recent study of British geology in the 1830s, Rudwick (1985) places Whewell among the 'élite'.

[52] K. Lyell (1881).

[53] Ruse (1979).

our survey of Whewell's scientific labours, we can and must properly consider his arguments at this point. Again, we shall see the significance of our connecting them, as Whewell's science and his philosophy intertwine.

We have seen how Whewell spoke of the *formal* part of a theory and of its *physical* part, the model being astronomy where Kepler provided the formal part (dealing with what actually happens) and Newton the physical part (dealing with the causes of what happens). In the *History of the Inductive Sciences* (1837a), Whewell took this taxonomy one step further, subdividing the physical part of astronomy into *dynamics*, dealing with the theoretical causes of change, and *physical* astronomy proper, where dynamics is applied to this world of ours and formal astronomy is actually deduced. This tripartite division was followed by Whewell when he turned to geology, and will guide our discussion.

Thus, we have *descriptive* or *phenomenal* geology, dealing with the actual facts) we have geological *dynamics*, where the theoretical causes of change are considered (rain, sleet, earthquakes, etc.); and we have *physical* geology, where dynamics is applied to this world of ours, in order to derive phenomena.

Of the first two sections, considered in their own right, Whewell had little to say. This was the domain of the full-time geologists, and Whewell was happy to praise Sedgwick's sterling work at the phenomenal level and Lyell's even more sterling work at the dynamical level. Predictably, Whewell cautioned that because geology is not yet fully formalized, it cannot be considered a fully mature science. But this was said in the spirit of commentary rather than criticism.[54] Then, with the ground cleared (if you will forgive the geological metaphor), Whewell felt he could turn to that which really interested him: physical geology.

First, we must seek out the fundamental idea appropriate to geology (and indeed to all historically based science, which Whewell labelled jointly the 'Palaetiological Sciences'). It is apparently that of *Historical Cause*—'In our present investigations, we consider the events which we contemplate, of whatever order they be, as forming a chain which is extended from the beginning of things down to the present time . . . ' Geological phenomena, therefore, must be considered 'the effect of what preceded the cause of what succeeded'.[55] Whewell was not entirely clear as to whether Historical Cause is truly a fundamental idea in its own right, or rather a conception of the more general (mechanical) idea of Cause. Perhaps there is no true answer, for the distinction between conception and idea is fuzzy.

What is important for our understanding is the fact that, in geology, Whewell identified a geological methodological rule stemming from the idea of historical cause which is the exact counterpart of a methodological rule stemming from the more general notion of cause: 'Effects are proportional to their Causes, and Causes are measured by their Effects.'[56] In geology, apparently,

[54] Whewell (1837*a*), iii. 549, 554.
[55] Id. (1840), ii. 112.
[56] Ibid. i. 171.

what this means is that you simply must infer causes which operated in the past from effects which we see about us today. The clue to past causes, as Whewell reiterated *ad nauseam*, is not present causes, but present effects.[57]

Against this theoretical background, Whewell felt able to discuss the two rival physical geological theories of the day—theories for which he himself coined the names 'Uniformitarianism' and 'Catastrophism'.[58] The uniformitarian, Charles Lyell, in his *Principles of Geology* in particular, was seen by Whewell to be pushing a 'steady state' theory of the earth, where everything remains much as it has always been—referring now to the inorganic world and recognizing that there will be fluctuations within limits—and where the causes of the past are much as they are today, both in nature and in intensity. The catastrophist, on the contrary, was characterized by Whewell as tending to see some sort of direction to geologic earth history, specifically from a much hotter state to the present temperate state, with occasional times of great upheaval, when the earth's surface was much changed. These great upheavals, 'catastrophies', were most probably naturally caused (that is, no need to postulate special divine interventions), but one could not presume that their causes were necessarily of a kind, let alone an intensity, similar to causes working today.[59]

There is no question but that theological motives were a significant factor in the divide between uniformitarians and catastrophists. Lyell (and Charles Darwin) were deists, seeing God's glory in the fact that He could create a world which, once started, would run indefinitely without need of (miraculous) intervention. Adam Sedgwick and other catastrophists were Christian theists. They believed in a God who is concerned always with His creation, and prepared to intervene as necessary, thus showing His infinite power and love.[60]

Whewell, the Christian clergyman, worshipped the God of the catastrophists, and it is therefore little wonder that he turned his back on Lyellian geology. However, Whewell was a professional scientist, and by 1830 there was no question of simply opposing scientific positions with theological dogma.

[57] Whewell (1837*a*), i. 615–16; id. (1840), ii. 117.

[58] Id. (1832), 126. Marxist thinkers constantly warn us of the dangers of reification, taking an arbitrary concept and raising it to the status of apparently real existence.One of the major ways in which reification occurs is through labelling. Give a dog a name, good, bad, or indifferent, and you are already on the way to making the brute real. Without saying Whewell's division was entirely arbitrary—certainly Lyell thought he himself was pushing a view distinctly different from that of other geologists like Buckland, and Sedgwick and others reacted as if this were just what Lyell was doing—I do suspect that Whewell's dichotomy did polarize geologists rather more than they might otherwise have chosen. Whewell's new terms stuck, both within and without the geological community, inclining people to think in terms of difference.

[59] Though catastrophes were probably naturally caused, there was a general tendency to link them with organic extinctions, which were in turn linked to the need for new creations, which (as we shall see) were most definitely *not* considered naturally caused. Perhaps this accounts for the use of the rather theory-laden term 'catastrophe', which certainly gives a theological flavour lacking in the upheavals supposed by Cuvier, from whom catastrophism was derived. Also, in 1830 Catastrophism was only just emerging from all sorts of identifications with Noah's Flood. See Ruse (1979).

[60] Ibid.; Rudwick (1972).

Uniformitarianism had to be shown inadequate or unnecessary, judged as science. And, since uniformitarianism was really staked more on philosophical factors than on brute scientific facts, Whewell felt himself pre-eminently qualified to argue the catastrophist case (and was so judged by his fellows, uniformitarian and catastrophist).

There was some empirical evidence pointing towards a directional earth history.[61] That which most impressed Whewell was the evidence pointing towards a cooling earth:

> When it is ascertained that in every part of the earth's crust the temperature increases as we descend below the surface, at first this fact seems to indicate a central heat: and a central heat naturally suggests an earlier state of the mass, in which it was incandescent, and from which it is now cooling.[62]

However, there had to be more said than just this. Such facts as these speak only against the Lyellian claims about earth's steady state. What about the key uniformitarian premise that no new causes or intensities are needed or permissible in geology? Here Whewell fell back on his thinking about historical cause. Past causes are to be inferred from present effects, not from present causes. If today's geology seems inexplicable by today's causes—and, relying on such phenomena as supplied by Sedgwick, that the comparatively recent elevation of the Alps could not have been effected by today's earthquakes, Whewell agreed that today's geology does indeed seem inexplicable by today's causes—then it is bad science and bad philosophy arbitrarily to insist on similar causes and similar intensities working throughout earth history. If the effects call for catastrophes, and Whewell thought they did, then catastrophism it must be.

No argument is needed to defend the claim that, at this point, Whewell's science was mixed with Whewell's philosophy. Indeed, here in geology they merge into one. And, significantly, those who opposed Whewell's conclusion, notably Lyell, recognized this fact and countered primarily on philosophical grounds. For instance, what about the question of simplicity, something which we know Whewell prized highly in a scientific theory? 'It is highly unphilosophical, it has been urged, to assume that the causes of the geological events of former times were of a different kind from causes now in action, if causes of this latter kind can in any way be made to explain the facts.'[63]

Whewell was not impressed. Simplicity is desirable, but not at any expense. We know that causes do sometimes get more intense—some volcanoes are stronger than others. Why then should we arbitrarily rule out yet greater

[61] Distinguish between an Earth history with direction and a progressionist Earth history. Most catastrophists, Sedgwick for instance, thought the direction of Earth history was one which showed progress, most particularly towards an earth-state fitted for humankind. This position was backed by a supposedly progressive fossil record—monad to man. Whewell would have agreed that the Earth progressed to a state ready for humans. But, as we shall learn, he pulled back from progressionist readings of the fossil record.

[62] Whewell (1840), ii. 130.

[63] Ibid. 126.

intensity, assuming the quite unwarranted premise that today's causes have never been more intense than they are now? In any case, if we are to defend Lyellian steady-statism, we must assume some kind of cyclical process to account for the fact that, indubitably, some parts of the globe are cooler today than they once were. Hot yesterday, cold today, hot tomorrow.[64] But, apart from the fact that uniformitarianism and catastrophism now start to merge, the pretence of simplicity has quite vanished: 'the hypothesis of uniformity . . . cannot possess any essential simplicity which, previous to inquiry, gives it a claim upon our assent superior to that of the opposite catastrophic hypothesis'.[65]

Yet was not Whewell being unfair to (uniformitarian) geology, making greater demands of it than he would of his own favoured astronomy? Such, at least, was the complaint of Lyell. 'Suppose it was recorded, on authority which you did not question, that at a certain period of the past, say 10,000 years ago, a remarkable inequality, or an unusual rise or fall of the tides took place, or a long suspension of tidal oscillations.' Surely, one would attempt to explain this phenomenon in terms of known causes and intensities. No new forces, of degree or kind, would be involved. Hence, concluded Lyell, one should do likewise in geology, and this points towards uniformitarianism and away from catastrophism.[66] As expected, Whewell disputed this conclusion. Causes must be inferred from effects, and if the effects (as in geology) point away from present causes, then so be it. In taking such a stance, it is clear that Whewell felt he had the deepest backing of his philosophy. Remember that, for Whewell, understanding in terms of causes is the aim of science, and understanding in terms of *verae causae* is the epitome of success. But, for Whewell, a *vera causa* was not necessarily something of which we have had past experience; rather, it is something lying at the heart of a consilience which explains all the facts. If catastrophes are needed to explain all the facts, then this was reason enough for Whewell to turn towards catastrophism.

All in all, therefore, Whewell felt that his science and his philosophy jointly pointed to the same geological conclusion, and although today we might have some qualms about the conclusion, we must surely agree as to the jointness.[67]

[64] Lyell did presume some kind of cyclical process, his 'grand theory of climate' (L. Wilson 1972).

[65] Whewell (1840), ii. 129.

[66] K. Lyell (1881), ii. 5.

[67] Today catastrophism is enjoying something of a renaissance. Certainly no one would deny the directionality of Earth history, from hot to cold, although its presumed time-span—4½ billion years—would have horrified Whewell as much as it would have delighted Lyell. Recently it has been argued that certain major past events, most particularly the demise of the dinosaurs, were caused by extra-terrestrial phenomena, specifically, impinging asteroids (Alvarez *et al.*, 1980). Interestingly, in fighting catastrophism, Lyell explicitly mentioned impinging extra-terrestrial objects as phenomena he wished to deny. Thus, today, catastrophism wins!

But of course it does not really, for the general methods of modern geologists, working from present to past, are Lyellian through and through (Ruse 1981). The moral, perhaps, is that suggested earlier, namely that Whewell did geology little service in thus polarizing uniformitarianism and catastrophism; although, as also noted earlier, given the vigour with which Lyell responded, Whewell was hardly solely responsible for the divide.

Biology

Finally, in this survey of Whewell's scientific thought, let me turn briefly to his ideas about what John Herschel highlighted as *the* most burning scientific issue of the day—the 'mystery of mysteries'.[68] I refer to the problem of the origins of organisms. We see about us a varied living world, and within the fossil record the story of many beings before us. Where did they come from?

Whewell, as you might expect, had very strong views on the subject, consolidating the position towards which, I believe, many were blindly groping.[69] Yet although Whewell's was a conservative position (indeed, he took the whole matter out of science, referring origins to miracles) it was far from an ill-thought-out dogmatic opposition to new ideas. Moreover, speaking generally, Whewell is a paradigmatic example of the troubled pre-evolutionists, as they found themselves inexorably forced towards that very conclusion they most abhorred. And, specifically, I suggest that Whewell's thinking was a great influence on the one who was utterly to destroy Whewell's world—Charles Darwin.

Whewell was certainly no literal Biblical Creationist (either of today's variety or the variety which existed in 1830). Moreover, like Lyell (although unlike virtually everyone else, uniformitarians as Herschel or catastrophists as Sedgwick), Whewell did not see a progressive unfurling (divine or otherwise) in the fossil record. Rather, he believed that organisms come into being, last for a time, and then usually go, at or around the time that other types of organism appear. There is no specific ordering, although humankind is a recent arrival on the scene.[70] Then what did cause new kinds of organism? In dealing with organic origins, we are still (as with inorganic geology) working with the past. Thus the idea of historical cause is still in operation. This means that the cause of origins—that which brought them into being in the first place—had to be something sufficient to make them as they are, and as they were (as represented in the fossil record). In the organic world, Whewell

[68] This phrase comes in a letter from Herschel to Lyell, 20 Feb. 1836. It was well known after it was quoted by Babbage (1838, p. 226) and was (significantly) used by Darwin at the beginning of *The Origin of Species*. Although *The Origin* was not published until 1859, it was in 1838 that the highly ambitious young scientist was giving major attention to the central problem of organic origins.

[69] W. Cannon (1960).

[70] The usual compromise by religious people such as Sedgwick was to allow that the Earth *per se* is old, but that humans appeared 6,000 years ago and that there were no creations thereafter. Thus one could have one's science and one's traditional time-span, as inferred from Biblical genealogies (Ruse 1979).

I am quite sure, incidentally, that Whewell was a non-progressionist for reasons which appealed also to Lyell—both saw organic progressionism as the slippery slope leading to evolutionism (Ruse 1976). Whewell's thinking on this point comes out clearly some twenty years later when he was involved in controversy denying the plurality of worlds (extra-terrestrial beings). Against Baden Powell, who claimed progression implies evolution implies laws implies extra-terrestrials, Whewell denied progression (Whewell 1853).

supposed that unexperienced catastrophes are needed. But (as we know), he made it quite clear that he did not think that these catastrophes would necessarily be of a non-natural (that is, miraculous) kind. In fact, he inclined to think they would be natural. However, in the case of organisms, Whewell flatly opted for miracles.

Part of Whewell's case for this stand was negative, depending on such facts as the inability of breeders to turn cows into horses. Whewell felt able to deny uniformitarian-type natural origins for new kinds of organism.[71] But this was not enough. Organic origins demand supernatural causes, 'successive acts of creation and extinction of species, out of the common course of nature, acts which therefore, we may properly call miraculous'.[72]

Was not Whewell grossly inconsistent in arguing as he did? Had not the clergyman swamped the scientist and philosopher? After all, it was Whewell who wrote, in a passage later made famous by Darwin's cheekily prefacing the *Origin of Species* with it, that events follow general laws, and are not a function of 'insulated interpositions of Divine Power'.[73] However, Whewell felt that he had to argue for miracles and he was prepared to argue also that his position was scientifically respectable. Science itself shows that science alone is not adequate!

The key to Whewell's position lies in a point uncovered in our discussion of his philosophy. In biology, although historical cause is crucial when dealing with the past, we must also be guided by that distincitve fundamental idea which applies when dealing with living (or once living) beings. This is the idea of Final Cause.[74] Organisms seem as if designed. Unlike inorganic objects, they are organized. They are teleologically directed towards ends, and must be thought of this way. Apply this point now to the problems of the past. How do we explain the origins of organisms, things which exhibit final causes? Blind laws—natural laws—lead to blind effects, to randomness. But causes must be adequate to effects. Hence we must invoke adequate causes, and these (for organisms) must be guided causes. Obviously, therefore, since organisms are not the result of human intelligence, they must be the result of divine intelligence. In short, we need miracles. And the point is that we have been driven to this conclusion by the best scientific reasoning. Science has shown that it cannot work alone; 'when we inquire whence [organisms] come into this our world, geology is silent. The mystery of creation is not within the range of her legitimate territory; she says nothing but she points upwards.'[75]

Whewell's argument was not that the miraculous origins of organisms break the laws of nature. Rather, they stand outside or beyond them. But, does God have His own laws, which He imposes upon Himself? Perhaps so, but, the early Whewell (that is, the Whewell of the 1830s) wanted to keep the mystery

[71] Whewell (1837*a*), iii. 575–9.

[72] Ibid. 574. Whewell did not allow that some extinction is natural. The unhappy story of the dodo was well known.

[73] Id. (1833*c*), 356.

[74] Id. (1840), ii. 78.

[75] Id. (1837*a*), iii. 588.

of mysteries clouded in mystery. God's creative power does not belong to 'what we are accustomed to speak of as the laws of nature . . .'[76] Science and humans can say nothing. Later, however, we see a shift in Whewell's position. In 1844, the Edinburgh publisher Robert Chambers published (anonymously) an evolutionary tract, *Vestiges of the Natural History of Creation*. Orthodox scientists like Sedgwick were appalled, and unorthodox scientists like Darwin, then finishing his own essay on evolutionism, prudently decided to delay publication. Joining in the general outcry, Whewell responded in about as insulting a manner as he could, namely by collecting already published anti-evolutionary passages from his earlier works, adding a stirring preface which did not even mention its target by name, and putting forth a slim volume with the revealing title, *Indications of the Creator*.[77]

However, although Whewell stood firmly against evolution, he did have to concede a move in the direction of that very position evolutionism epitomizes—the introduction of organisms by a natural, law-bound process. Chambers may have left Whewell unmoved, but Whewell's good friend, the anatomist Richard Owen, was by then pushing his Anglicized version of transcendentalism, where all organisms are believed to be modelled on one of a number of fundamental blueprints—'archetypes'.[78] This position did influence Whewell.

From Owen's viewpoint, archetypes neatly solved a problem pressing on any perspective on organisms which takes adaptation (linked by Owen, like Whewell and all the others, to final cause and God's design) as fundamental. How does one explain the apparently useless isomorphisms between organisms—the skeletal similarities between forelimb of man, foreleg of horse, wings of bird and bat, flipper of whale, paw of mole, and more? These appendages are all used for different functions, and yet they are linked through common form. They exhibit what Owen labelled 'homologies'. Owen was therefore happy to agree with those European thinkers who argued that the whole of nature is formed according to certain basic ground-plans, and he put much effort into making out the particular nature of a number of key archetypes.

Whewell was keenly aware of the problems faced by comparative anatomists like Owen, and for this reason alone he would have welcomed archetypes. But, for Whewell the philosopher and educator, archetypes were even more important than they were for Whewell the scientist and natural theologian. After the writing of the *History* and the *Philosophy*, Whewell turned increasingly from Kantianism to Platonism. The reasons for this switch were varied, although undoubtedly a major factor was the importance Plato accords to mathematics in the education of the young, a position strongly endorsed by the conservative Whewell in the face of the university reformers of his day. But be that as it may, Whewell began to regard his fundamental ideas less as Kantian conditions of the understanding, and more as Platonic Forms or Ideas, transcendental entities with a real existence. Owen's archetype theory was, therefore, manna

[76] Id. (1832), 125.
[77] The second edition did identify *Vestiges* as its target.
[78] Owen (1848).

from heaven (to use a metaphor Whewell would have appreciated). In organisms, one could actually see the participation of the things of this world of change in the eternal entities of the Real World, to put matters in Platonic terms.

But there was a price which had to be paid. No longer could Whewell deny any knowledge of God's creative practices. Whewell the Platonist was also Whewell the Christian, and so he had now to agree that we have some insight into those very rules or laws by which God binds Himself in the creative act. Referring to archetypes, Whewell allowed that one cannot 'doubt that this spectacle of analogies and resemblances implies the existence of laws in the mind of the creator, according to which he has proceeded in the work of Creation'. At once, he tried to cover himself, adding that people of his persuasion

> cannot go onwards, with the writer of the 'Vestiges', from resemblance to sequence, and from sequence to causation. They cannot venture to say that animals followed one another upon the earth in the order of their anatomical resemblances; and that their anatomical differences grew out of each other naturally by general laws.[79]

However, one might be forgiven for thinking that, whatever he might have said in excuse or reservation, Whewell's admissions about God's laws were the thin end of a very large wedge, a wedge which would finally open the door to full-blown evolutionism. Certainly, we find people like Owen eventually being pushed into an evolutionary stance.[80] Moreover, we find that many simply refused (or were unable) to sense the subtleties of a position like Whewell's, and merely took it as a step on the path to a fully natural understanding of organic origins.

Concluding our story of Whewell's entanglement with the problem of organic creation, let me note that Darwin (who had known Whewell very well, when an undergraduate at Cambridge and later, in the late 1830s, through the Geological Society) sent his old friend a copy of the *Origin* when it was published. Whewell accepted it graciously, regretting however that he was unlikely to accept its conclusions. And in 1864, in a preface to a final (7th) edition of his Bridgewater Treatise, Whewell again affirmed his preference for miracles—and, as he had done nearly twenty years before against the *Vestiges*, refused to identify his target by name!

Whewell therefore set his face firmly against the major scientific advance of his age. And yet, science itself was to have the last laugh on Whewell, making use of him despite his own ardent counter-efforts. Darwin's *Origin* was no mere compilation of facts, hoping by sheer force to make the case for evolutionary descent. Rather, it was one long, deliberate, and carefully constructed argument, designed to persuade the reader that, although no one could enter a time-machine and fly back to the past, present evidence points

[79] Whewell (1846), 13.

[80] Owen was never quite sure whether to excoriate Darwin and his followers as disciples of Satan, or to sue for plagiarism on the grounds that he had thought of it all first.

overwhelmingly to a shared origin of transmutation, from simple forms, by natural processes. In short, Darwin wanted to persuade the reader that evolution, what he called 'descent with modification', through his own particular causal mechanism of natural selection, was a *vera causa*.[81]

And what was the criterion of *vera causa*-hood that the young Darwin, working on his ideas in the late 1830s, seized upon? None other than that of his friend and advisor, William Whewell! Thus, Darwin was led to make a consilience of inductions the focus of his theorizing, and it is this that structures the *Origin*. From evolution through selection follow the effects of the organic world: instincts, homologies, geographical distributions, fossil remains, and much more. Conversely, these effects convince one that, although unseen, Darwin's central mechanism is a cause of the highest kind, a *vera causa*. For all that he argued otherwise, Whewell lives on today in the major unifying theory of biology.[82]

Conclusion

Whewell was not a great scientist. That, he candidly admitted himself.[83] His major work, on the tides, although worthy, was essentially a failure. At least, it was if one judges it by those inductive leaps that he detailed in his *History* and analysed with such care in his *Philosophy*. And elsewhere in science, he either failed to pursue insights or (as in the case of biology) looked firmly backwards rather than forwards.

Yet there has been much gained from our rapid survey of Whewell's scientific endeavours. First, through the work and interests of one man we have had an almost unique insight into the science of Britain in the second quarter of the nineteenth century.[84] Second, although Whewell may not have achieved true greatness in his science, there is that in his work which engages interest in its own right, taking one well beyond the ordinary. I refer particularly to Whewell's prescient insights in mineralogy and in political economy. He was, quite genuinely, much ahead of his time.

Third, as we have seen, Whewell influenced others. Sometimes the influence may not have been entirely to the good. One thinks, for instance, of the way in which Whewell helped polarize geological theorizing. But such influence existed, nevertheless. Sometimes the influence was significant and important, even to the present day. The case of Darwin and the *Origin of Species* has just been discussed. The central biological consilience is no less crucial to today's evolutionists than it was in 1859.

81 Ruse (1979).

82 Ruse (1973); id. (1982). There is much more to the story of Darwin's methodology than I have been able to mention here. For more details see Ruse (1975*a*) and Hodge's contribution to this volume.

83 Todhunter (1876), ii. 286.

84 I say 'almost unique' because Herschel was almost as catholic as Whewell in his interests.

And fourth, and most importantly, by looking at Whewell the scientist we throw light on Whewell the historian and philosopher of science, and conversely. Whewell may not have been a great scientist, but I would argue that, as a commentator on science, he did achieve greatness. Hence, we should cherish any understanding that can be gained in this quarter. Whewell claimed that his philosophy guided his science, and that his science provided grist for his philosophy. Our survey has surely shown that this is true. In all of the areas of science in which he worked, Whewell tried out his philosophical ideas. Conversely, his scientific achievements were taken straight up as examples for his philosophizing. The symbiosis was one of strength and weakness, at least judged from our vantage point, a century and a half later. In respects, Whewell's philosophy drove his science forward. One thinks, for instance, of Whewell's work on mineralogical classification, and of his use of formal methods in political economy. In respects, Whewell's philosophy pulled his science back. One thinks of Whewell's doubts about the existence of subsensory particles, and of his opposition to naturalistic explanations of organic origins.

But which ever way Whewell's science was tugged, it was his philosophy that was tied to the other end of the rope. Whewell, although a man of many parts, would have wanted us to see his work as an integrated whole. We can now surely agree that he was right in such a wish, and that in attempting such an integrated vision we have paid homage to him and brought insight to ourselves.

5. *Passing on the Torch: Whewell's Philosophy and the Principles of English University Education*

by PERRY WILLIAMS

AS Master of Trinity College, William Whewell was the most powerful person in the University of Cambridge at a time when university education was a major political issue. The industrialization and commercial expansion of the early nineteenth century had elevated new groups to rival the old established rulers of society: Dissenters challenged the Church of England, manufacturers challenged the landed gentry, radicals challenged the traditional politics of Whigs and Tories; and in these battles, the universities were disputed territory. People now thought of universities as having a national role, as institutions which might help to create (or preserve) the form of society they wanted to see; thus their struggle for different forms of society became also a struggle for different forms of university education. Whewell was a strong contender in this struggle, and this chapter will be about the form of education which he thought a university should supply, the role which he believed universities played in creating a nation's future, and the effect on society he was trying to produce through his influence at the University of Cambridge.

But this chapter will also be about his philosophy. Like Geoffrey Cantor and Richard Yeo, I take the view that Whewell was trying to establish a unified account of the nature of knowledge, embracing both the physical sciences, represented in his *History* and *Philosophy of the Inductive Sciences*, and the moral sciences, represented in his *Elements of Morality* and his teaching as Professor of Moral Philosophy. Whewell's philosophy of knowledge is relevant here because he saw a very close connection between these works and his most important educational writings, *Principles of English University Education* (1837) and *Liberal Education* (1845). On all their title pages he placed a

I would like to thank Simon Schaffer for helping and supporting my Whewell studies beginning with my undergraduate dissertation in 1980; Andrew Cunningham for teaching me how to do history and encouraging me to toughen the argument of this article; and Sue Morgan for sharing her insights with me (see n. 5). I would also like to thank Harvey Becher, Menachem Fisch, and Richard Yeo for helpful conversations and criticisms.

distinctive motif, specially commissioned for the *History* in 1837,[1] which showed a flaming torch (representing knowledge) being passed on from hand to hand. At one level, the connection between these works was logical and thematic, his philosophy of knowledge forming the basis of his educational writings; indeed he claimed that the argument of the *Principles* was directly suggested to him by his work on the *History of the Inductive Sciences*.[2] Even taking the connection at just this level, then, in order to understand why Whewell advocated the form of university education that he did, we also need to understand why he developed his philosophy of knowledge. But once we have done this, it will emerge that the connection between these works was closer still, for, I shall argue, Whewell's development and propagation of a philosophy of knowledge and his involvement in the university reform controversy represented a single enterprise with a single purpose.

Some aspects of the connection between Whewell's educational and philosophical thought have been studied before. Martha McMackin Garland, for instance, has related the ideal of liberal education expounded by Whewell and others in the Cambridge Network to their unified view of truth, in which physical, moral, and theological knowledge were harmoniously combined.[3] But my account of the *purpose* of Whewell's unified project may be more unexpected. The question of what Whewell was trying to do in the educational sphere has been in general answered only superficially; Garland is content with the answer 'improvement' or 'modernization', as though improvement was all of one kind. As we shall see in the first section of this chapter, Whewell's ideas about what constituted improvement were in total contradiction with those of radical reformers. We will also see that there was more consistency between the 'improvement' policy of Whewell's youth and the unqualified conservatism of his later years than is generally realized; the apparent inconsistency arises not because Whewell became rigid and 'ossified' with age,[4] but because we have misinterpreted what he was trying to do through such 'improvements' as encouraging the increased teaching of the sciences and attacking the teaching of Paley's Benthamite philosophy.

The second section of this chapter will be about what Whewell was trying to do through his philosophy of knowledge. Inevitably there will be some overlap with the chapters of Menachem Fisch and Richard Yeo, but I have found it necessary to give an additional account here. For whereas Fisch stresses the technical and philosophical aspects of Whewell's philosophy of knowledge and Yeo focuses on its relation to contemporary intellectual debate, I will be following a line of investigation started by Sue Morgan[5] and arguing that

[1] Todhunter (1876), i. 118.

[2] Whewell (1837), 1.

[3] Garland (1980), esp. 39–50; S. Cannon (1978), ch. 2. Becher (1975) provides a much more sophisticated analysis of the philosophical and theological basis of Whewell's ideal of liberal education; this paper is unfortunately rather difficult to obtain.

[4] Garland (1980), 40.

[5] Morgan (1985). I am indebted to Morgan particularly for drawing my attention to the significance of Jones's attack on Ricardo.

Whewell's underlying aims were practical and political. Following this, in the third section of this chapter, we will be able to see that his constant purpose, in the whole series of projects which culminated in the *History* and *Philosophy of the Inductive Sciences*, then in the policies laid down in the *Principles of English University Education* which he pursued throughout the rest of his life, was to counteract the spreading influence of utilitarianism, atheism, and radicalism by bringing people to appreciate the true nature of knowledge—'[giving] a right and wholesome turn to men's minds', as he put it.[6] His aim, in short, was to make subversive thought unthinkable.

This presentation of Whewell's work as part of a social and political conflict may come as a surprise. The cultural dominance which the members of the Cambridge Network achieved has tended, as Jack Morrell and Arnold Thackray observe, to 'blind us to the reality that they . . . were but a competing faction within early Victorian culture';[7] and nostalgia for their harmonious and unified world of learning has tended to make us forget that the harmony existed only within the Network (if there). We have not expected, for example, to find deep conflict between Whewell and university reformers outside Cambridge. Yet, as we shall see in the next section, there was a fundamental opposition between Whewell and those who set the university reform movement going in the 1830s, whose aims went well beyond university reform to the promotion of wider social and political change.

I

'This is the age of reform', wrote William Hamilton in 1831, at the start of an *Edinburgh Review* article on the University of Oxford. 'Next in importance to our religious and political establishments, are the foundations for public education; and having now seriously engaged in a reform of "the constitution, the envy of surrounding nations," the time cannot be distant for a reform in the schools and universities which have hardly avoided their contempt.'[8] This extremely aggressive article followed an *Edinburgh Review* tradition of hostility towards the two English universities, running back to the early years of the journal when John Playfair and Sydney Smith condemned Oxford and Cambridge for their antiquated and useless curricula.[9] Since that time, these Edinburgh Whigs had formed an alliance with leading Midlands Nonconformists and London disciples of Jeremy Bentham, whose political aims were similar in as much as they required the wholesale reform of old institutions—most obviously Parliament, but also the English universities. These Whigs, Dissenters, and Radicals saw Oxford and Cambridge as currently operating in the service of the old ruling classes. For one thing, the main subjects of study—classical

[6] Whewell to Hare, 25 Dec. 1833, Todhunter (1876), ii. 175.

[7] Morrell and Thackray (1981), 21.

[8] [Hamilton] (1831), 384.

[9] [Playfair] (1807–8); [Smith] (1809–10).

literature, Aristotelian logic, and the mathematics of Euclid and Newton—were determined by gentlemanly ideas of polite manners rather than the practical needs of future statesmen and professionals. For another, religious tests and compulsory attendance at chapel effectively excluded those who were not Anglicans from studying there at all, while the college fellowship system was effectively a branch of the career structure of the Church of England rather than a teaching organization. Nor was it likely, with their rigid and oligarchic constitutions, that the universities would reform themselves to meet the needs of the new commercial industrial classes.[10]

When these Whigs, Dissenters, and Radicals attacked Oxford and Cambridge, they were looking to the universities not only of Scotland, but of post-Revolutionary France and the German states, especially those founded on the model of the new University of Berlin. These were State rather than Church institutions, being state-supported and (supposedly) operating for the benefit of the state as a whole, rather than one particular class. Their curriculum was modern and taught by specialist professors, each an expert in his own subject, unlike the English universities, where the main teaching was provided by clergyman tutors with pretensions to universal competence. The new university in London (later University College, London), which the Whigs, Dissenters, and Radicals set up in 1827 to break the Oxbridge monopoly, was designed on Scottish and German lines: it was to teach useful knowledge, such as political economy, the natural sciences, medicine, and modern languages; the teaching was to be professorial, with complete *Lehrnfreiheit* (freedom for the student to attend lectures as he chose), and no religious restrictions of any kind.[11] This 'godless institution' was sufficiently threatening to Tories and High Anglicans to goad a number of them into founding their own rival King's College, where 'instruction in the doctrines and duties of Christianity as taught by the Church of England should be for ever combined with other branches of useful education'.[12] But by that time, the reformers had gone beyond setting up alternative institutions and were in a position to conduct an assault on Oxford and Cambridge themselves.

Two things made it seem likely that they would succeed. The first was the passage of the 1832 Reform Bill and the subsequent formation of a Parliament better disposed towards reform in general. The second was the increasing support for some degree of university reform at least from Whigs at the universities themselves: not merely sometime graduates, like Hamilton, but men still working at or closely associated with the universities. At Cambridge, these included Whewell's friend Adam Sedgwick; his assistant tutor Connop Thirlwall; and George Pryme, Professor of Political Economy and MP for the town of Cambridge since the 1832 election. The Dissenters' demand for

[10] Tillyard (1913), chs. 2–3; Cardwell (1957), 25–33, 51–9; [Hamilton] (1831, 1831*a*); [Hogg?] (1831); [Anon.] (1831).

[11] [Brougham] (1825); [Macaulay? Brougham?] (1825–6).

[12] Charter of King's College, quoted in University of London (1912), 10.

unrestricted admission and graduation was supported by these three in particular, with Pryme and Sedgwick organizing a petition to Parliament, while Thirlwall published a pamphlet defending their action, for which he was expelled from his tutorship by Christopher Wordsworth, the Tory Master of Trinity. The parliamentary campaign for the admission of Dissenters was led by Lord Radnor, who in 1834 actually succeeded in passing an appropriate Bill through the House of Commons, though unsurprisingly it was overturned by the Bishops in the House of Lords.[13] In 1837, Radnor raised the issue of university reform again, this time with a Bill for a Royal Commission to investigate the statutes of the Oxford and Cambridge colleges. Like Hamilton in 1831, he pressed for reform on legal and historical grounds: originally, the argument ran, the teaching and government of the universities had been in the hands of all their graduates, but in the seventeenth century, at the instigation of Archbishop Laud, the college Heads had seized control and allowed or caused the professorial chairs to decay into sinecures, replacing them by a college tutorial system; hence the colleges were now acting in violation of the original statutes.[14]

It was at the end of 1837, while the Master and Senior Fellows of Trinity were hastily revising their statutes so as to reconcile them with current practice and thus destroy the most powerful argument of the reformers,[15] that Whewell brought out his book *On the Principles of English University Education*. He had already published a pamphlet during the Dissenters controversy, condemning Thirlwall for having criticized his own College in public and for denying the value of compulsory chapel.[16] In 1835 he had published a pamphlet giving his *Thoughts on the Study of Mathematics*, which received a hostile notice from William Hamilton in the *Edinburgh Review*, who regarded it as a very inadequate justification for the exclusive concentration on mathematics at Cambridge.[17] His new book, which included the *Thoughts* and his reply to Hamilton as an appendix, was essentially an argument against change. In the three sections of the book, Whewell maintained the necessity of continuing the 'English' system of tutorial teaching in subjects such as mathematics and classics; the priority, during most of a student's career, of college teaching and examinations over those of the university or private turors; and the importance of a college-based 'discipline system' of constant control and supervision and compulsory attendance at lectures and chapel. In short, in all respects except one—he supported increased teaching of the natural sciences—Whewell contradicted the demands of the reformers.

[13] Thirlwall (1834); [Anon.] (1834). See also [Wilson] (1834); [W. Hamilton] (1834); Winstanley (1940), 83–96, 73–8; Garland (1980) 73–7.

[14] Hansard, Parliamentary Debates, xxxvii (1837), 1001–43; id. xxxviii (1837), 509–30, 658–75. [W. Hamilton] (1831), 384–427. Masters of Arts still took an oath to teach daily in the schools for 5 years, Doctors for 2 years: Winstanley (1940), 249. See also Kemble (1837).

[15] Winstanley (1940), 186–90, 95.

[16] Whewell (1834*d*).

[17] W. Hamilton (1836).

Whewell was no crass reactionary though, stubbornly clinging to the past. He presented new and intellectually sophisticated grounds for the maintenance of the status quo at the English universities, based on a new ideal of what universities were for. He retained as basic the principle that the purpose of a university was to provide a 'liberal education', traditionally the education of gentlemen who did not need to earn a living, as distinct from a professional or a practical education. This at once invalidated any kind of argument from utility, for instance that universities should teach modern languages and the physical sciences because they were useful for business and industry; such narrow and specialized training might appropriately follow a university education, but was not to form a basic part of it. As he was later to write, 'the object of a liberal education is, not to make men eminently learned or profound in some one department, but to educe all the faculties by which man shares in the highest thoughts and feelings of his species'.[18] Yet this passage also shows how Whewell had altered the concept of liberal education. Whereas in the eighteenth century a gentleman's education had been directed towards the acquisition of taste and polite manners,[19] a liberal education for Whewell now meant the cultivation of the basic faculties of the human mind; its purpose was to 'draw forth and unfold a portion of our common human nature', 'to connect a man's mind with the general mind of the human race'.[20]

Whewell was not of course the only person to redefine the concept of liberal education and the role of the university in this way,[21] but it had a special meaning for him because of its foundation in his philosophy of knowledge. He had already published enough of this in the *History of the Inductive Sciences* of 1837 to be able to invoke it explicitly in the *Principles of English University Education* later that year:

> By a history of each of the sciences in succession, I have proved, I hope satisfactorily, that their progress depends upon the distinctness of certain fundamental ideas; and that these ideas, being first clearly brought into view by the genius of great discoverers, become afterwards the inheritance of all who thoroughly acquire the knowledge which is thus made accessible.[22]

All knowledge in the sciences, according to Whewell, was based on Fundamental Ideas—certain general relations by which sense impressions were organized. The physical sciences, he argued in the *History* and *Philosophy*, were based on Ideas such as Space, Time, Number, Cause, Externality, Media, Likeness, and Polarity, and he believed that there were Fundamental Ideas in morals and politics too; later, in the *Elements of Morality*, he was to enumerate these as the Ideas of Humanity, Justice, Truth, Purity, and Order. The significance of these Ideas was that, once held distinctly in the mind, they enabled people to see the principles of a science as universally and necessarily true. The history of the sciences was thus a history of the progressive intuition of these Ideas: now

[18] Whewell (1845), 107–8. [19] Rothblatt (1976). [20] Whewell (1845), 6–7.
[21] Rothblatt (1976); Garland (1980). [22] Whewell (1837), 20.

that the Idea of Force (or Cause of change of motion) was distinctly possessed, for instance, it was impossible to conceive of Newton's First Law of Motion as false, although it was once believed to be untrue.

This distinctness of Ideas, generally diffused throughout society, Whewell identified with civilization. As with 'liberal education', he was one of those who altered the meaning of the word at this time; it was no longer the process of acquiring polite manners or 'civility', but the state of a nation's culture, the general level of development of the higher faculties of the mind.[23] Whewell saw universities as having a vital role to play in the development of this culture, since they educated men of the upper class, who were to become leaders of the community, and who would therefore affect the mental character of the nation.[24] The role of the universities was thus to act as centres of learning from which distinct Ideas would diffuse out across the country. Such was Whewell's hope for the future progress of civilization:

> One idea after another, of those which constitute the basis of science, becomes distinct, first in the minds of discoverers, then in the minds of all cultured men, till a general clearness of thought illuminates the land; and thus the torch of knowledge is handed forwards, thousands upon thousands lighting their lamps at it as it passes on; while still from time to time some new Prometheus catches a fresh light from heaven, to spread abroad among men in like manner.[25]

It was this promise that he represented in his title-page motif.

But if the university's job was to cultivate distinct Ideas in its students, Whewell argued, there were certain implications. First, it was necessary that the chief subjects taught should be ones in which 'principles having clear evidence and stable certainty . . . form the basis of our knowledge'.[26] Secondly, it was necessary that they be taught in a manner which Whewell called 'practical': the student was not to be allowed to remain a passive listener but made to take an active part, repeatedly producing his own knowledge for practice and correction.[27] The mathematics teaching of college tutors provided the model; 'the pupil is required to prove orally a proposition in Euclid or in Mechanics, and is interrogated as to the connexion of the steps of the proof, or as to the application of the proposition to a particular case'.[28] Though the pupil would no doubt already be able to see the self-evidence of the axioms of mathematics, practice at deductive inference from them would fix and develop the Fundamental Ideas on which they were based. 'Other persons may have the idea of space, as well as the geometer;—the idea of force and matter, as well as the mechanician; but these ideas shine with a clearer and steadier light in the minds of those who constantly work by such lamps, and therefore, carefully tend and trim them.'[29] By contrast, a subject such as philosophy, particularly prized at the German universities, was unsuitable for basic

[23] Rothblatt (1976), ch. 2; Williams (1976), s.v. 'Civilisation'.
[24] Whewell (1845), 1–2; (1850*d*), 17–18.
[25] Id. (1837), 27.
[26] Ibid. 9.
[27] Ibid. 5–8.
[28] Id. (1845), 110.
[29] Id. (1837), 16.

university education; its foundations were constantly changing, which allowed only 'speculative' teaching, as through the passively received lectures of a professor.[30] Whewell confirmed the importance of basic educational subjects for science and civilization by a brief historical survey, which showed that '*the progress of science corresponds to the time of practical teaching; the stationary, or retrograde, period of science, is the period when philosophy was the instrument of education*'.[31]

He used analogous arguments to show the necessity of classical study forming part of a liberal education, in order to transmit that culture of the mind related to the expression of thoughts and feelings[32] (he was later to speak of a Faculty of Language, or reflex thought).[33] What he was particularly concerned to contradict in the *Principles* was the claim that classics of modern European literature should replace those of ancient Greece and Rome. He did this by an appeal to continuity. Through their examples of poetry, eloquence, history, criticism, grammar, and etymology, the Greek and Latin classics had been the base on which the language and literature of later ages rested; hence a knowledge of them was necessary to inherit that accumulation of mental wealth. They were the common intellectual estate of all educated people, to cut oneself off from which would render it impossible to understand even modern authors. 'In what a condition should we be, if our connexion with the past were snapped;—if Greek and Latin were forgotten? . . . The mental connexion of the present nations with each other, as well as with the past, would thus be destroyed. What would this be but a retrograde movement in civilisation?'[34]

In the second section of *Principles*, Whewell set out his view of the proper relationship of college and university teaching. Here again he contradicted the demands of the reformers, for he maintained that the 'office of the Colleges' was of the highest importance to the cause of education in our Universities';[35] while other elements might play a part, they could not be allowed to interfere with the basic system of direct, practical instruction by college tutors. University examinations, he suggested, should not disturb students' work before their third year, since study directed towards an examination tended to become mere cramming. Similarly, the lectures of university professors might be of benefit to students in the later stages of their education, but only after they had first received a clear appreciation of basic principles; if English professors cancelled their lectures or drew small audiences only, it was because they were of lesser significance in English university education, not because they were lazy or corrupt, as reformers claimed. Private tutors too might have a role in university education, he thought, but only if under the direction of a college tutor, for such private coaches typically provided only cramming without real education or moral guidance.[36]

[30] Whewell (1837), 9–11. [31] Ibid. 22, emphasis in original. [32] Ibid. 31–8.
[33] Id. (1845), 8–11. On reflex thought, see also e.g. id. (1845*b*), i. 159.
[34] Id. (1837), 37–8. [35] Ibid. 61–2. [36] Ibid. 54–80.

The moral side of a liberal education was the subject of the third section of the *Principles*, on discipline. 'We must . . . *educate*, not *instruct* merely: we must infuse a sense of moral and religious responsibility, as well as mere knowledge: we must form the principles of conduct as well as the intellect.'[37] For the cultivation of the student's moral as well as intellectual Ideas, Whewell looked to the influence of the colleges, each a little world in which, as he quoted, the statutes decreed 'that all inferiors behave themselves towards their superiors in a submiss and reverent manner:—the Scholars towards the Bachelors, the Bachelors towards the Masters of Arts, these towards Bachelors and Doctors of Divinity, and all towards the Master as the supreme governor, and also towards the eight Seniors as fathers and leaders.'[38] In particular he looked to the tutors, to whom it mainly fell to administer punishments—not in retribution, but to mould a student's attitudes and behaviour[39]—and to enforce attendance at college lectures and chapel. He was well aware that they could compel attendance only, and not attention or devotion, yet he maintained that the compulsion had value: that by compelling the forms they could support the substance and so help establish beneficial habits of mind.[40] Once again, Whewell contradicted the demands of the reformers, both supporting the college system as a necessary part of liberal education, and rejecting totally the *Lehrnfreiheit* associated with the universities of France, Germany, and London.

Overall then, the *Principles* presented a new and well-founded philosophy of liberal education that was essentially opposed to reform. That was the way it was understood by Charles Lyell, a Whig with a long-standing interest in university reform,[41] who regarded it as 'one of the most authoritative & able defences of "things as they are" '.[42] In his 1845 *Travels in North America*, while attacking Oxford and Cambridge with the now customary historico-legal argument, he singled out Whewell's *Principles* for criticism: since he wanted to see more scope for professorial teaching of the moral, political, and physical sciences, he objected in particular to its defence of the virtual monopoly of classics, mathematics, and tutorial teaching.[43] Whewell was hurt by this attack, which he felt to be unfair; by this time, he was contemplating ways of building up professors' audiences and expanding teaching of the sciences, and he thought he could see these things advocated in the *Principles*. He wrote to Lyell and forced him to see some justice in this; he had, after all, agreed that there was *some* role for professorial teaching and that a degree of insight into the progressive sciences was an essential part of liberal education. Nevertheless, Lyell in his reply maintained his conviction that 'in its general scope & spirit, it is a panegyric of the peculiar machinery of tuition, & plan of study then & now adopted in Oxford & Cambridge, & by no means calculated to lessen the

[37] Ibid. 81. [38] Ibid. 83–4. [39] Ibid. 93–102.
[40] Ibid. 102–27. [41] See Lyell (1827).
[42] Lyell to Whewell, 17 Apr. 1847, Whewell Papers, Trinity College Library, Cambridge University, Add. MS a. 216[57].
[43] Lyell (1845), i. 303–7.

prejudices of those who admire & uphold that system'. Furthermore, he could not agree with some of Whewell's maxims about the teaching of the sciences, which he considered 'as peculiarly opposed to those University reforms for which I was contending in my Travels'.[44]

It is important to realize that although Whewell supported teaching of mathematics and the sciences, he only supported teaching of a particular kind—the kind validated by his philosophy of knowledge. This is most evident in his 1835 pamphlet *Thoughts on the Study of Mathematics*, which was not, as Hamilton had implied in his hostile review, primarily a vindication of mathematical study, being aimed not so much at reformers outside the university as at Whewell's teaching colleagues inside it. A common objection to the exclusive study of mathematics was that it would lead a student to expect the same precision and certainty of principles in other subjects and so unfit the mind for dealing with other kinds of truth, but Whewell paused only briefly to rebut this. For him, of course, truth in other subjects *was* essentially similar to truth in mathematics, so this danger did not arise—so long as the mathematics was taught properly.

> If mathematics be taught in such a manner that its foundations appear to be laid in arbitrary definitions, without any corresponding act of the mind:—or if its first principles be represented as borrowed from experience, in such a manner that the whole science is empirical only;—or if it be held forth as the highest perfection of the science to reduce our knowledge to extremely general propositions and processes, in which all particular cases are included;—so studied, it may, I conceive, unfit the mind for dealing with other kinds of truth.[45]

These were tendencies which he could see becoming more prevalent in mathematics teaching at Cambridge,[46] and the pamphlet was an attempt to identify them to his colleagues, to show their philosophical and educational unsoundness, and so extirpate them. In each case the tendency was to hinder elementary students from achieving a firm grasp of first principles—that is to say, distinct Fundamental Ideas. In the first case, the principles were made out to be arbitrary definitions; in the second case they were taught as experimental discoveries, as contingently instead of necessarily true; and in the third case, they were confused by being presented through the generalities of analytical mathematics.

This last case requires some explanation, since Whewell is usually regarded as having been a champion of the new analytical mathematics of Laplace and

[44] Whewell to Lyell, 31 Mar. 1847; Lyell to Whewell, 17 Apr. 1847, Whewell Papers, Add. MS a. 216^{56-7}.

[45] Whewell (1837), 171.

[46] Whewell named no individual, but referred only to 'a peculiar school of mathematical speculation' (Whewell (1835*a*), 150), and 'recent mathematical literature . . . by authors who have recently passed through their under-graduate career' (Whewell (1835*a*), 176). From Harvey Becher's work on mathematics teaching at this time, I would guess that John Hymers was one of those he had in mind (Becher (1981), 21–2). The only other work Whewell refers to is *Geometry without Axioms* by Peronnet Thompson, one of the 'Philosophical Radicals'.

Lagrange. Indeed he was one of those responsible for introducing it into Cambridge in the 1810s, though Harvey Becher has argued that even then he saw analysis only as a set of techniques for solving physical problems, whereas the French mathematicians and British analysts such as Woodhouse and Peacock wanted it developed as part of pure mathematics.[47] At all events, Whewell reversed his position somewhat in the 1830s. As a young graduate, he had looked forward to a time when the algebraical reasoning of analysis should have totally replaced the geometrical arguments of Newton;[48] as a 41-year-old tutor in 1835, he opposed the prominence of analytical mathematics in the Cambridge curriculum on the grounds that the abstractions of algebra led to 'the obliteration of all the peculiar fundamental conceptions which belong to the different parts of mathematics'.[49] The analyst could obtain his answer just as well if he had forgotten, or did not know, whether his *x* represented an angle, a distance, a force, or a mass, so there was no encouragement to the intuition of Fundamental Ideas.[50] Whewell was not, of course, attacking the study of analysis by those aiming to become professional mathematicians; he saw it as a useful, even an essential, tool, but it was not to be studied before the basic mathematical Ideas had been made distinct. He believed that it could not take the place of the geometrical reasoning of Newton in the university mathematics curriculum, and he repeatedly fought the promoters of analysis on this issue.[51]

Whewell's policy on the teaching of the sciences, and what parts of them should be taught and how, was similarly underwritten by his philosophy. The *Thoughts on the Study of Mathematics* concluded with a recommendation for introducing Newtonian mechanics and hydrostatics into the elementary curriculum of the 'poll men'—the majority of students who took the less demanding examinations for an Ordinary Degree; he was able to put this into effect the next year (1836) as a member of the tripos reform syndicate, and he supported the course with a new textbook, *The Mechanical Euclid* (1837).[52] The rationale for this extension of the curriculum was clear. As he later wrote in the *Philosophy*, in the section on intellectual education:

> It is undoubtedly in the highest degree desirable that all great advances in science should become the common property of all cultivated men. And this can only be done by introducing into the course of a liberal education such studies as unfold and fix in men's minds the fundamental ideas upon which the new-discovered truths rest.[53]

[47] Enros (1981, 1983); Grattan-Guinness (1985). Enros argues that the role of the Analytical Society has been overrated; for the traditional story, see Dubbey (1963). For Whewell's role, see Becher (1981).

[48] See his remark about people soon '[only reading] a few propositions of Newton as a matter of curiosity', Whewell to Herschel, 1 Nov. 1818, Todhunter (1876), ii. 30.

[49] Whewell (1835*a*), 171.

[50] Id. (1845), 44–6.

[51] In general on analytical mathematics, see id. (1845), 38–77. For his defence of Newtonian methods against his colleagues, see id. (1832*c*), pp. vi–viii.

[52] Id. (1835*a*), 180–1; id. (1837), 16–17, 29 n.; id. (1845), 190, 34–5.

[53] Id. (1847), ii. 363–4.

In other words, the torchlight of distinct Fundamental Ideas was to be spread ever wider. But this recommendation extended only to sciences which had become as stable and permanent as Mechanics and Hydrostatics. For he went on to say:

> No Ideas are suited to become the elements of elementary education, till they have not only become perfectly distinct and fixed in the minds of the leading cultivators of the science to which they belong; but till they have been so for some considerable period. . . . Hence it was perhaps reasonable that a century should elapse from the time of Galileo before the rigorous teaching of mechanics became a general element of intellectual training; and the doctrine of universal gravitation was hardly ripe for such an employment till the end of the last century.[54]

It was this passage in particular which made Lyell think that Whewell, despite his general support for teaching of the sciences, was not a real friend of reform. 'These doctrines', he wrote,

> are admirably calculated to make our Universities the last strongholds into which antiquated methods of teaching and thinking will retreat & where they will make a stand long after they have been fairly driven out of the open country. . . . The regular well-paid standing army of University teachers availing themselves of your maxims will go on for a hundred years teaching what their ancestors taught . . .[55]

In fact, Whewell was not in a position to be strict with these maxims; there were other pressures on him, in particular the need to reverse the decline in lecture attendance suffered by all professors except those of mathematics and divinity—a problem of which, as a professor himself, he was acutely aware. It was probably this, as much as any desire that students should 'become acquainted with some portion of this body of accumulated and imperishable knowledge',[56] that motivated him in bringing forward the new regulations of 1848, which made attendance at one course of professors' lectures obligatory for all poll men.[57] The same set of regulations gave the lectures of certain professors another important role: that of providing the instruction for two new triposes in Natural Sciences and Moral Sciences—a system he had advocated in his *Liberal Education* of 1845. But Whewell had not diverged from his 1837 *Principles* as much as it might seem, for this professorial teaching in the sciences was emphatically only a supplement to the basic liberal education of mathematics and classics.[58] The latter were what he now called,

[54] Whewell (1847), ii. 373.

[55] Lyell to Whewell, 17 Apr. 1847, Whewell Papers, Add. MS a. 216[57].

[56] Whewell (1845), 17.

[57] Whewell had already made several efforts to boost lecture attendance. See Winstanley (1940), 178–81; Becher (1986), 80–4.

[58] Whewell's original proposal was that the new triposes, like the Classical Tripos, should be open only to those who had reached Honours standard in mathematics. In the event, he was content that they should be open to those who had only reached Ordinary Degree standard. He did not approve of the Classical Tripos being favoured in this way, however. See Winstanley (1940), 198–213; Whewell (1845), 223–7; id. (1850*d*), 4–23.

using the terminology of Coleridge and Peel,[59] 'Permanent Studies', 'those portions of knowledge which have long ago taken their permanent shape'. Subjects which were still developing, such as modern literature, analytical mathematics, and the sciences, were counted as 'Progressive Studies'; these he held to be desirable also, but they were only to follow the Permanent Studies and never to take their place.[60] His chief concern in *Liberal Education* was not in fact the advocacy of new Progressive Studies, but the strengthening and stabilizing of the Permanent. Much of the book was devoted to his plan for establishing a basic, standardized mathematics course, overseen by a Board of Mathematical Studies, which would be the preliminary to any study for Honours; he considered the existing course insufficiently fixed and definite and too much in neglect of the essential permanent principles.[61]

Whewell's other main purpose in writing *Liberal Education* was once again to defend the University from radical reform. Radnor's Bill of 1837 had come to nothing since the Whig administration was weak and reluctant to take it up, but by 1845 Whewell could see 'tolerably plain indications that the old Universities are not to expect a continuance of the protection they have been accustomed to receive at the hands of Government',[62] so he thought it important to state once again the principles which had to be maintained. He reissued the *Liberal Education* in 1850 and 1852 with additional sections to bring the discussion up to date with current events, for his fears had turned out to be fully justified: despite the university's efforts to pre-empt parliamentary interference by revising its own statutes, major reform packages were being introduced under the direction of the Royal Commission of 1850–2 and the Statutory Commission of 1856–9. Now that he was Master of the biggest and most important of the colleges, Whewell was heavily involved in the negotiations, becoming notable for his vigorous defence of the established order. On the syndicate for revising the university statutes, he maintained that the Vice-Chancellor should retain his absolute right of veto, to avoid him having to bring before the university 'any project, however crude and mischievous'.[63] Later he defended the reputation and authority of the College Heads during the 'democratic frenzy', as he called it, of a popular assault on their domination of university government.[64] He came into conflict with the members of the Royal Commission over their proposal to fund a generous expansion of university teaching through a tax on the colleges, which he called 'a violation of the rules of property so strong, as to belong only to times of revolution and confiscation'.[65] When the Statutory Commissioners declared their intention to

[59] Coleridge (1830), 16–21, 33–4 etc. See Schaffer, Ch. 8 in this volume.

[60] Whewell (1845), 21–2. See also ibid. 95–117, 163, etc.; id. (1837), 42–6, 133–4.

[61] Id. (1845), 196–7, 207–9.

[62] Winstanley (1940), 187–8. Whewell to J. C. Hare, 12 Aug. 1845, quoted in Winstanley (1940), 198.

[63] Ibid. 238–42. Whewell (1852*c*), 70.

[64] Winstanley (1940), 42–57; letter to his wife, 12 May 1855, Stair-Douglas (1882), 440.

[65] Winstanley (1940), 259–63, 267; Garland (1980), 130.

limit the tenure of college fellowships and to open them to married men and laymen, Whewell led the opposition from the colleges in general and Trinity in particular.[66] He was also opposed to the democratization of college government, in particular the introduction of annual meetings of all the Fellows, down to the most junior,[67] but here he was forced to admit defeat.

'A very sad season of my College life,' he wrote in anticipation of these new meetings, 'to have the College pulled in pieces and ruined by a set of school-boys.'[68] The whole quarter-century of his Mastership was coloured by this belief that the lower academic orders were generally irresponsible and not to be trusted. As Vice-Chancellor, he made himself unpopular by suppressing undergraduate rowdyism with a vehemence which alarmed even his friends,[69] but could surely have been anticipated from his statement that the position of pupil was incompatible with such behaviour as attending political meetings or shouting.[70] Trinity Fellows too were made to feel the weight of his authority, for he banned them from smoking in their Combination Room and refused to allow them to have private keys to the College.[71] Other incidents contributed to his general reputation as an arrogant and overbearing bully: for instance, his rehanging of the paintings in the Fitzwilliam Museum without consulting the other members of the governing syndicate, which led to their unanimous resignation in protest; or his prestige battle with the Assize Judges over whether or not they had the right to accommodation in the Master's Lodge.[72] Whewell the Master, as he was known by the Cambridge generation of Maxwell, William Thomson, and Henry Sidgwick, was well represented in his statue, sculpted soon after his death, which displayed to good effect that massive, powerful body which had once caused a prize fighter to remark what a man the ring had lost when they made him a parson.[73] Taut and alert, seated as if in judgement,[74] Whewell still glares down imperiously on visitors to the Trinity College chapel.

Whewell's authoritarianism was not a feature of his last years only; his powers as Master and the rigidity and dogmatism of old age simply made it more conspicuous. His philosophy of knowledge validated and encouraged his belief in absolute authority, both intellectual and moral; for if true knowledge lay in the intuitive appreciation of Fundamental Ideas as real and definite as a flaming torch, then those who possessed the torch, whose Ideas were more distinct, were fully entitled to disregard the misguided opinions of others. Whewell's philosophy of knowledge, in other words, justified his insistence on

66 Winstanley (1940), 339–72; Whewell (1857*a*).
67 Winstanley (1940), 352–3; id. (1947), 240–1.
68 Letter to Kate Marshall, 30 Oct. 1857, Stair-Douglas (1881), 497.
69 Stair-Douglas (1881), 275–6, 283–4, 285–7.
70 Whewell (1837), 130–1.
71 Robson (1964), 168–76.
72 Winstanley (1940), ch. 9; id. (1947), ch. 2.
73 *DNB*, s.v. 'Whewell', 460.
74 For a contemporary description, see Stephens (1894), 84.

a strongly hierarchical relationship between teachers and pupils, and seniors and juniors; and we have already seen how it underlay his defence of the college system (in which authority in teaching and discipline could be most easily maintained) and the priority of subjects to which his philosophy of knowledge applied (and hence which upheld the teacher–pupil hierarchy most strongly), such as mathematics (properly taught), the better established parts of the physical sciences, and to a lesser extent classics. In short, his philosophy of knowledge justified all aspects of his overall opposition to radical university reform.

I do not wish to imply that Whewell's educational politics was simply a working out of the logical implications of his philosophy of knowledge. Such an account would be wholly inadequate; there is no logic which *compels* anyone to apply a philosophy of knowledge to educational questions. (Modern commentators on Whewell's philosophy certainly do not do so.) Yet it does seem that Whewell's philosophy of knowledge preceded his involvement in the university reform controversy, coming earlier in time, and in some way giving rise to it. According to Whewell, the views he expressed in the *Principles of English University Education* were suggested to him by 'a long and somewhat laborious course of researches on the principles and history of science'[75]—in other words, the writing of the *History* and *Philosophy of the Inductive Sciences*. Why should that have suggested writing the *Principles* to him? The common passing-on-the-torch motif on the title pages gives us a clue. What it implies is that Whewell saw his philosophical writings and his educational writings as a single project with a single message and a single purpose. His involvement in educational politics would thus have been a simple continuation of the work which he had begun in developing and propagating his philosophy of knowledge, which therefore naturally justified his educational views. But if this is the case, then to understand fully what Whewell was trying to do in his educational politics, we must also understand what he was trying to do in his philosophy of knowledge. And this will be the subject of the next section.

II

'My Induction', was how Whewell referred to the *History* and *Philosophy of the Inductive Sciences* while he was writing them.[76] Induction was a subject on which he had been especially working since 1830, when he first began to write a book

> *to give the description and rules of the inductive philosophy*. To shew in what it peculiarly consists: how far the conception of it originally promulgated [by Bacon] was a true one; what are its precepts, its formulae, its prohibitions; what its [prominence?], its

[75] Whewell (1837), 1.

[76] Whewell to Jones, 6 Oct. 1834, 2 May 1835, 11 Nov. 1835, Todhunter (1876), ii. 192–3, 210–11, 230.

grounds[,] its prospects. But more especially . . . to shew how this method has been exhibited and exemplified since it was first delivered . . . [77]

Whewell's conception of Induction at this time was not the one which we normally associate with him. When he began this book (which he never completed), he thought of Induction as a simple Baconian ascent from observed phenomena to general principles, and it was not until 1834 that he began arguing that coming to see phenomena as connected by a general principle depended on the intuition of some Fundamental Idea. But although he had changed his conception of Induction, it remained the central theme of the *History* and *Philosophy of the Inductive Sciences*; there he presented all the history of a science which was not an Induction as being either a Prelude or a Sequel to one, for according to his philosophy of knowledge it was only with an Induction that any wholly new knowledge was acquired.

Why was Whewell interested in Induction? Why did he change his conception of it? What was he trying to achieve by writing these books? The answer which Menachem Fisch provides in his chapter is that Whewell was trying to solve a particular philosophical problem, that of explaining how 'excellent science', as Fisch calls it, could come about. The answer which Richard Yeo provides in his 1979 article and his chapter in this book is that Whewell was trying 'to dissociate science from Utilitarianism and empiricist philosophy' and '[confirm] Christian values'.[78] These accounts accord with the one I am about to give in many respects (for example, the date at which Whewell introduced the Fundamental Ideas into his philosophy, the importance of his opposition to Utilitarianism and empiricism, and his defence of Christianity), and I do not want to deny their truth or validity, as far as they go. But I shall argue here that such philosophical problems as the nature of induction and the acquisition of knowledge in the sciences were of concern to Whewell only because of certain contemporary problems which confronted him during the 1820s and 1830s, and that these problems were not just academic matters of intellectual debate but serious social and political threats which promised to destroy civilization as he knew it.

In those years, there was a dramatic rise in the following of that group of men now generally known as the Philosophic Radicals. Through their periodical the *Westminster Review* and through enthusiastic and eloquent new disciples such as Charles Austin, the utilitarian philosophy of Jeremy Bentham, James Mill, and David Ricardo spread rapidly amongst the younger generation, especially the more intellectual undergraduates; in 1825, Mill joked that his essays were 'the textbooks of the young men . . . at Cambridge'.[79] Utilitarianism's attraction lay in its wide-ranging intellectual power, and its radical potential as

[77] Whewell Papers, R18 17[12]. 28 June 1830 is the date at the start of the book, and the draft preface is the first item. The remainder of the contents may of course be of a later date.

[78] Yeo (1979), 502–3.

[79] Mill to McCulloch, 18 Aug. 1825, quoted in Bain (1882), 292. See Thomas (1979), 451–2, 135, etc.

a solvent of received wisdom and a critique of the established order. Just as the phenomena of the physical world could all be reduced to Newton's laws, so the phenomena of the moral world could be reduced to laws of human nature. All questions of ethics, law, and politics were to be decided by appeal to principles such as the principle of association, attributed to Locke, that ideas in the mind were nothing but the sensations of experience, written onto a blank slate and connected together by association; or the principle, due to Bentham, that human beings are governed in their actions only by pleasure and pain, and that therefore the only criterion for whether an action or a social institution was good or bad was whether it tended to increase or decrease the sum total of human happiness; or the population principle of Malthus, that an unrestrained population will grow faster than the available food supply. There was disagreement amongst utilitarians as to the precise political programme which all this implied, but it certainly included sweeping measures of economic reform, legal reform, parliamentary reform (perhaps even total democracy), and educational reform.[80]

Whewell was hostile to utilitarianism for a number of reasons, the most important being that the utilitarians portrayed the world in a way which ignored or denied the existence of God. People were portrayed as governed entirely by selfish pleasure-seeking, society as divided by divergent class interests, and the misery of starvation as inevitable; the world of the utilitarians was materialistic and lacking in benevolent design. We can infer that this was repugnant to him from his reaction to atheism in the natural sciences. Since his youth, he had been strongly committed to what we might call their devotional aspect, in that he was accustomed to study of the natural world leading to contemplation of its divine Creator, His power, wisdom, and benevolence.[81] Probably the initial attraction of analytical mathematics for Whewell was that it provided a way of doing this better; the programme of Newton could be fulfilled,[82] and the motions of the solar system, down to the finest detail, could be shown to follow from just a few laws of dynamics, the empirical discovery of which showed the Creator's design. French analytical mathematicians, however, presented the laws as pure definitions or self-evident truths,[83] thus making the solar system self-sustaining, even (if one accepted Laplace's 'nebular hypothesis') self-creating: a world from which God was absent. Whewell was moved to fight this interpretation; in a number of articles and prefaces he argued that the

[80] Ibid., especially ch. 3 on James Mill. See also Halevy (1928); Collini, Winch, and Burrow (1983).

[81] See e.g. Whewell to his sister Ann, 1 Jan. 1822, Stair-Douglas (1881), 72–3.

[82] 'The student who feels a proper admiration for the system of the *Principia*, ought to look forwards to the complete development of it in the *Mechanique Celeste*, as the ulterior subject of his labours.' Whewell (1823), p. v.

[83] 'It appears as if their attachment to the forms and processes of pure analysis . . . had given them a disrelish for the more physical and inductive part of the reasoning, and made them comparatively indifferent as to the manner in which they arrive at that part of the subject where the machinery of analysis begins to work.' Whewell (1819), p. vi.

French mathematicians' view of the laws of dynamics was logically fallacious,[84] and in a private notebook in 1825 he proved to his own satisfaction that the discovery of laws of nature did not make it unnecessary to think of God, for they were no power or agency in themselves but only the rules by which the supreme Agent operated.[85] But Whewell's devotional use of the sciences was threatened from the other side by High Churchmen like his friend Hugh James Rose, who saw mathematics and experimental philosophy no less than utilitarianism as the products of a worldly and materialistic age.[86] (See Yeo's chapter, Section I.) In letters to Rose and in a series of university sermons in 1827, Whewell defended the sciences against the charge of atheism, claiming that, pursued correctly, the physical and moral sciences provided 'food for consolatory and admiring contemplation' of God's power and goodness.[87] It was only when pursued by men 'with unenlightened and unenlivened minds',[88] that they produced abominations like Laplacean physics and utilitarianism;[89] and this was the only reason that 'about the modern physical and moral sciences, as they are often presented to us, there hangs an atmosphere chilling and deadening to religious feeling'.[90]

The second reason for Whewell's hostility towards utilitarianism was that it was 'mischievous', or as we might say, subversive. Not only did it provide the lower classes with reason to hope for political power and the overthrow of the established order, it positively encouraged young people to disregard traditional wisdom without troubling to understand it, and to criticize and judge their elders. Again, we can infer that this was disturbing to Whewell from his reaction when students seemed to show the same impatience and conceitedness in their mathematical study. Here it was a case of them neglecting their elementary work and rushing on to the more abstruse parts of advanced mathematics, attempting to answer difficult, high-scoring questions in the Tripos when they did not have a proper understanding of mathematical basics. In 1827, on a syndicate for reviewing the Tripos regulations, Whewell used his influence to encourage better habits: the syndicate confined questions on higher mathematics to the last two days of the exam, and introduced for the first time printed question papers, which would make it clear to future students that a mastery of elementary mathematics was needed to do well on the first two days.[91] Apparently Whewell saw analytical mathematics as a special temptation for young mathematicians; the algebraical techniques were

[84] Whewell (1828*b*); id. (1819), pp. v–vi; id. (1832*d*), pp. ix–xii; id. (1834*c*).

[85] Todhunter (1876), i. 360–5.

[86] Rose (1826), 12. Whewell to Rose, 19 Nov. 1826, 12 Dec. 1826, Todhunter (1876), ii. 75–9.

[87] Whewell Papers, R17[13], fo. 3 reverse.

[88] Ibid. fo. 14.

[89] The subjects of his second sermon (Whewell Papers, R6 17[14]), and, as originally intended, his fourth (R6 17[17]). A passage of the latter is quoted in Todhunter (1876), i. 330–1. In the event, Whewell gave a different sermon in place of this, probably because Jones had complained that it would steal the thunder of his forthcoming book; see Todhunter (1876), ii. 79–83.

[90] Whewell Papers, R6 17[13], fo. 13.

[91] Rouse Ball (1899), 117–37, 187–219; Whewell (1845), 188; Becher (1981), 4–14.

powerful and exciting to use, and it was all too easy for a student to think himself superior for knowing some analytical trick which his teachers did not.[92] In his first textbooks, he had been trying to make the new analytical methods more popular and more available, but by the late 1820s he came to see analytical mathematics as something to be discouraged rather than encouraged amongst undergraduates; the revisions which he made to his textbooks in 1832 therefore included partitioning off analytical methods from the elementary material, integrating the argument of his Dynamics books more closely with Newton's *Principia*, and arguing strongly for the retention of geometrical methods.[93] He was by now in his late thirties, and unable to sympathize with a younger generation which rejected Newtonian mathematics and took up utilitarianism. 'You may depend upon it,' he wrote to his college contemporary Richard Jones in 1828, 'that we lived exactly at the right time for catching all the better influences of the university and that there have been nobody half so wise and enlightened in the later generations'.[94] Though this was a tongue-in-cheek remark over a minor issue, it nevertheless testifies to Whewell's sense of alienation from the younger generation, and his anxiety about the bad influences working on them.

The events of the early 1830s must have confirmed Whewell's worst fears for the consequences of the spread of utilitarianism. The 1830 revolution in France had been led by middle-class radicals and supported by mass unrest, and for a while it seemed that the same explosive mixture might produce a revolution in England too. His friend Jones, a curate in Sussex, found himself in the middle of uprisings and rick-burnings in the autumn of 1830, and though Whewell was more distant from the rioting, he was almost as much alarmed by what seemed like 'a break up of society', and one which seemed certain to spread.[95] The uprisings of 1831, after the House of Lords blocked the Reform Bill, must have been even more frightening; jails were stormed and the prisoners released, and the houses of peers and bishops were burned. By the end of 1832, after the Reform Bill had been passed, though it was clear that there would be no revolution, yet the radicals had won their way sufficiently to ensure that Reform was going to remain a political issue for the rest of the decade.

While all this was happening, Whewell came to see 'Induction' as a possible answer to the spread of atheism, radicalism, and utilitarianism. His friend Jones had already adopted it as his chief weapon in his fight against the political economy of the utilitarian David Ricardo. For Jones, as for Whewell at this time, 'Induction' was part of the method of enquiry laid down in the *Novum*

92 See Whewell (1845), 72–3; id. (1835*a*), 175–7.

93 Whewell (1833*e*), p. ix. *Analytical Statics* (1833*d*), was effectively the analytical sections of the first edition, carved out and made into a separate book. The revised *Dynamics* extended over three volumes (1832*d*, 1832*e*, 1834*e*). See Whewell to Wilkinson, 2 Oct. 1832, Todhunter (1876), ii. 147.

94 Whewell to Jones, 4 Dec. 1828, Whewell Papers, Add. MS c. 51^{58}.

95 Jones to Whewell, Oct.–Nov. 1830, 21 Oct. 1831, Whewell Papers, Add. MS c. 52^{17-19}, c. 52^{41}; Whewell to Jones, 16 Nov. 1830, Whewell Papers, Add. MS c. 51^{91}.

Organum of Francis Bacon, which they had avidly read and discussed as undergraduates: a method of *induction* from observations to concepts or laws, followed by *deduction* of novel consequences: of reasoning first *to* general principles and then *from* them.[96] Jones's argument was that Ricardo's theory of Rent had been established by deduction from general principles which had not been first established inductively, and which investigation could show to be not generally true; the conclusions of Ricardo's theory, for instance that the interest of landlords was opposed to the interest of every other class in society, were therefore fallacious.[97] The project which Jones set for himself was to construct a properly *inductive* political economy, based on a comprehensive collection of observations (the first book was to be an investigation of five different classes of rent in some twenty different countries from Scotland to China), which would simultaneously refute and replace that of Ricardo and his followers. Whewell was 'quite satisfied . . . of the truth of Jones's general views'[98] and became an enthusiastic supporter of his project, chivvying him on with reminders of his duty 'to throttle the shortwinded, rotten pseudo-political economists' and obtaining for him a printing contract with the Cambridge University Press.[99] (See Becher's chapter, and Yeo's chapter, Section II.) Whewell himself began working on a complementary project on *deductive* political economy, expressing general economic principles in mathematical form to enable clearer calculation of their deductive consequences.[100] Despite having no inductively established principles to work from, he was still able, in two papers to the Cambridge Philosophical Society in 1828 and 1831, to demonstrate his mathematical technique, and to make clear the limitations of Ricardo's principles, the deductive errors he had made, and the existence of alternatives.[101] Whewell had more than one eye to the possible influence of these papers on undergraduates. 'I do hope', he wrote to Jones, 'that [they] will prevent my friends the young mathematicians from looking with too profound and unquestioning an awe upon the crabbed dogmatizing of the Millians.'[102] If they properly understood the nature of induction, and its distinction from deduction, they would inevitably see the fallacies in the economics of Ricardo and Mill.

But Induction was for Whewell more than a way of avoiding logical fallacies: it was also a way of avoiding irreligion and intellectual conceitedness. True knowledge and its pursuit, he believed, was necessarily concordant with

96 Jones (1859), Prefatory Notice by Whewell, pp. xix–xx.

97 Jones (1831), pp. xiii–xiv, xxxi–xxxix. Jones never completed any of the other three volumes. See Checkland (1951); Goldman (1983).

98 Whewell to Herschel, 23 Nov. 1827, Todhunter (1876), ii. 86.

99 Whewell to Jones, 10 Sept. 1827, Whewell Papers, Add. MS c. 51[41]. Letters between Whewell and Jones, 1827–9, Whewell Papers, Add. MS c. 51[43–56, 65]; c. 52[15].

100 Whewell to Jones, 5 Mar. 1829, 4 July 1830, Whewell Papers, Add. MS c. 51[62, 85].

101 1829 and 1831 are when these papers were first read to the Cambridge Philosophical Society: Whewell (1830, 1831*b*). Hollander (1983) makes some convenient long quotes but I think misunderstands Whewell's political economy. See chapters by Becher and Yeo in this volume.

102 Whewell to Jones, 2 Apr. 1829, Whewell Papers, Add. MS c. 51[63].

Christian belief and Christian morality, as he saw them. It was on these grounds that he defended the physical and moral sciences against the charge of atheism in his 1827 sermons, arguing:

> that in instances when these reasonings and conclusions have breathed a spirit at variance with godliness, they have been equally at variance with sound philosophy and a right exercise of the natural understanding. That having *sometimes* left their inferences half-formed & imperfect, and sometimes having extended their assumptions & assertions far beyond the scanty *foundations* of their knowledge, [men] have *thus*, & no otherwise, sunk down to doctrines repugnant to the perfections and revealed will of God.[103]

We should be clear that Whewell was not claiming that certain doctrines were false *because* they conflicted with the Christian religion, but that such a conflict was an indication that they could be shown to be false *on independent grounds*. What was theologically wrong or morally wrong was also philosophically wrong. Conversely, the philosophically correct method of inquiry was necessarily theologically and morally correct. The main message of the sermons was in fact a moral one.

> If *others* have forgotten or neglected, let *us* ever remember, the temper and disposition in which it becomes us, not as believers and Xtians only, but even as the disciples of a rational prudence, and a sound philosophy, to conduct our enquiries. Let us recollect that not boldness and acuteness only, but patience and composure of mind are requisite. . . . Let us press upon all whose minds spring forwards with eagerness in the paths of speculative research, this sobriety of inference and intellectual modesty. Let us inculcate a self-vigilance; a mistrust of any rapid and dazzling views which may seem suddenly to flash upon us, and which may promise to spare all further time & labour in our researches:—above all a mistrust of [many?] peremptory dogmas and maxims, which are current in the world, as such always have been current; and by which the most difficult questions are at once decided, and all further doubt stigmatized as folly.[104]

Faced, then, with the manifold errors of the contemporary sciences, Whewell formed the conviction that to follow the proper method of inquiry was to be continually mindful of God, and to be humble and respectful of tradition.

Now we can see why Whewell would have been goaded by the events of the early 1830s into embarking on a project to explain the nature of Induction to people in general. If he could only make them understand Induction in the same way that he did, then the credibility of utilitarianism, and other forms of atheistic radicalism, would be destroyed. In 1830, therefore, he began work on a book which would not only give a philosophical account of Induction but provide historical illustrations of how it worked, and to this end he began to collect material for 'inductive histories' of chemistry, geology, magnetism, and optics.[105] This book was never completed, but Whewell used the work for it in

[103] Whewell Papers, R17 13, fo. 14, emphasis in original. [104] Ibid., fos. 20–1.

[105] Whewell Papers, R18 17[12]. Also R18 9[14], pp. 17–18; Whewell to Jones, 26 Dec. 1830, 9 May 1831, Whewell Papers, Add. MS c. 51[94, 105].

a number of other writings in the early 1830s, which can also be considered as part of the Induction project.

In 1832 he produced a new textbook which, unlike the others, was intended not just for good mathematicians but for poll men, the mass of students, and perhaps also for the general educated public outside the university. Here he set out to present the inductive as well as the deductive side of mechanics through 'a sketch of the history of the leading principles of the science'. By showing 'through how many attempts, and after how many efforts of the most intelligent speculators, every one of these doctrines has been reduced to its final simplicity and certainty', he tried to make clear that the principles were dependent on experience and not, as the French presented them, arbitrary definitions or self-evident truths.[106] Some of the 'inductive history' material found its way into his review of John Herschel's *Preliminary Discourse on the Study of Natural Philosophy* of 1831. Herschel, as one of the undergraduate circle of Bacon-readers, had presented a very Baconian account of method in the sciences, which gave Whewell his own chance to promote a proper understanding of the concept of induction by both definition and historical example, distinguishing it carefully from deduction and anticipation (an error which was 'far, even in our own times, from being obsolete'). He also found space for a few jibes at utilitarian philosophers and the radical William Cobbett, and an approving quotation of Herschel's claim that the study of natural philosophy did not lead to doubt of revealed religion but rather the reverse ('one of the finest essays on the moral conduct of the intellect which has ever been produced').[107]

Jones meanwhile, having finished his book on rent, was working along parallel lines, investigating the etymology of the term 'induction' since Aristotle, and writing a short popular exposition of the concept for the benefit of a pupil.[108] The pupil was in danger of being corrupted by the utilitarian Nassau Senior's contribution to Richard Whately's *Elements of Logic*, which contained some objectionable remarks about political economy being founded on self-evident propositions, and that there should therefore be as little disagreement amongst economists as amongst mathematicians;[109] Jones's popular exposition of induction was intended as an antidote. 'What I am about', he wrote to Whewell, 'falls in very well with your own notions. I think it may be useful to young persons and teach them not to talk and decide but to be modest and hold their tongues.'[110] And indeed, though Whewell doubted how effective an exposition of induction could be without examples, Jones's aim was one of which he wholeheartedly approved. He hoped

> that by and bye the whole world will think [Senior's notions] as nonsensical as we do. I shall be glad to see your speculations about induction when they are finished, for among

[106] Whewell (1832*c*), p. v.
[107] Herschel (1830); Whewell (1831*a*), 400, 406.
[108] Jones to Whewell, 24 Feb.–21 Mar. 1831, Whewell Papers, Add. MS c. 52^{20-31}.
[109] Whately (1827), 309–10.
[110] Jones to Whewell, 28 Feb. 1831, Whewell Papers, Add. MS c. 52^{22}.

other questions it is certainly an important one how the true faith can best be propagated. I have done what I could in my review of Herschel. I do not know whether you looked at it in that view, but I intended it to be as good an attempt as I could make to get *the people* into a right way of thinking about induction.[111]

The point of their Induction project was public education; they were writing so as to have an effect on the way people thought. The 'right way of thinking about induction' which they wanted to inculcate was perhaps most clearly expressed at this time in the Bridgewater Treatise on astronomical natural theology, which Whewell was commissioned to write in September 1831, no doubt on the strength of his 1827 sermons.[112] The part of the book which Whewell himself regarded as most important and distinctive was the two later chapters on 'Inductive' and 'Deductive Habits of Mind'.[113] 'The impression produced on men's minds by discovering laws of nature' was devotional, he argued; the experience of seeing the connection in a heap of facts led the discoverer to contemplate the designing Intelligence. If this induction was forgotten, 'the impression produced on men's minds by tracing the consequences of ascertained laws' would be that the laws were self-evident, for one would never be led to think of any Intelligence beyond them. This explained the atheism associated with Laplace and other second-rank deductive mathematicians, whereas the truly great mathematicians such as Galileo, Newton, Boyle, and Pascal, who had made inductions and actually discovered something new, were all pious men.[114] Thinking about induction in this way, as well as making evident the logical weakness of utilitarianism, would encourage a devotional attitude towards natural knowledge, humility towards its discoveries and possessors, and a respect for tradition. It was these associations which Induction had for Whewell when he informed a friend at the end of 1833 that he was planning to write 'a philosophy, such as shall really give a right and wholesome turn to men's minds'.[115]

In one major respect, however, the philosophy which Whewell had in mind was different from that which he had been expounding since 1830. Some time in 1833 he had, for the first time, begun to conceive of Induction as he was to present it in the *Philosophy of the Inductive Sciences*: as involving the intuition of a Fundamental Idea.[116] Why did Whewell introduce the Fundamental Ideas into his account of Induction? Can we explain this change as being demanded by the internal logic of his theory, or as the 'influence' of Kant? I believe that the account I have just given of what Whewell was doing in the preceding years offers the possibility of a better explanation. Through the various writings of

[111] Whewell to Jones, Feb. 1831, Todhunter (1876), ii. 115.

[112] Bishop Blomfield to Whewell, 4 Sept. 1830, Whewell Papers, Add. MS a. 201[44]. Whewell to Jones, 9 May, 28 July, 1 Oct. 1831, Whewell Papers, Add. MS c. 51[105, 111, 114].

[113] See Whewell to Lord Monteagle, 4 Nov. 1853, Stair-Douglas (1881), 427.

[114] Whewell (1833*c*), chs. 5–6.

[115] Whewell to Hare, 25 Dec. 1833, Todhunter (1876), ii. 175.

[116] Compare Whewell (1832*d*), 32–5 with id. (1834*c*) and id. (1834*e*), pp. iv–xv.

the Induction project he had been trying to explode the credibility of atheistic and radical philosophies, in particular utilitarianism: what happened in 1833 was that he saw a way of doing this better.

Probably his inspiration came during discussions with his friend Adam Sedgwick over the revision of the latter's recent Commemoration Sermon for publication.[117] Sedgwick was trying to 'teach . . . the junior Members of the University to think correctly on the more important branches of academic study',[118] which meant recognizing that mathematics and classics cultivated the higher faculties and encouraged religious devotion, and that the attention paid to the philosophies of Locke and Paley was wrong and dangerous. The attack on Locke and Paley was no academic issue;[119] although their philosophies had been respectable enough in the eighteenth century when the university adopted them, by the 1830s they were being regularly cited by radicals in support of utilitarian philosophy.[120] The utilitarians claimed that basic moral ideas were only arbitrary and accidental associations of experiences, and that therefore the only justification of actions was expediency, the promotion of well-being for oneself and others; Sedgwick denied this and claimed that morality was not a matter of what was *expedient* but of what was *right*. The mind *did* contain ideas that were not derived solely from experience: they were innate, but latent at birth and brought out by experience. The mind at birth was not, as Locke had maintained, just a blank sheet, but a sheet 'already touched by a celestial hand', and the glorious pattern emerging on its being plunged into surrounding colours was the result of design, not accident.[121]

This part of Sedgwick's *Discourse* was important to Whewell; he was to quote it later.[122] What it would have shown him was that the utilitarians could be attacked not just on their methodology, for failing to be inductive, but on their epistemology, their basic view of the nature of knowledge and its acquisition. This would be a deeper and more powerful kind of attack; all their radical criticism of traditional ideas and institutions lost its plausibility if one believed that there existed innate or emergent ideas, which were somehow fundamental and fixed. This realization would explain why Whewell now began to pursue a new line of work, trying to identify Ideas of this kind in the physical and moral sciences. His reading of Kant would have helped him see Fundamental Ideas in the physical sciences, Ideas which became distinct in people's minds through the process of Induction, and which enabled them to see the principles of each science to be universally and necessarily true (something which experience alone could not explain).[123] And he would have

117 Sedgwick (1833), pp. v–vi. Clark and Hughes (eds.) (1890), i. 404.

118 Sedgwick (1833), p. 102, n.F.

119 Garland (1980), Ch. 4, presents 'the reaction against Paley and the Benthamites' as the abandonment of an out-moded 18th-century doctrine, as though there were no Benthamites in the 1830s. S. Cannon's account is more subtle (1978), 59.

120 Halevy (1928), 23.

121 Sedgwick (1833), 46.

122 He quoted it in his preface to Mackintosh (1862), 36.

123 Whewell to Herschel, 22 Apr. 1841, Todhunter (1876), ii. 298; aphorisms 14, 15 in Whewell (1847), ii. 445.

seen how he could extend the old Induction project so as to give people an awareness not only of the nature of Induction but also of the Fundamental Ideas at the basis of the physical sciences.

It was in this mood, then, that he wrote his famous letter to Jones outlining his plan for a three-book work: a *History*, 'to point out how each of the important advances [in the sciences] was made';[124] a *Philosophy*, to counter 'the doctrines now most widely prevalent respecting the foundations and nature of truth' and so 'check the prevalence of grave and pernicious error';[125] and a *Prospects of Inductive Science*, to investigate 'the question of the possibility and method of applying Inductive processes . . . to other than material sciences; as philology, art, politics and morals'.[126] The *History*, when it appeared, was dedicated to Herschel, with whom he shared the project of explaining to '*the people*' to concept of Induction;[127] the *Philosophy* he dedicated to Sedgwick, with whom he shared the project of refuting 'the fallacies of the ultra-Lockian school',[128] and the *Prospects* he surely intended to dedicate to Jones.

That the *Prospects* was never written is indicative of how Whewell shifted the ground of his opposition to utilitarianism. This part of the plan derived from his collaboration with Jones, when the emphasis had been on the empirical, data-gathering side of induction; he left it behind as he became more interested in the Fundamental Ideas, an interest with which Jones could not sympathize. The decisive point came in 1835 when Whewell, searching for arguments to use against Paley, turned to Mackintosh's *Dissertation on the Progress of Ethical Philosophy*, and from it caught 'a glimpse . . . of the inductive history of ethics', something which he had 'long been wishing and struggling for'.[129] Following up that glimpse, in 1837 he preached four sermons 'on the foundations of morals' and the following year took up the Knightbridge Chair of Moral Philosophy, converting it from a sinecure to an active teaching post. The analogy between the physical and moral sciences was now complete; while publishing the *History of the Inductive Sciences* and putting the finishing touches to the *Philosophy* Whewell could teach from his Chair that in the moral sciences too there was knowledge that was certain and stable, based on the same kind of Fundamental Ideas. (See Yeo's chapter, second half of Section III.)

By 1837, then, as Becher and Yeo observe, Whewell had largely shifted his attention 'from mathematics to moral philosophy', and had developed the basic philosophy of knowledge which (with modifications) he was to promote

[124] Id. (1857), i. 3.
[125] Id. (1847), i. 5–6.
[126] Id. to Jones, 27 July 1834, Whewell Papers, Add. MS c. 51^{173}.
[127] Interestingly, the dedication was signed as coming from Jones's London house, not Trinity College; Todhunter (1876), i. 101.
[128] Whewell (1847), i. iii.
[129] Whewell to Jones, 9 May 1835, Todhunter (1876), ii. 212. For Whewell's earlier seeking after that glimpse, see his Induction notebooks dating probably from Dec. 1833, e.g. Whewell Papers, R18 17^8, p. 4; R18 17^6, pp. 71–3.

for the rest of his life. What I have tried to show here is that his *History* and *Philosophy of the Inductive Sciences* and teaching of moral philosophy were the latest in a whole series of projects, including his collaboration with Jones in political economy, his 1827 sermons, and the various writings of the Induction project, all of which had the same general aim. That aim, conceived in response to the alarming spread of utilitarianism, atheism, and radicalism, was one of public education: 'to give a right and wholesome turn to men's minds', to give 'the people' a correct understanding of the nature of knowledge and its acquisition, so that they would see as clearly as Whewell did the false foundations of these doctrines and not be led astray. His opposition to these deeply threatening movements should not be seen as the 'context' of his philosophy of knowledge, nor 'background', nor 'external factors', for it was none of these things. It was the whole point of Whewell's work; it was the reason why he thought it worth while to spend most of a decade in writing two massive books on the history and philosophy of the inductive sciences.

III

We are now in a better position to understand Whewell's new interest in university education in the 1830s and what he was trying to do through his subsequent involvement in the university reform controversy. In two parallel publications of 1835,[130] he extended the scope of his crusade against the spread of utilitarianism, atheism, and radicalism, by attacking not just unsound views of the nature of knowledge but also forms of education which encouraged them. The first publication was a new edition of Mackintosh's *Dissertation* for Cambridge students, in which he inserted a preface explaining its moral importance and directing the work to the youth of England.[131] Since Sedgwick's sermon and subsequent *Discourse*, he had been uncomfortably aware of the fact that by teaching undergraduates the philosophy of Paley, which denied the existence of innate moral ideas, the university was virtually teaching them to be utilitarians. Whewell hoped that Mackintosh's *Dissertation* would be a suitable antidote to Paley: as he wrote to Jones, it contained 'arguments against the utilitarian view' which would hopefully correct 'some prevalent errors', and it provided a 'glimpse' at least of a moral philosophy based on Ideas.[132] His second publication of that year was his *Thoughts on the Study of Mathematics*. In this, as we have seen, he on the one hand attacked forms of mathematics teaching which impeded appreciation of Fundamental Ideas, and on the other recommended the general teaching of elementary mechanics and hydrostatics

[130] Whewell twice mentions the two projects in successive sentences in his letters to Jones, 9 May 1835, 26 May 1835, Todhunter (1876), ii. 212, 213–14.

[131] Whewell to Jones, 26 May 1835, Todhunter (1876), ii. 213–14; preface to Mackintosh (1862), 46.

[132] Whewell to Jones, 9 May 1835, 21 Aug. 1835, Todhunter (1876), ii. 212, 222–3.

'to dissipate the impression, that mathematical reasoning is applicable only to such abstractions as space and number'.[133] What Whewell was doing in both these publications was attempting to shape university education into an instrument by which the new generation would receive and respect the stable, permanent truths, both moral and physical, handed down by their teachers; he tried to curtail forms of teaching which interfered with this, and encourage forms of teaching which promoted it.

But by 1837, proposals for the radical reform of Oxford and Cambridge along the lines of the French and German universities or University College, London, seemed to pose a bigger threat to the younger generation's sense of permanent, godly truth than any current form of teaching. As we saw in the first section, the proposals included the replacement of tutors' practical teaching by professors' speculative teaching, and abandoning the teaching of subjects in which 'the old truths will always be true', such as mathematics, for the teaching of subjects in which there is 'revolution after revolution', such as philosophy. Whewell devoted a whole chapter of *Principles* to pointing out the moral consequences of this. Study in which 'there is nothing old, nothing stable, nothing certain' could only turn an impressionable student into a 'restless speculator, . . . destitute not only of . . . clear ideas, and . . . habits of exact thought, . . . but devoid also of that steady belief in the permanent nature and value of . . . truth, which is an essential virtue of the understanding'. Equally importantly, in the study of mathematics the scholar was 'led to entertain a docile and confiding disposition towards his [necessarily superior] instructor', whereas by appealing to his critical spirit through the study of philosophy he would feel himself 'placed in the situation of an equal and a judge, with respect to his professor', and, probably passing through phases of allegiance to several different teachers, could 'hardly fail to look upon them with a self-complacent levity, which involves little of respect'. Whewell believed that the 'vehement and general hostility to the existing institutions of their country' reputedly shown by the young university men of Germany and France was a natural consequence of 'an education which invokes the critical spirit'.[134]

In *Principles* again, Whewell pointed out the moral and social consequences of adopting the free system of *Lehrnfreiheit* associated with the universities of France, Germany, and London, in which there was no control over a student's conduct or his attendance at lectures or chapel. The consequences were that the students '[received] with the utmost indignation all attempts to subject them to any kind of rule,' and '[assumed] a right to interfere in the appointment of professors and similar matters'; one could see this from the behaviour of London medical students, the recent unrest in one of the Paris schools, and the many occasions on which German students had shown their indignation 'by organized migrations, or by putting some particular University under the

[133] Whewell (1835*a*), 179.

[134] Id. (1837), 46–53.

interdict of the clubs of the young men'. Whewell had no doubt 'that the tendency of the free system, if introduced into the English Universities, would be to corrupt the character, and deprave the manners of the students'.[135]

The overall object of the book was thus to set out, in opposition to the systems of the French and German universities and the upstart University College, London, Whewell's interpretation of the principles of *English* university education, 'the general Principles on which the work of the English Universities *is* and *must be* conducted' (my emphasis).[136] That work, as we saw, he considered to be one of 'liberal education', which he identified with the transmission of stable and permanent truth in the form of distinct Fundamental Ideas, which constituted civilization itself; and for this work he deemed the college system, the traditional curriculum, and compulsory Anglican chapel to be essential. Whewell believed that if this were generally understood, there would be no desire for radical university reform, which amounted to the destruction of the existing system, but only the moderate reform and improvement of the kind for which England was noted.

> England has constantly impelled the progress of thought and institutions in Europe, while, at the same time, she has held back from the extravagances and atrocities to which the progressive impulse has urged more unbalanced nations. The bold and vigorous metaphysicians of England first put in motion the speculative movement of modern Germany; but England refuses to follow the wild onward whirl of system after system, to which this movement has led. English teachers of political freedom, and the free institutions of England, called up the spirit which has broken the bonds of more than one despotism; but England (thank God!) was never hurried into the democratic madness of her nearest neighbour.

England had had a Reformation without the abolition of an established Church, a revolution without the destruction of her monarchy, and a reform of Parliament without a fatal wrench to her constitution: reason for confidence, Whewell said, that she would not disastrously remodel her universities on some foreign or imaginary plan. But just to be sure, he concluded with a warning to 'all whom it may concern' (a clear reference to the expected Royal Commission) that the proper spirit in which to undertake reform was one 'not of hatred, but of reverence for the past; not of contempt, but of gratitude towards our predecessors'. For however much confidence he might express in the 'practical sense and fixed sobriety' of the English, it was the horrible possibility that they might follow 'more unbalanced nations' into 'extravagances and atrocities' and 'democratic madness' that had made his book necessary in the first place.[137]

We can now see that the educational principles which Whewell conceived in the 1830s and pursued throughout the rest of his life had the same aim as his philosophy of knowledge: to immunize people against the influence of utilitari-

[135] Whewell (1837), 129–31. See Jarausch (1974).
[136] Ibid. 3.
[137] Ibid. 138–40.

anism, atheism, and radicalism. Whewell's means of achieving this were also the same: by bringing people to appreciate the existence of stable, permanent, divine truths in both the natural and moral world; in other words, the Fundamental Ideas. We can thus see why he used the motif of a flaming torch being passed on from one hand to another on the title pages of all his works on philosophy and education: they had a common purpose and a common message; the accompanying motto, '*lampadia ekhontes diadosousin allelois*' (carrying torches they pass them on to one another), he took, significantly, from Plato's *Republic*, a work at once about politics, moral philosophy, the nature of knowledge, and education; a work which he was later to translate.[138] Like Plato and the German philosophers, Whewell used light to represent true knowledge, the distinct Fundamental Ideas, which actively gave meaning to experience like a torch casting light. The symbol itself was a denial of the utilitarian or Lockean view of knowledge. Even more importantly, the torch was being *passed on*; the pupil was not making fire for himself or taking it from heaven (only a great genius or discoverer could do that) but receiving it from his master with gratitude and reverence. The image perfectly sums up the message Whewell was trying to convey, and the result he was trying to achieve.

How does the account which I have presented here affect our view of Whewell? As far as the history of education is concerned, it is clear that he did not, as is often said, start out as a reformer and end up as a conservative. From the 1830s, he consistently opposed radical university reform and even moderate reforms which seemed too close to Benthamism for comfort, resulting in an estrangement from Adam Sedgwick and his other erstwhile friends of a Whig persuasion.[139] His 'reforms' or 'improvements', as they have seemed, acquire another character now that we know why he supported them. He attacked the teaching of Paley's moral philosophy not (*pace* Garland)[140] because utilitarianism was an outmoded eighteenth-century doctrine, but because it was an all too live and threatening nineteenth-century movement. He advocated the increased teaching of the physical sciences not because they were useful but because they were true; because their study would discourage utilitarianism, encourage devotion to God, and promote social stability.

It is part of our usual view of Whewell that he was a defender of the natural sciences, in contrast to High Churchers who believed, on the strength of continental evidence, that their pursuit encouraged materialism, atheism, and questioning of the established order.[141] We can now see that while Whewell supported the natural sciences, he feared their pursuit for the same reasons as the High Churchers. Unlike them, however, he believed that the power and

[138] *Republic*, I 328 a. 3. In his translation of this work, Whewell presented Plato as basically a moral philosopher, (1861), p. vi.

[139] See his letters to Jones, 23 Mar. 1834, Whewell Papers, Add. MS c. 51[164]; 26 May 1835, Todhunter (1876), ii. 214.

[140] Garland (1980), ch. 4.

[141] Andrew Cunningham and I intend to publish soon on the changing form of the pursuit of natural knowledge in this period, under the title *The Invention of Science*.

sophistication of the new continental sciences had to be acknowledged, and so he sought to ensure that they were brought back into the service of God and the established social order. Through his philosophy of knowledge, he tried to limit the importance of the new developments, particularly in astronomy and geology, by demonstrating the epistemological basis of the sciences—Fundamental Ideas, which, when appreciated, affirmed the wisdom of God and the permanence of knowledge.

To see Whewell's philosophy in this way may shed some light on his well-known dispute with John Stuart Mill (discussed in detail in Buchdahl's chapter). Although the dispute was conducted in philosophical terms, it inevitably had a political significance, and Mill as well as Whewell understood it as such.[142] 'The tendency of his efforts', wrote Mill, 'is to shape the whole of philosophy, physical [and] moral, into a form adapted to serve as a support and a justifiction to any opinions which happen to be established.'[143] A radical such as Mill, forced continually to contend against established ideas, was bound to be hostile to a philosophy such as Whewell's which reinforced them,[144] in the same way that he was bound to be hostile to institutions such as the old English universities, in which 'no thought can find place, except that which can reconcile itself with orthodoxy', and to educational writers like Whewell who believed that 'the proper object of an intellectual education is not to strengthen and cultivate the intellect, but to make sure of its adopting certain conclusions'.[145] Mill was not quite accurate in this accusation, for in fact Whewell believed these two things to be identical: strengthening and cultivation of the intellect, for him, meant the intuition of Fundamental Ideas, which inescapably involved the adoption of certain conclusions, indeed coming to see them as self-evident. But for Mill, of course, intellectual education was synonymous with intellectual freedom.

In the 1860s and 1870s, utilitarian followers of John Stuart Mill such as Henry Sidgwick and Henry Jackson began to be successful in implementing the very kind of university reform which Whewell had feared would '[put in] serious and extensive jeopardy the interests of the civilisation of England and of the world'.[146] Religious tests were ended for college fellows, new Triposes were introduced with a corresponding decrease in emphasis on the traditional subjects, and centralized university teaching became increasingly the norm. In these decades,the intellectual world, like the political world, was remade after the fashion of liberalism. A new generation of professional men of science, such as Thomas Henry Huxley and John Tyndall, proclaimed the separation, even the antagonism, of science and religion, and across the range of new academic specialisms, from biblical criticism to anthropology, there were the

[142] For an early reaction of Whewell to Mill, see his letters to Hare, 15 and 30 Oct. 1838, Todhunter (1876), ii. 270–3. He observes that despite Mill's criticism of Bentham, he is a writer of the *London and Westminster Review*, whose objects are 'to destroy the church and democratize the nation'.

[143] Mill (1852), 453.

[144] Id. (1873), 273–4.

[145] Id. (1852), 451–2.

[146] Whewell (1837), 53.

same attempts to subject received ideas to rational criticism and set up the subject on a purely secular basis.[147] We can see, therefore, that the fear which Whewell expressed in the 1830s was not illusory; civilization as he conceived it—the passing on of Ideas that were divine, permanent, and fundamental to the human mind—was indeed brought to an end.

There is a convention of taking 1859, the date of the publication of the *Origin of Species*, as marking this change: the 'demise of the truth complex', the 'fragmentation of the common context'.[148] Thus the story that Whewell banned the *Origin* from Trinity College library, whether true or not, is entirely appropriate;[149] it represented an intellectual and a social change which he was bound to oppose. But what we have seen here is that Whewell did not encounter this threat for the first time in 1859; he recognized it in the 1830s, in the utilitarianism, the atheistic sciences, and the radical politics of the time. The purpose of his philosophy of knowledge and his ideal of liberal education was to 'pass on the torch', and so prevent these movements coming to power; for he believed, correctly, that if they did so, it would mean the end of civilization as he knew it.

[147] Winstanley (1940), ch. 6; Annan (1955); Harvie (1976); Kent (1978); Turner (1978); Barton (1976).

[148] S. Cannon (1978), ch. 9; Young (1985), ch. 5.

[149] The original source for this is F. Darwin (1877), iii. 261, but the story is probably apocryphal.

6. *Indications of a Creator: Whewell as Apologist and Priest*

by JOHN HEDLEY BROOKE

SO confident was William Whewell of the durability of natural theology that in his *History of the Inductive Sciences* he would describe a hymn to the Creator as the 'perpetual song' of the temple of science.[1] However misplaced that confidence proved to be, it stands in sharp contrast to the perceptions of those historians and philosophers who have treated natural theology, with its repertoire of design arguments, as a force already spent by the close of the eighteenth century. Those who have expressed astonishment that William Paley, the noted Anglican apologist, could write with such jaunty assurance after Hume had so devastated the foundations of natural religion, must find it even more astonishing that hymns to a Creator remained a prominent feature of scientific culture, in Britain at least, until Whewell's generation had passed away. How does one account for the longevity of a tradition which, whilst not confined to Britain, was nevertheless so conspicuous here that observers have sometimes spoken of a peculiarly British propensity to proclaim a holy alliance between science and religion?

That alliance has often been described, rarely explained. I make no pretence to do so in this essay,[2] but it may be useful, by way of introduction, to construct an eclectic model, the various elements of which may help to make the viability of natural theology at least comprehensible:

1. Natural theology could retain a function so long as it remained useful in the promotion of science itself. Appeals to the wisdom of the Creator might help to appease those who saw in science a threat to traditional religious values. In so far as the natural sciences could supply arguments for divine Providence, so those arguments became a useful resource when urging the moral worth of a scientific education.[3]

2. A quite different role for natural theology was preserved wherever a rational defence of Christianity was perceived as a priority in rebutting the allegations of sceptics, scoffers, and deists. The *Astronomical Discourses* of Thomas Chalmers had, as one of their functions, the protection of Christianity against the jibes of Tom Paine and others who had exploited a plurality of worlds to ridicule

[1] Whewell (1857), iii. 393.
[2] An attempt at a more systematic explanation will be found in Brooke (1985).
[3] Brooke (1979), 46; Shapin and Barnes (1977).

conservative images of creation and redemption. Those of Chalmers's congregation who felt shrivelled into insignificance by the vastness of the universe were reassured that God's care for them was no more compromised by the existence of other mansions in the heavens than it was by the teeming millions of His creatures revealed by the microscope.[4]

3. There was also a continuing role for natural theology, in so far as it proved able to mediate between different religious professions. The divisiveness of religious dogmas could always be mitigated a little where agreement could be reached on certain basic propositions that were considered susceptible of rational demonstration. Recent scholarship has stressed this mediating role in the particular context of emerging scientific societies, such as the Royal Society in the 1660s and the British Association for the Advancement of Science in the 1830s, where it was expedient to play down political and religious differences in defence of a common interest: the cultivation of science and of a propitious public image.[5]

4. An interest in natural theology was always liable to surface in those who had a commitment to both science and religion—for the simple reason that it forged an intellectual bond between two interests that might otherwise have remained unrelated. This unifying role within the psyche of particular individuals would tend to continue so long as it remained relatively common for practitioners of science either to be clerics, to be actively involved in ecclesiastical affairs, or to retain the faith in which they had been nurtured.

5. An eclectic model for the durability of natural theology would also have to take into account its potency as a political resource. There is no simple equation in this matter (one thinks of Joseph Priestley, both natural theologian and political radical). But it was not difficult to extract a reassuringly conservative message from the divine dispensation as disclosed in nature. A fixed natural order was readily projected on the social order, and urgently so when social stability appeared to be in jeopardy.

6. Natural theology could also have a continuing philosophical function, wherever reason and its powers were celebrated in attempts to suppress both religious enthusiasm and religious scepticism. This particular function has been thrown into the limelight by Barbara Shapiro's correlation between science and 'latitudinarianism' in late seventeenth-century England, and by the 'Broad Church' which Morrell and Thackray have located among the leaders of the British Association.[6]

7. Finally—though usually of primary interest to historians of science—natural theology was liable to be sustained as long as it could supply regulative principles for the interpretation of the organic world, that is, until such time as a non-teleological principle could be invested with plausibility. It is true that,

[4] Chalmers (1817); Cairns (1956); Rice (1971).
[5] Wood (1980), 19; Brooke (1979); Morrell and Thackray (1981), 224–45.
[6] Shapiro (1968); id. (1969); Morrell and Thackray (1981), 230, 225–6.

in Germany, the philosophy of Kant helped physiologists to detach their teleological principles from a physico-theological base, but that was a refinement that gained only isolated support in pre-Darwinian England.[7]

The striking feature of this model is that it might have been manufactured with Whewell in mind. He would seem to be an almost perfect exemplar of each of the seven elements identified:

1. A recurrent role for natural theology in the promotion of science could be illustrated by sermons he preached in Cambridge during 1827. In a series of university sermons, he sought to elevate the importance of science by persuading his congregation that they had nothing to fear from natural laws. Laws, after all, presupposed a Legislator. In the last analysis, the ordinances of nature were motions of the divine Will, 'rules of his operation and production'.[8] There was nothing to fear from enquiries into natural order, for 'how can this ordering and reasoning be conceived except we suppose an Intelligent Ruler employed in deducing, from the laws, the facts . . . from which we, reversing this course of deduction, can, to those containing principles ascend again'.[9] That science could now explain events once attributed to divine intervention was cause for celebration, not mistrust.

2. The value of natural theology in exposing the irrationality of irreligion can be illustrated from the same set of sermons. The threat of irreligion was a pressing one for Whewell because it had become the fashion, among men of science themselves, to interpret the conclusions of celestial dynamics, for example, in a less pious manner than had his great predecessor at Trinity, Isaac Newton. There had been of late a certain coldness of feeling among students of nature, which Whewell likened to the cold courtesy shown to another human being one wished to shrug off.[10] By defining the spirit in which the sciences should be pursued, natural theology could raise the temperature of religious feeling and awareness. His eye was fixed on Laplace, his mind on the damage done by philosophers who presumed to '*unlearn* the necessity for the interfering mind and the *amending* hand which their greatest teacher looked to'.[11] If irreligious men could find new solace in a self-regulating solar system, natural theology, as Whewell expounded it in his Bridgewater Treatise, robbed them of it: 'It would be as if the savage, who had marvelled at the steady working of the steam-engine, should cease to consider it a work of art, as soon as the self-regulating part of the mechanism had been explained to him.'[12]

3. The enduring role of natural theology as a mediating agent is thrown into relief by Whewell's performance at the meeting of the British Association in

[7] Lenoir (1982), 1–53.

[8] Whewell, Sermon, 11 Feb. 1827, Whewell Papers, Trinity College Library, Cambridge University, R6 17^{14}.

[9] Ibid.

[10] Whewell, Sermon, Feb. 1827, Whewell Papers, R6 17^{13}.

[11] Whewell, Sermon, 11 Feb. 1827, Whewell Papers, R6 17^{14}.

[12] Whewell (1839*a*), 350.

1841. There was no longer any great risk, Whewell announced, of 'deviating into literary, or metaphysical, or theological discussions'.[13] But an implication of his address was that a consensus on the conclusions of natural theology provided the best means of avoiding wrangles over religious doctrine in an association dedicated to science. Thus he immediately recalled the 'elevated strains of religious reflection' that had been heard from his predecessors.

4. On the unifying role of natural theology Whewell would, again, appear the perfect exemplar. He has proved a friend to those social historians of science who have seen in his career that union of clerical and scientific interests which, by the next generation, would have largely disappeared.[14] His particular style of natural theology has been taken to epitomize a scientific ethos which, redolent of Oxbridge privilege and aristocratic pretension, would soon be challenged by the more plebeian and secular ideology of T. H. Huxley.[15] Certainly, as both scientist and priest, Whewell did seek to unify his interests. It was a merit of final causes that they could, in his own word, 'connect' the conclusions of science with religious concerns.[16]

5. As for natural theology as a political resource, Whewell may never have descended to the platform on which William Buckland preached his Easter sermon in 1848: 'equality of mind or body, or of worldly condition, is as inconsistent with the order of nature as with the moral laws of God'.[17] But he did associate political instability with those countries which had failed to bring science and religion into union, where superstition had given way to the prevalence of irreligion.[18] His recommendations on the principles of university education had connotations which were manifestly anti-revolutionary. He was not enamoured of the German system, at least in the teaching of philosophy. Those revolutions in thought, by which Hegel had superseded Schelling who had superseded Fichte who had superseded Kant, had led not only to a 'dire shedding of ink' but also to a serious unsettling of tender minds:

> In Germany and France, we are told that there prevails among the young men of the universities a vehement and general hostility to the existing institutions of their country . . . Such a consequence may naturally flow from an education which invokes the critical spirit, and invites it to employ itself on the comparison between the realities of society and the dreams of system makers.[19]

A knowledge of science, its history, its progression towards fundamental truths, its being most successfully prosecuted by men of religious integrity (which he argued in his *History*), could, by contrast, provide a sheet-anchor. It could help to sustain that respect for the divine Mind which Whewell considered a prerequisite of respect for human authority.

[13] Whewell (1842*b*), p. xxxiv.
[14] Turner (1978), 356–76.
[15] Desmond (1982); id. (1985), 50.
[16] Whewell (1839*a*), 355.
[17] Cited by Gordon (1894), 243–5.
[18] Whewell, Sermon, Sept. 1841, Whewell Papers, R6 17[49].
[19] Whewell (1837), 48–52; Brooke (1986).

6. As for the celebration of reason, who else among nineteenth-century natural theologians developed such a strong epistemological argument for the belief that one could gain at least a partial access to the divine Mind?[20] Whewell was deeply committed to a 'promised *land of knowledge*, a land offered to us no less bounteously than the richer Canaan of our salvation'.[21] It was, after all, the divine gift of reason which differentiated man from animal. The human mind had its faculties for the discovery of truth. However much they might be misdirected they were not to be mistrusted.[22] As Richard Yeo has so succinctly put it, for Whewell 'man's ability to know the world was the measure of his special place in creation'.[23]

7. Who, finally, was more explicit than Whewell in specifying a teleological reading of nature as a *sine qua non* of right reasoning in the biological sphere? Reviewing the history of physiology, he declared the merit of Cuvier to consist 'not in seeing that an animal cannot exist without combining all the conditions of its existence; but in perceiving that this truth may be taken as a guide in our researches concerning animals'.[24] Hence Whewell's disquiet when Geoffroy St Hilaire proposed an alternative principle, that of unity of plan, on the basis of which teleological reasoning might be proscribed. One of Whewell's responses to that threat was distinctly pragmatic: even those who opined that no manifestation of means apparently used for ends in nature could prove the existence of design in the Author of nature 'still cannot resist the necessity which compels men to assume, in organized beings, the existence of an end'.[25] Kant's philosophy of biology was, therefore, of value to Whewell in that it could be wheeled out against St Hilaire. As exemplar of our model Whewell fits the bill to perfection because, despite availing himself of Kantian principles, he knowingly dissociated himself from the agnostic implications of Kant's critique of physico-theology. He was aware of Kant's subtlety but at the same time implied that the critique was refutable.[26]

A polyvalent model to account for the longevity of British natural theology has apparently been corroborated by reference to one of the most learned of Cambridge polymaths. Merely to confirm a preconceived model, however, might—with some justice—be considered a somewhat sterile enterprise. But there is a way forward and it is provided by Whewell himself. There is a fertile theme in the extent to which he may be said to be a typical defender of physico-theology. The more closely one studies his theological works the more apparent it becomes that he pursued a distinctive style of natural theology, heavily qualified by considerations drawn from what he believed to be the fundamentals of a Christian and pastoral theology. There is, in his capacity as priest, another side to Whewell in which natural theology is underplayed in so far as it exploits design in the physical universe.

[20] Yeo (1979).
[21] Whewell, Sermon, Feb. 1827, Whewell Papers, R6 17[13].
[22] Ibid.
[23] Yeo (1979), 497.
[24] Whewell (1857), iii. 390.
[25] Ibid. iii. 386.
[26] Ibid.

Hints of this other side of Whewell are discernible in his published works and they have not gone unnoticed. Observers have sometimes detected a sense of discomfiture in the Bridgewater Treatise, as if the assignment he had accepted committed him to stronger claims than, in his better judgement, he might have wished to approve. Not that there was any lack of conviction behind his central argument that the mutual adaptation of the laws of nature (more than the minutiae of specific 'designs') testified to divine Wisdom. Nor was he reticent in proclaiming that advances in the sciences could lead up to a belief in a most wise Maker. But to testify to or lead up to is some way from a formal demonstration. Whewell's reserve may be seen in the frequency with which he adopted the interrogative form: 'Can the lenses of the eye, for instance, be formed and adjusted with an exact suitableness to their refractive powers, while there is in the agency which has framed them, no consciousness of the laws of light?'[27] His discomfiture might be seen in his frequent recourse to the double negative: 'It will be difficult or impossible to exclude from our conception of this wonderful system, the idea of a harmonizing, a preserving, a contriving, an intending Mind, of a Wisdom, Power, and Goodness far exceeding the limits of our thought.'[28] In certain convolutions, the interrogative and the double negative were fused: 'What are the habits of thought to which it can appear possible that this could take place without design, intention, intelligence, purpose, knowledge?'[29] Then there is the hint, in the concluding chapter of the Bridgewater Treatise that the argument from design would probably be persuasive only to those who had cultivated the right habit of thought. The perception of 'design', he suggested, had to be closer to the premise than to the conclusions of the argument.[30] In his *Philosophy of the Inductive Sciences* he was to present the point more sharply: the idea of a First Cause was 'not extracted from the phenomena, but assumed in order that the phenomena may become intelligible to the mind'.[31]

In his published sermons, especially those on the foundation of morals, the inadequacy of physico-theology is given special emphasis. In November 1837 he delivered an exegesis of Romans 1: 20 in which he noted that his text had often been understood to focus on God's power and wisdom as evidenced in creation. But he was dissatisfied with that interpretation: 'It was not alone the power and skill of a divine artificer which the Apostle held to be discoverable in the constitution of the created world, but the holiness of a divine lawgiver, the justice of a divine judge.'[32] The limitations of a purely physico-theology were duly underlined: 'Brutes, as well as man, are wonderfully made; and show forth the wisdom and skill of Him who framed them; but they suggest nothing as to a *moral* author of the world.'[33] Where, he asked, are we to find anything which discloses a holy as well as a wise Creator? His answer, following the

[27] Such passages occur throughout the Bridgewater Treatise; see, especially, Whewell (1839*a*), 141.

[28] Ibid. 13–14.

[29] Ibid. 146.

[30] Ibid. 344.

[31] Cited by Yeo (1979), 502.

[32] Whewell (1837*d*), 16.

[33] Ibid. 18.

author of Romans 2: 14–15, was in the human conscience. When Whewell was preaching to the godly, rather than to the British Association, his theology was far more Pauline than references to his Broad Churchmanship might imply. 'Our great and important practical business', he would say, is 'to cherish the sense of our own demerits, and a reliance upon a righteousness better than our own'.[34] Whewell may not have discounted arguments from design, but the fundamental datum for an earnest natural theology was the existence of the human conscience.

Whewell's published works on moral philosophy supply further evidence that his most cherished apologia was not a physico-theology of the type popularized by Paley. If there was a thread running through the early sermons and his later ethical writings, it was his critique of utilitarian ethics. If Whewell is to be placed in context, it has to be recognized that the main thrust of his natural theology was directed against the calculus of pleasures and pains deriving from Bentham and Paley. His central preoccupation was how to defend an educable conscience as the ultimate reference for morality against a utilitarianism which, in its implications, failed (so Whewell believed) to distinguish humanity from the brutes. The question, for Bentham, had been not 'can animals reason?', nor 'can they speak?', but 'can they suffer?'[35] Seizing that very contention, Whewell retaliated that we are to be humane to them because we are human, not because we and they alike feel *animal* pleasures.[36] When considering his resistance to theories of organic evolution, it must be borne in mind that any conception which might dissolve the discontinuity between man and beast would help to sanction an ethical system he had spent years trying to eradicate.

The arguments employed by Whewell to eliminate the Benthamite heresy and to reinstate an 'internal principle' in the soul, whereby right and wrong could be differentiated from gain or loss, are too numerous and involved to be considered here. In any case, Whewell's moral philosophy has been skilfully and sympathetically analysed by Schneewind.[37] A few of his moves are worth noting however, in that they reveal a mind convinced that the greatest threat to Christianity came not from advances in the sciences but from a philosophy which, denying that actions could be inherently right or wrong, assessed them merely by the balance sheet of their consequences. One of his objections was purely practical: how could one possibly calculate all the consequences of every act, right down to the last ripple?[38] Moreover, to weigh a summation of pleasure against a summation of pain was too crude a form of accountancy since it rolled into one different types of pain and evil which, when properly distinguished, were not additive. For Whewell, the evil of the experience of pain was quite different from the evil of sin. By blurring such distinctions, a

[34] Ibid. 49.
[35] On the secular implications of Bentham's emphasis, see Thomas (1984), 176.
[36] Whewell (1852*a*), 223–4.
[37] Schneewind (1968), 108–41.
[38] Whewell, Sermon, March 1852, Whewell Papers, R6 17[72].

utilitarian calculus could have the baleful consequence of countenancing a sinful act if it supposedly led to a net gain of pleasure.[39] The happiness produced by actions, Whewell argued, depends on their morality, not vice versa.

In his *Lectures on the History of Moral Philosophy in England*, he objected to Bentham's system on the ground that it could not correct popular opinion. By constantly juxtaposing the *popular* with the *moral* sanction, Bentham could not determine what was moral without first knowing what was popular, since popularity supplied one of the pleasures by which the morality of the action would be assessed. Whewell was not impressed: 'if murder, sensuality, falsehood, oppression, be in any cases popular, this popularity tends to make them virtues, for it gives them the *reward* of virtue'.[40] Finally, for our purposes, there was that recurrent motif which reflected Whewell's high sense of man's uniqueness in creation: 'The morality which depends upon the increase of pleasure alone would make it our duty to increase the pleasures of pigs or of geese rather than those of men, if we were sure that the pleasure we could give *them* were greater than the pleasures of men'.[41] Whewell's position, by contrast, is that the faculty of conscience, however 'grievously obscured and perverted in man's present condition', is darkened but not destroyed.[42] It ever speaks 'the language of a region which lies above the tumult and dissonance of appetite and passion.'[43] With Joseph Butler he asserted the natural supremacy of conscience whilst at the same time retrieving that notion from caricature. In his Preface to *Butler's Three Sermons on Human Nature* he emphasized that neither Butler's contention, nor his own, was 'that each individual person has such a faculty, which is the proper judge or standard of his actions; that if he conform his conduct to his conscience, he must act rightly'.[44] This was a travesty and open to obvious objections based on cultural variation. Rather, conscience was a faculty, as reason was a faculty—a 'power by exercising which we may come to discern truths, not a repository of truth already collected in visible shape'.[45]

Whewell's analogy between reason and conscience shows how erroneous it would be to ascribe to him the belief that mankind was blessed with a 'moral sense' akin to the senses of sight and hearing. That common mode of speech was to trivialize human judgement.[46] Because the conscience had to be enlightened and instructed before it could become a safe guide, it was described in his *Elements of Morality* as 'only a subordinate and intermediate authority'.[47] Since it was never fully formed, but always in the course of formation, there were dangers in appealing to it to justify particular actions. It has been said, Whewell observed, 'that if I *talk* of my humility, I lose it; something of the same kind may be said of conscience'.[48] Whewell's rejoinder

[39] Whewell, Sermon, March 1852, Whewell Papers, R6 17[72].
[40] Whewell (1852*a*), 218. [41] Ibid. 223. [42] Whewell (1837*d*), 24.
[43] Ibid. 25. [44] Whewell (1848*a*), p. xiii. [45] Ibid., p. xiv.
[46] Ibid., p. xxv. [47] Whewell (1845*b*), i. 238. [48] Ibid. 240.

to the utilitarians was not, then, that conscience was the 'Voice of God', but that it could become such if, and only if, it was informed by moral and religious instruction. He wanted to be able to say that 'to disobey the commands and prohibitions of conscience, under any circumstances, is utterly immoral', without admitting that obedience to conscience was a sufficient condition to make an act moral.[49] There was no highway code implanted in the human heart. But there was a divinely given faculty which, during the history of the human race, had allowed moral principles to be unfolded, much as Whewell believed the principles of the natural sciences to be unfolded by the constant interplay between theory and fact in which the contribution of the mind was paramount.[50]

This minor excursion into the published works on morality is not the digression it may appear. It serves to underline two points which are fundamental to any appraisal of Whewell's natural theology. Firstly, however ingenious (and ingenuous) his recourse to physico-theology in the Bridgewater, it was 'the consideration of God's workmanship as seen in our souls'[51] which he most highly prized. Secondly, his understanding of how the conscience was to be educated was deeply informed by his familiarity with the New Testament epistles, and especially the letter to the Romans. Ultimately there was no simple division between 'natural' and 'revealed' religion as far as Whewell was concerned. In defending his conviction that 'right and wrong are peculiar qualities, and no mere modifications of pleasure and pain', he was looking for a sense of 'duty' which would be consistent with the 'plainest teaching of revelation'.[52] And one of the things that was recognized in revelation was that even when men knew what was required of them, they were incapable of performing it. Deviation from the good was a characteristic of the human will. From such a thraldom, philosophy was powerless to rescue.[53] As Butler had argued before him, it was only through revelation that one learned of the deliverance made possible through Christ. Whewell's congregations were not spared references to the Atonement: 'That perfect obedience which the law demands, but demands in vain, leads us to lean upon the satisfaction of the Cross'.[54] There were limits to Whewell's rationalism. In the refining of conscience, 'reason is not enough . . . ; it is religion which gives the hope and reality of a power of enlightenment'.[55]

By insisting that reason was not enough, Whewell implied that any theology exclusively based on reason would be deficient. There were limits to the value of a purely natural theology, conspicuously so wherever considerations drawn from physico-theology threatened to override cherished Christian principles. His essay *Of the Plurality of Worlds* provides the most vivid illustration from

[49] Ibid.

[50] Whewell exploited an analogy between the exercise of the cognitive and moral faculties on many occasions; e.g. in Whewell (1837), 35, 41. See also Schneewind (1968), 119–20.

[51] Whewell (1849), p. ix.

[52] Whewell (1837*d*), 31–3.

[53] Ibid. 62–3.

[54] Ibid. 86.

[55] Whewell (1845*b*), i. 238.

the corpus of his published works. Whewell's adversary, David Brewster, clung to the popular dogma that extraterrestrial life was rife in the universe. For Brewster, as for many exponents of a conventional physico-theology, a plurality of worlds was not so much a feature of natural theology as a desideratum. To make sense of the myriads of stars invisible to the naked eye it had been common, from the end of the seventeenth century, to confer meaning upon them by making them subservient to the needs of other beings on other planets similar to our own. If the celestial regions were destitute of life, Brewster argued, it would be as if a railway entrepreneur were to send a train all over Europe, with all its carriages illuminated but only one occupied.[56] All sense of the economy of creation would be lost. But Whewell's God was not made in the image of a railway king. The politics, the ethics of creation mattered more than the economics. Whewell was considered eccentric for wishing to depopulate the heavens, but there can be little doubt that his motives had much to do with the transcendental value he attached to the uniqueness of man as an intellectual and moral being. I have argued elsewhere[57] that Whewell's strategy may have been, in part, a response to popular theories of organic evolution, epitomized by the anonymous *Vestiges of the Natural History of Creation*. The author of that work, Robert Chambers, had exploited a plurality of worlds in at least two ways which Whewell found offensive. Not only did the existence of life on other worlds lend credibility to naturalistic processes of evolution (where similar physico-chemical conditions prevailed, analogous processes would be induced); it was also a useful vehicle for turning a doctrine of creation into a reductio ad absurdum. It was surely inconceivable, Chambers had argued, that God had, as it were, visited each world in turn, conjuring its inhabitants into existence.[58] Surely He would have preferred to establish the laws by which life would have naturally developed? Placing his argument in the discourse of natural theology, Chambers could capitalize on precedents which had given greater weight to God's foresight than to his freedom. His offence, as Whewell perceived it, was to confer a spurious respectability on a doctrine of mind which allowed human sagacity to be prefigured in the dog or elephant. Whewell, in other words, distanced himself from the discourse of natural theology as sustained by Brewster and as revamped by Chambers. As Michael Crowe has recently suggested, his antipathy to extraterrestrial life probably had its deepest roots in a theology of Incarnation which he had gradually come to appreciate was difficult, if not impossible, to translate to other spheres of intelligent life.[59] Brewster's lament was that by admitting abortive features of creation, Whewell appeared to be evacuating natural theology.[60] Whewell's lament was that Brewster appeared not even to have considered whether there might be religious grounds for believing the unique position of man.[61]

[56] Brewster (1854), 13 and *passim*.
[57] Brooke (1977), 221–86.
[58] Chambers (1844), 161.
[59] Crowe (1986), 281; see also Yeo (1979), 505–11.
[60] Brooke (1977), 282.
[61] Whewell (1855).

These observations, culled from his published works, provide a justification for raising a question which has not, to my knowledge, been addressed in any detail: did Whewell see any demonstrative force in the argument from design? Given the primacy of his moral theology, it would also seem reasonable to ask what value he was prepared to place on appeals to the natural world in the context of religious edification. These interrelated questions are to be pursued in the remainder of this essay. They are important not only for clarifying Whewell's position, but also because the answers may throw further light on the continuing survival of natural theology itself. Might it not be argued that it proved so viable precisely because it did not have demonstrative pretensions? Where the argument from design was presented as a *proof* of God's existence and attributes, it was clearly vulnerable to the Humean and Kantian critiques. But perhaps the argument could function in various non-demonstrative modes, thereby evading the strictest censure? Could it not, for example, serve to corroborate a pre-existing faith? Could it not help to excite a sense of awe and wonder in a devotional context? Could it not be joined with other components of a natural theology to prepare a common ground for dialogue with other faiths or internecine traditions? Could it not help to assuage doubts occasioned by specific scientific discoveries? If so, if its principal functions *within* a religious communion were not demonstrative, then arguments of the Humean type may hit the wrong, or at least a largely irrelevant, target.[62]

To test such a hypothesis it becomes essential to examine how appeals to nature did in fact operate in a devotional context. Whewell makes the perfect case-study since he was both preacher and philosopher. One hesitates before invoking his sermons, yet they become a necessary resource if the argument is to be developed. They also constitute a rich deposit since there are numerous unpublished sermons in the Trinity College archive. What follows is based on a sample of about forty of these spanning the period 1827–59. Exciting they are not. On the status, style, and limitations of natural theology, however, they are peculiarly revealing.

The Subordinate Status of Natural Theology

What, in this world, is it important for man to know? Whewell's answer to that question involved a stratification of priority. It was far more important to be aware of one's spiritual condition than engage in a purely intellectual exercise concerning the existence and attributes of God. For Whewell, there was a fundamental sense in which knowledge of God derived from natural theology was of subordinate status. It would be surprising if it were otherwise, but

[62] In raising these questions, I am delighted to express my debt to my colleague in the Religious Studies Department at Lancaster, John Clayton, who has been investigating the non-demonstrative functions of the traditional theistic proofs; Clayton (1986), 724–84.

Whewell made the point with unusual clarity and vigour. The proudest treasures of human knowledge

> are not here an allowed object of our thoughts, except in subordination and relation to that more rich and perfect knowledge, which connects us with our Heavenly Teacher. The end of our wisdom is to make us wise to salvation;—the summit of our attainments to know our fallen nature and the provision made for its restoration by our Father and Saviour.[63]

This is typical of many such statements and it is associated with Whewell's willingness to differentiate between different kinds of truth. He explicitly denounced the view that 'all substantial and enduring truth, *may* be, and *should* be, conformed to that type which has been but approved for the physical sciences'.[64] Natural theology was not a passport to those higher truths which were only found in revelation. How, through natural theology, could one learn of the Fatherhood of God, or that the ultimate goal of human existence was union with Him? Neither the intellect nor the will sufficed to bring men closer to God. Both, Whewell declared, are defective. To the intellect must be added faith.[65]

These are the predictable outlines of any Anglican sermon—except that Whewell's message, time and again, was more deeply inspired by, and steeped in, a Pauline vocabulary than a routine performance would require. That God was in Christ reconciling the world to Himself, that we may be joint heirs with Christ, were propositions that no philosophy could establish.[66] They nevertheless embodied knowledge which Whewell described as 'the highest, the proper occupation of man's mind in the most consummate exercise of his intellect'.[67] The subjugation of natural theology was quite explicit. Indeed Whewell's Pauline theology of hope was almost its antithesis: 'We are as St Paul says (Rom. 8: 24) to be saved by Hope. Not by Hope of that which can be seen by the natural eye or the rational power alone: Hope directed towards that which is so seen, is not the Christian's hope.'[68] As Whewell's exegesis of Romans 3: 8 makes plain, one could not follow St Paul with eyes left trained on the works of creation. It was the purpose of the epistle to the Romans, Whewell affirmed, to extend the view from the Law to the Gospel, from the unrighteousness of man to the righteousness by faith in Jesus Christ, 'from the work of Creation to the work of Redemption'.[69]

The shift of focus from creation to redemption is such a recurrent theme in Whewell's preaching that the image of a broad-church rationalist may conceal more than it reveals. He repeatedly stressed the 'perpetual dissonance' between man's passions and his sense of duty.[70] Whilst his evangelism in Trinity

[63] Whewell, Sermon, Feb. 1827, Whewell Papers R6 17[13]. [64] Ibid.
[65] Each of these points was developed in a Sermon dated 12 Nov. 1848, Whewell Papers, R6 17[59].
[66] Ibid. [67] Ibid. [68] Ibid.
[69] Whewell, Sermon, Mar. 1852, Whewell Papers, R6 17[72].
[70] Whewell, Sermon, 6 Feb. 1853, ibid., R6 17[74].

Chapel hardly bordered on the melodramatic, his theology had a pronounced evangelical content. It is certainly not to be supposed, as have some historians of science, that the 'broad-church' label is non-problematic. The legerdemain by which respect for the scientific enterprise has sometimes been incorporated as a defining characteristic of a 'broad-church' mentality cannot be countenanced. It is of no value to describe Michael Faraday as a Broad-Churchman when, as Geoffrey Cantor has shown, his respect was for a style of science informed by the evangelical values of the Sandemanian sect to which he belonged.[71] Evangelical motifs in the theology of Adam Sedgwick have similarly led Paul Marston to question whether the broad-church tag does anything other than obscure the religious commitment of Whewell's colleague at Trinity.[72] But what of that other illustrious colleague, Julius Hare, whom Whewell described as his 'forerunner' and to whom his sermons on the foundations of morals were dedicated? Was Hare not the archetypal Broad-Churchman, and Whewell also by association?

Emphatically not, according to F. D. Maurice who, in a spirited critique of the category, set out to rescue Hare from the blight of an imprecise label.[73] Maurice rejoiced that the Anglican Church was already broad enough to comprehend such disparate spirits as Hare and Richard Whately, but the many divergences between them (and between others allocated to the Broad Church) seemed to

> go some way towards proving, first, that a Broad Church party, such as has been dreamed of, is impossible; and secondly, that if it were possible it would be unnecessary, seeing that a body has existed here for about a thousand years, which is considerably more inclusive than the new creation could ever become.[74]

The burden of Maurice's critique was, however, that a figure such as Hare simply did not conform to the connotations which the term 'Broad Church' had acquired. He had not conspired to form a school consisting of restless spirits and Germanophiles. Nor did his theology display that 'mild poco-curante' character that might be associated with the attempts of a littérateur to remove the rough edges of the Church. As Maurice could not fail to note, Hare's intensive study of Martin Luther had led him to, if anything, a fuller proclamation of the doctrine of justification by faith than to 'guarded statements' or 'middle ways'.[75] True, Hare had been deeply critical of the party spirit infecting the Church, but a theology, and a personal example, which transcended party lines did not have to be lukewarm. What Maurice wished to say of Hare might, with some justice, be said of Whewell. The way of party was the easy way, the broad way in Christ's sense. The more difficult was that strait and narrow way which sought no facile reconciliation between competing schools of thought, but rather sought to persuade the clergy of different parties that they were united by higher bonds than those of school. No less than Hare did

[71] Cantor (1985).
[72] Marston (1984).
[73] Maurice (1874).
[74] Ibid., p. xix.
[75] Ibid., pp. xlviii–xlix.

Whewell proclaim the gospel of justification by faith; no less perhaps did he see that German biblical scholarship could lead to a deeper understanding of how God spoke to men than was enshrined in the newly conventional notion of verbal inerrancy. In so far as Whewell concurred with Hare, he concurred with one who, as Maurice succinctly concluded, was friend and counsellor of those who valued depth, not breadth, in their theology.[76]

If there were more evangelical elements in Whewell's theology than is commonly supposed, at least two other implications deserve notice. Firstly it may become easier to understand why he was attracted to and absorbed by the history of science. In 1827 he had posed the question as to how it was possible for men of science to pursue their research without religious feeling. His answer had been that there must be 'something defective and partial in the conduct of their faculties, something vicious and mistaken in their trains of research'.[77] Hence the attraction of a history of science which would expose 'cases where discoveries and researches, in themselves, it may be, containing somewhat of truth and reality, had been extended to assertions far too wide and general for the compass of their scanty and narrow foundations, and had *thus*, and no otherwise, led to doctrines repugnant to the . . . revealed will of God'.[78] What he wanted to show was that science pursued irreligiously was liable to go off the rails. There is no concealing the fact that in his early historical writings he was out to demonstrate that the greatest scientists had been Christians whose discoveries had been made possible very largely through preconceptions of order, design, and purpose derived from their piety.[79] The second implication is simply this: how nature is interpreted depends on prior beliefs and principles of reasoning. A prior reverence will produce a reverent reading. And this would suggest a serious qualification to the scope of natural theology. Belief in a good and wise Being might be made stronger and more vivid by contemplating the combinations of the natural world, but could a natural theology alone produce that prior reverence? The content of the sermons suggests not.

The Confirmatory Rather than Demonstrative Function of Natural Theology

Whenever scientific knowledge was invoked in his sermons, Whewell spoke of the *confirmations* of faith.[80] Scientific pursuits could 'feed and elevate . . . devotion when it exists'.[81] From a pastoral angle attempts at rational inference to design were inappropriate. Members of his congregation would be invited to

[76] Ibid., p. lxxxix.

[77] Whewell, Sermon, Feb. 1827, Whewell Papers, R6 17[13]. [78] Ibid.

[79] Whewell (1839*a*), 308. [80] Whewell, Sermon, Feb. 1827, Whewell Papers, R6 17[13].

[81] Ibid. Without wishing to assimilate Whewell's theology to that of the evangelicals, his references to the confirmatory role of natural theology would certainly be compatible with theirs. See Rosman (1984), 44–7.

render a reason for the faith within them, but what Whewell required was a route via the heart to a settled habit of mind.[82] Only when the proper habit of thought was established would the inference from design to Designer carry conviction. Whewell was adamant that a discourse of theistic *proof* was alien to his intentions, even when dealing with critics of the faith.[83] In contexts where he declared the universe to be 'full of symmetry and order and beauty and life' he would still insist that it was not his purpose to offer 'new proofs of religious truths' or to remove 'the objections of those who seek in their doctrines an exercise of the reason only without ever referring to their heart and conscience'.[84] That the laws of nature implied a Legislator was not so much an argument for confounding the sceptic as the expression of an appropriate *feeling*:

> We ought to feel that when we are following . . . the traces of Law and Order in the natural world, we have thus offered to us the means of confirming and enlightening in our minds the general notions of Law and Order, which applied in the moral world, are necessary to instruct our consciences and to elevate our nature. We ought to feel that all the Laws of the Universe are so woven together that we cannot separate or disentangle them, and are necessarily by any thread drawn to the very highest point: the Judge and Purifier of all . . . [85]

The vocabulary of confirmation is quite explicit.

That Whewell was prepared to use the term 'regulative principle'[86] in a theological, as well as a scientific, context helps to underline the sense in which he believed propositions referring to final causes were not so much inference from facts as presuppositions of a fruitful interpretation of nature. This was, of course, the point on which he would insist in his *Philosophy of Discovery*: the Idea of final cause 'is not deduced from any special facts, but is assumed as a law governing all facts in organic nature, directing the researches and interpreting the observations of physiologists'.[87] A theistic philosophy could explain why an idea supplied by the mind was so extensively applicable to the organic world, but a theistic explanation was not the same as a theistic proof. References to proof do occur in the *Philosophy of Discovery*, but, even there, considerations come into play which suggest that they must be read with circumspection. The context in which the word 'proofs' occurs was one in which Whewell was tracing the epistemological consequences of a prior theism, rather than one in which he leaped directly from the applicability of fundamental ideas to the existence of God. His enterprise was not one of proving the attributes of God from a study of nature. Indeed one focus of his discussion was what *little* could be known of the Mind of God through the mediation of fundamental ideas.[88]

[82] Whewell, Sermon, 11 Feb. 1827, Whewell Papers, R6 17[14].
[83] Id. Sermon, 20 Feb. 1827, ibid. R6 17[16]. [84] Ibid.
[85] Id., Sermon, Sept. 1841, ibid., R6 17[49].
[86] Id., Sermon, n.d. 1843, ibid., R6 17[55].
[87] Whewell (1860), 369–70.
[88] Ibid. 373–4, 377–8.

That natural theology had a corroborative rather than a demonstrative role surely does have implications for any analysis of its durability. There is a sense in which it could be sheltered from a Humean attack by the modesty of its pretensions. Whewell, moreover, had additional reasons for preferring modesty.

The Limitation of Natural Theology by the Limitation of Human Nature

In a devotional context Whewell's question was how the unregenerate could be brought to the state in which they would accept the forgiveness and Fatherhood of God, with all that that implied. Experience showed that it was through what Paul had called, in 1 Corinthians 1: 21, the foolishness of preaching (and therefore ultimately through God's Spirit) rather than through rational argument.[89] In purely pragmatic terms rational 'proofs' were disappointing in the effects they achieved. Human nature was so 'mixed and imperfect' that 'in influencing men's lives reason leads to an effect far less than, man being a rational creature, we might have expected from it'.[90] Whewell knew only too well that 'demonstration produces a result far short of its claims'. And this was true even of those 'proofs of our duty from our immortality with all their seeming vigour'.[91] Left to themselves they were barren. It is difficult to see how he could have regarded arguments from design as any more fruitful. Demonstration was demoted. But worship 'brings forth consequences far beyond its pretensions'.[92] By stressing the efficacy of apparently inadequate and extraneous forces in bringing men to God, Whewell was able to emphasize the divine initiative more than if it were simply a question of rational demonstration. Agents without intrinsic efficacy could more readily be understood as instruments of the Deity 'subservient either to the ordinary or the extraordinary operations of his Holy Spirit'.[93]

The impliction would seem to be that even if a natural theology could be erected its value would be severely limited by the very essence of human nature. The limitation was expressed in another way when Whewell assigned natural theology to a category of 'speculative religion'. This had the effect of removing it from the main body of Christian doctrine and making it less relevant to a principally devotional religion. There was no complete demarcation, for Whewell could refer to advances in knowledge providing 'new food for our contemplative devotion'.[94] But the devotion is once again pre-existent. He explicitly denied that advances in knowledge could bring men nearer to God as His *servants*.[95] Even within the category of 'speculative religion' there was the problem that 'the attributes of the Deity and his creation have purposes beyond the reach of our theories'.[96] Since the principal objective of the

[89] Whewell, Sermon, n.d. 1843, Whewell Papers, R6 17^{55}.
[90] Ibid. [91] Ibid. [92] Ibid. [93] Ibid.
[94] Id., Sermon, Feb. 1827, ibid., R6 17^{13}.
[95] Ibid. [96] Ibid.

Christian was progressive sanctification, the domain of speculative religion was quite narrow for 'no outward or speculative reason can tell us what these better tendencies are'.[97]

True religion, for Whewell, was a matter of experience more than a matter of speculation. That which had to be experienced was the power of God working in the human heart. '*There is no other way*', he would exclaim, the emphasis his own.[98] Accordingly, if the propositions of a speculative religion remained in that category, they were not merely of limited value, they were of no value: 'they are as though they were not'.[99] Something more was always needed to 'convert a belief of the reason into a motive of the heart'.[100]

The implication would seem to be quite straightforward. Even if the existence of God could be proved via natural theology, the proof would be valueless since what was required was an 'active and living appropriation of the belief in God'.[101] There was nothing original or peculiar in this. Whewell was simply reaffirming that Christian doctrines were of the form that they entailed belief *in* something, not merely belief *that* such and such was the case. But it was sufficient to make him suspicious of allusions to design which slipped too readily off the tongue. When the British Association blew into Plymouth in 1841, Whewell had words of admonition for his scientific brethren:

> It is not enough for us to say the works of nature are the works of God, and therefore in studying nature we are rendering homage to God. This is indeed little better than a mere form of words. For how do we give to God the service of our thought when we pursue the objects which amuse or interest our minds without thinking of Him at all?[102]

Nor, he continued, was it enough to say that the works of nature

> impress us with wonder and admiration, and that there is in these feelings a deep religious import:—for what religion is there or what value, in any mere admiration of nature, if it in no way connects with reverence and love for a God who sees and hears us, our Redeemer and Comforter . . . the Lord *our* God.[103]

The problem could only be overcome when philosophy was, as it were, fused with piety: 'we must make our philosophy and our piety, not only reconcilable, but identical'.[104] Morrell and Thackray have claimed that, gathered together in the British Association, the liberal Anglicans worshipped at the shrine of science rather than that of the Christian God. This was undoubtedly a *tendency*, of which Whewell was evidently aware. But it is an inadequate characterization since Whewell himself, who is after all one of their prime exemplars, denied the disjunction which Morrell and Thackray affirm.

[97] Id., Sermon, 20 Feb. 1827, ibid., R6 17[16].
[98] Ibid.
[99] Id., Sermon, n.d. 1843, ibid., R6 17[55].
[100] Ibid.
[101] Id., Sermon, 11 May 1851, ibid., R6 17[68].
[102] Id., Sermon, Sept. 1841, ibid., R6 17[49].
[103] Ibid.
[104] Ibid.

A Circumscribed Role for Natural Theology: The Raising of Questions, not Answers

From what has been said so far, it might be supposed that natural theology was entirely denuded of value. But this would be to go too far. Like Thomas Chalmers, Whewell seems to suggest that natural theology can be a useful vehicle for raising questions that can only be satisfactorily answered in terms of a Christian theism.[105] The interrogative style in the Bridgewater Treatise is replicated in the sermons. Instead of pitting the design argument against the materialist, Whewell simply asks: 'who appended to matter these properties?'.[106] Who, he asks, created this matter? Who 'framed and compacted this skeleton of a world?'. There is even a superficial resemblance to the pre-eminent process theologian of the twentieth century, Charles Hartshorne, in that Whewell would appeal to the coherence of the universe in preference to more primitive notions of design: who, he asked, gave the parts of the world 'that coherence by which it is prevented from crumbling into nothing?'.[107]

The severe subordination of a physico-theology is nevertheless thrown into further relief by the realization that the really critical questions, for Whewell, emanate from the moral sphere. That men did not live by the Moral Law was a ground for enquiring what consequences the divine Mind had annexed to the violation of His laws. Considering how dimly *we* see what these consequences are, how much greater must be the divine wisdom. This was a train of thought which found its way into the *Philosophy of Discovery*:

> He who sees all these consequences and has provided for the establishment of His Law in the whole history of the human race, must be supposed to be infinitely elevated above man in wisdom;—more even in virtue of this aspect of His nature, than in virtue of that which is derived from the contemplation of the universe.[108]

The existence of physical law, knowable by the human mind, certainly raised critical questions. Whewell did not, however, suggest that the answer could only be in terms of *creatio ex nihilo*. Again in the *Philosophy of Discovery* he observed that 'the laws of the universe discoverable by the human mind, are the laws of the divine Mind, whether or not there was a time when these laws first came into operation, or first produced the world which we see'.[109] He was prepared to adduce scientific evidence, especially from geology, which seemed to point towards a beginning, but no extrapolation from the evidence could be 'followed *up to* a beginning'.[110] That it was impossible to come to any conclusion by natural reason on the subject of Creation was yet 'another evidence of the infinitely limited nature of the human mind, when compared

[105] Chalmers (1833), ii. 282–93.
[106] Whewell, Sermon, 11 Feb. 1827, Whewell Papers, R6 17[14].
[107] Ibid. See also Hartshorne (1964).
[108] Whewell (1860), 381.
[109] Ibid. 382.
[110] Ibid. 383.

with the Creative or Constitutive Divine Mind'.[111] By allowing natural theology to raise questions rather than answer them, Whewell effectively—though whether intentionally it is difficult to know—side-stepped another of the Humean objections. He was not in the business of reversing a chronological sequence of cause and effect until it eventuated in an uncaused cause.

Indications of a Creator: The Suggestive Role of Natural Theology

To suggest that Whewell restricted natural theology entirely to the raising of questions would be less than accurate since, especially in a devotional context, he did admit that a study of nature could excite and intensify a sense of awe and wonder. But such a response, he suggested, was more like the response to a symphony than to a reasoned argument. A 'hidden delight', a 'grateful admiration' could be evoked, but the question of proof need not arise.[112] The sermon in which the analogy with a symphony was pursued was the very one in which he dissociated himself from a discourse of proof. An analysis of Whewell's vocabulary in the presentation of natural theology points to the conclusion that reflections on nature rarely had more than a suggestive role: they pointed to possibilities; they proved nothing. When he responded to *Vestiges* by republishing material under the title *Indications of the Creator* his choice of words was itself indicative of a degree of restraint. Even when he developed his epistemological argument he was not in the habit of claiming that human reason, with its powers duly unfolded, did more than *point to* its origin.[113] The implication was that men were to be nudged towards richer perceptions for their life; not rationally propelled.[114]

In this nudging process the main consideration might not be what science had established, but rather what science, by its very nature, would never explain. A question raised in one of his sermons was why the appearances of nature—a starlit sky or the effusive glow of the rising sun—have such a palpable effect on the human spirit. The impression, he insisted, was one 'which no mere laws of material nature can account for—which no science can explain'.[115] Why certain human emotions should correspond with the different seasons was another example of the inexplicable. Science might explain the occurrence of a storm, but not that 'gloom of the spirit which the tempest brings'. The purpose of his argument was to establish that the objects of the natural world 'have a place and an operation in the world of spirits'. In constructing a critique of deism (which was his ultimate objective here) his principal recourse was not to that philosophy of nature which Hooykaas has described as semi-deism.[116] He was not trying to create space for an absentee

[111] Ibid. [112] Id., Sermon, 20 Feb. 1827, Whewell Papers, R6 17[16].
[113] Id., Sermon, Feb. 1850, ibid., R6 17[64].
[114] Id., Sermon, n.d. 1843, ibid., R6 17[55].
[115] Id., Sermon, 15 Oct. (?) 1850, ibid., R6 17[66]. [116] Hooykaas (1959).

god to intervene from time to time. It is more that he wished to invest the whole material universe with spiritual significance. No argument from design informed by scientific knowledge could meet that requirement. Committed to a dualistic ontology, his appeal was to a divine agency which, directly or indirectly, worked on the human spirit. In this respect the limitations of natural theology were associated with the limitations of science itself: 'as no natural causes, no material agencies, no physical laws, can explain such an operation, we have still, in spite of all the discoveries of natural science, a wide and varied region left to us, in which God's work produces upon our souls, impressions mysterious and inexplicable.'[117] However broad his churchmanship, however rationalist his theology, it did not preclude the admission of angels in the mind.[118]

The Veiling of Natural Theology

A science-based natural theology had further limitations in the sense that an advancing knowledge of nature could generate new mysteries in which God's designs and purposes were veiled in obscurity. There was no simple antithesis between knowledge and ignorance in the apprehension of nature: 'as we discover new laws the laws of these laws become more and more mysterious and lose themselves in the darkness which surrounds the throne of light'.[119] Whewell had at least two examples in mind. The more one learned about other worlds the more their purpose became shrouded in mist. The more one learned of the geological succession the greater the number of unanswered questions. In part, Whewell's argument was designed to disperse the hubris of scientific arrogance: 'strangely framed indeed must their minds be who can find in science aught to make them proud'.[120] But the implication was that science, in so far as it fed into a physico-theology, did so as much by multiplying mystery as by removing it, as much by revealing our presumption in claiming to know God's purposes as by revealing His ends. In a sermon of 1827 there had been an almost Cartesian ban on teleological argument: the purpose in these things 'suits His perceptions and thoughts, not ours'.[121] Nor was this *entirely* out of line with what later appeared in the *Philosophy of Discovery*. Whilst it was of fundamental importance to Whewell that the Idea of final cause mediated between the divine and human mind, he also acknowledged that 'it would be presumptuous to suppose that we can enter into [God's purposes] to any but a very small degree'.[122]

During his career advances were made in the life sciences which threatened the teleological argument in the form in which Paley, for example, had presented it. We have already had occasion to refer to the doctrine of unity of

[117] Whewell, Sermon, 15 Oct. (?) 1850, Whewell Papers, R6 17[66].
[118] Ibid.
[119] Id., Sermon, 20 Feb. 1827, ibid., R6 17[16].
[120] Ibid. [121] Ibid. [122] Whewell (1860), 370.

plan by which St Hilaire disposed of a teleological reference. Whewell illustrated the threat, drawing on an example discussed by Richard Owen:

> that the ossification of the head originates in several centres, and thus in its early stages admits of compression, has been pointed out as a provision to facilitate the birth of viviparous animals; but our view of this provision is disturbed, when we find that the same mode of the formation of the bony framework takes place in animals which are born from an egg . . . In this way, the admission of a new view as to unity of plan will almost necessarily displace or modify some of the old views respecting final causes.[123]

Whewell was clearly grateful to Owen for having defined the way forward: design could still be affirmed 'in the modification of the plan in special ways for special purposes'.[124] Whilst this was fine in principle, the integration of homological and teleological considerations nevertheless disturbed the clarity of focus by which ends had been seen in means. In fact Whewell admitted as much in his *Philosophy of Discovery*: 'And with regard to the *Design* which we see in the organs of living things, though we can confidently say we see it, how obscurely is it shown, and how much is our view of it disturbed by other laws and analogies.'[125]

Natural Theology and the Supernatural

Whewell's principal strategy against deism was not so much to create space for a semi-deism as to spiritualize space itself. Having said that, however, he did in his early sermons regard science as an aid to religion precisely because it showed that 'by no impulse of causes known to us' had the Earth become an abode for man.[126] Whilst reassuring his congregations that they had nothing to fear from the extension of natural law, he himself drew up the drawbridge on certain issues, and notably when human origins became the *explanandum*.[127] He was careful to insist in his published works that he was not foreclosing discussion. It was rather that naturalistic accounts of the sequence of organic forms had been tried and found wanting. In the sermons, as one might expect, he was less reserved in arguing for divine fiat. It might even be said that the argument from design played second fiddle to the argument for fiat. The intellectual justification for invoking the supernatural was not, however, derived from scientific ignorance alone. Rather, as in Butler's *Analogy of Religion*, the principal consideration was that the Creation and the Destruction of the world were works of 'supernatural interposition'.[128] Looking backwards

[123] Id. (1857), iii. 559.
[124] Ibid. 560. See also Brooke (1977), 282; Yeo (1979), 508–9.
[125] Whewell (1860), 378.
[126] Whewell, Sermon, 18 Feb. 1827, Whewell Papers, R6 17[15].
[127] Whewell (1838*a*), 23.
[128] Whewell, Sermon, 18 Feb. 1827, Whewell Papers, R6 17[15].

or forwards there were ultimate precedents. In his early sermons, as he explored a series of coincidences between the conclusions of Scripture and the conclusions of natural science, he saw no reason to play down the notion of supernatural power. In his more apocalyptic remarks one can see how a theological catastrophism almost certainly did predispose him against a complete continuity of secondary causes.[129]

That any natural theology had to show deference to the supernatural power of God was made clear in a sermon on the book of Job. Whewell's text referred to the intervention of Jehovah following the efforts of Job and his friends to rationalize suffering. In that intervention God's power was contrasted with man's nothingness. Moreover, nature testified to that power. But the point of the story was that it was a power so transcendent that it was inappropriate for man to question God's moral government. In so far as natural theology could establish the unfathomable *power* of God, so it put constraints on the further use of reason. The message Whewell extracted was that it is incongruous and indefensible for man to judge the rules of God's government by his reason only.[130] It is not always appreciated that a minimal natural theology, in which the power alone of God's attributes was deduced, could actually militate against a more ambitious theodicy.

Practical Defects of a Physico-Theology

Whewell's argument that laws presupposed an Agent, whilst a useful vehicle for reassuring the faithful, had a deficiency of which he was fully aware. He was, in fact, so sensible of it that it would be surprising if he had pressed the argument further. The difficulty was that the more accustomed one grew to the regularities of nature, the greater the danger of spiritual indifference. He made the point most forcibly in a sermon preached in 1849, after Cambridge had been smitten with cholera. In such times of trouble theological rationalization did come to the fore. During the cholera epidemic of 1832, it had still been common for 'low' churchmen and methodists to see divine judgement visiting inebriates and reprobates with stiletto precision.[131] But since the Master of Jesus had recently died, that was hardly an option open to Whewell, even if he had wanted it. Instead he constructed a theodicy which gives us particular insight into the relations between providence and natural law as he envisaged them. Whewell's point was that 'the very frequency and regularity of the occurrences which *should* speak to us of God's providence produces upon us a contrary effect'.[132] Hence Whewell's rationale for the pestilence. It was an 'extraordinary' visitation designed to heighten a waning sense of dependence: 'If we forget Him in times of plenty, is it not most fit that He should send upon

[129] Whewell, Sermon, 18 Feb. 1827, Whewell Papers, R6 17^{15}.

[130] Id., Sermon, 20 Feb. 1827, ibid., R6 17^{16}.

[131] Morris (1976), ch. 7.

[132] Whewell (1849*a*), 10.

us times of scarceness? If we forget Him in times of health, is it not most fit that He should send upon us times of sickness?'[133] In a nutshell, how could the argument from laws to Legislator have coercive power when even Christians themselves were only too prone to take the course of nature as a 'matter of course'?[134]

A second practical defect arose from the fact that to the *natural* eye of man, the presence of disorder, struggle, pain, and suffering spoke not of harmony but of dissonance. That there was some ulterior design was therefore more a matter of faith than of reason. Whewell was not unhappy to dwell on the dissonance because it allowed him to emphasize that the most immediate experience of disharmony was within man himself. As part of the common religious experience of mankind, a need had been felt for a 'purity and strength greater than [that] found belonging to human nature'.[135] And it was this which led men to 'look to some dispensation of divine mercy as yet unseen'.[136] In Whewell's judgement, the predisposition to believe in a God of mercy arose not from some facile harmony urged by physico-theology, but from the dissonance in man's moral experience. In an important sense Whewell built on, rather than denied, that groaning of creation which Hume had sought to exploit. The moral experience of mankind was reflected in the 'inarticulate voice of nature'.[137] The religious impulse had arisen from the *combination* of pain and gratification. In both the moral sphere and in the department of physical nature there was a sense of derangement which no cheerful reference to the bounty of nature could dispel.

In Whewell's reflections on Romans 8: 28 one finds a quite sensitive account of how religious beliefs had been formed long before the Christian dispensation. But he interpreted the darker side of nature in a manner quite different from Hume—a manner which rendered it in some measure immune to Hume's objections. There was no physico-theological argument of the kind that Hume had expelled, at least in the sermons. For Whewell the issue had never been 'how much suffering would there have to be in the world before a teleological argument for beneficent design would be destroyed?' The fundamental issue, which in large measure determined how nature should be read, was the moral, not the scientific, experience of mankind.

Conclusion

It hardly needs to be said that the hortatory style and the devotional import of sermons make them problematic as sources. A persona may—usually does—modulate in tune with audience and perceived propriety. I therefore make no claim to have uncovered the quintessential Whewell. His preaching was, nevertheless, an important and regular aspect of his life. To neglect its content

[133] Ibid., 11–12. [134] Whewell, Sermon, 9 July 1854, Whewell Papers, R6 17[67].
[135] Id., Sermon, 6 Feb. 1853, ibid., R6 17[74]. [136] Ibid. [137] Ibid.

would be to neglect one of the radii of Whewell's prolific mind. To the last he maintained that *something* of God could be learned from nature: the fundamental ideas on which science was based (the ideas of space, time, substance, force, matter, and final cause) did give access to the divine Mind.[138] But I hope this excursion into Trinity Chapel and St Mary's Church may help to explain why the heading of his next paragraph was 'Though but Little'.[139]

I have also suggested that one of the reasons why natural theology survived during Whewell's lifetime was that it did not always claim too much. Within a religious community it often functioned without demonstrative pretensions and, as Whewell's sermons certainly show, there could be theological (as well as philosophical) reasons for moderating its claims. By the time Whewell's *Philosophy of Discovery* was published, Darwin's *Origin of Species* was already making its mark. To juxtapose those two books would, for some historians, be to juxtapose the principal representatives of two incommensurable paradigms.[140] As if to anticipate such an interpretation, Darwin, at a critical juncture in his intellectual development, had emended that passage in which Whewell wrote: 'the use of every organ has been discovered by starting from the assumption that it must have *some* use'. In his own copy of the *History of the Inductive Sciences*, Darwin 'corrected' the sentence to read: 'the relation of every organ has been discovered by starting from the assumption that it must have *some* relation'.[141]

In the light of the claim that Darwin destroyed the argument from design it may be instructive to reflect on how much was left to be destroyed in 1859. Whewell's wrangle with Brewster over extraterrestrial life suggests that natural theology was already a house divided against itself.[142] In the spheres of anatomy and physiology, morphological arguments, as we have seen, were already obscuring the teleological. Even Whewell's epistemological argument was running into trouble. Facing criticism from Lewes and Mill, he did own to one difficulty. The test of axiomatic truths was that the contrary of them had to be inconceivable. And yet 'persons, till they have in some measure studied the subject, do not see this inconceivableness'.[143] He could not repudiate the view that the generality of mankind would not feel the axioms of chemistry, or even of mechanics, to be necessary truths. Not that this damaged his position, but it conferred upon his natural theology a peculiar élitism. From the late seventeenth century one of the attractions of the argument from design had been its ostensible appeal to the generality of mankind. In his Preface to *The Wisdom of God Manifested in the Works of the Creation*, John Ray had observed that 'you may hear illiterate persons of the lowest rank of the commonality affirming, that they need no proof of the Being of a God, for that every pile of grass, or ear

[138] Whewell (1860), 377. [139] Ibid. 378.

[140] See e.g. Gillespie (1979), chs. 1, 2.

[141] See Manier (1978), 51–5. For a more detailed study of the complex interactions between Darwin's emerging theory and the legacy from natural theology, see Brooke (1985).

[142] I have argued this thesis at greater depth in Brooke (1977).

[143] Whewell (1860), 338.

of corn, sufficiently proves that'.[144] In Whewell's system the democracy had gone. Indeed a cynic might say that, in the absence of revelation, it was only through reading Whewell that one could gain access to the divine Mind, for through that medium alone would the uninitiated learn what the 'fundamental ideas' were.

[144] Cited by Brooke (1974), 13.

7. *William Whewell's Philosophy of Knowledge and its Reception*

by RICHARD R. YEO

DURING Whewell's lifetime references to his polymathic aspirations coloured most assessments of his achievements. Like Sydney Smith's well-known aphorism, these characterizations often carried negative judgements concerning Whewell's metascientific deliberations.[1] Even before embarking on his vast history and philosophy of the inductive sciences Whewell himself anticipated the opinions of 'people [who] bid us beware of the Demon of universal knowledge', although he freely confessed to George Morland in 1815 about 'certain yearnings after the whole circle of the sciences, certain ecstatic aspirations after universal knowledge, certain indefinite desires to approximate to something like omniscience'.[2] Thirty years later he referred to the Greek philosopher Erastosthenes whose activites embraced philosophy, geometry, astronomy, and poetry and, clearly with his own case in mind, observed that: 'It is seldom that one person attempts to master so many subjects without incurring the charge, and perhaps the danger of being superficial.'[3]

Whewell believed that the reception of his major works was influenced by perceptions about their ambitious intentions. When trying to decline an invitation to serve as President of the British Association he told Roderick Murchison that 'My *History* and *Philosophy* of Science are disqualifications, not qualifications, for being put at the head of the scientific world.' Writing to Lord Northampton on the same matter he remarked:

> Nobody knows better than I do how little I have done in science; . . . Of my *History of Science* the principal notice taken by men of science has been of a hostile kind; and I do not think that any practical cultivators of special sciences will feel any deference for a person who has presumed to speculate about them all.[4]

I would like to thank John Brooke, Menachem Fisch, and Simon Schaffer for their most helpful comments on drafts of this chapter.

[1] See the quotation from Smith in Becher's chapter, this volume. Smith is quoted in Stephen (1885–90), xx. 1361. See ibid. 1370, for a report of the quip about Whewell's assumption in rejecting the plurality of worlds: 'through all infinity there was nothing so great as the master of Trinity'. Cf. Herschel (1867–8), pp. li–lxi; Anon. (1866); H. Carlyle (1882), 142–4; Anon. (1881), 137–9.

[2] Quoted in Todhunter (1876), ii. 8, 10.

[3] Todhunter (1876), i. 35–6; also cited with comment in Maxwell (1890), ii. 528.

[4] Whewell to R. Murchison, 18 Sept. 1840, Todhunter (1876), ii. 286; Whewell to Lord Northampton, 5 Oct. 1840, ibid. 293.

Even before the *History* appeared, Charles Lyell admitted regretting that Whewell had not concentrated his powers 'on some one department of physical science and become a giant in that'. But he also added that it was now clear that 'you have given a greater impulse to the advancement of science among us by being a Universalist'.[5] Similarly, not long after his death, some commentators looked back on Whewell's efforts as a valuable corrective to the narrow specialization they observed in contemporary scientific life. Although critical of the way in which Whewell failed to build on the work of the *History* and *Philosophy*, his biographer, Isaac Todhunter, emphasized the current importance of the programme they embodied:

He [Whewell] did not despair of the fortunes of knowledge; he held that real connexions must subsist among the apparent infinite mass of details, and he encouraged investigators to seek for such general ideas in all their peculiar occupations. The lessons which he explicitly taught and implicitly suggested seem peculiarly useful at the present day when every pursuit is specialized.[6]

Since Whewell's work has largely been approached with reference to his idealist epistemology and his views on methodology (and then mainly in relation to his dispute with J. S. Mill), it is worth recovering what he saw as the broad scope of his project: the attempt to consider 'the whole circle of the sciences' in order to establish a philosophy of knowledge embracing not only the natural, but the moral sciences.

In his *Autobiography*, Charles Darwin recalls praising Whewell's *History* in the presence of the naturalist Robert Brown, and receiving the reply, 'Yes, I suppose that he has read the prefaces of very many books.'[7] Whatever the justice of this remark, it is useful for us to read the prefaces of Whewell's books, for in the introductory passages of both the *History* and the *Philosophy* he makes some interesting remarks about the nature of his intellectual project and, in doing so, reveals some of the tensions within it. In the first of these works Whewell noted that its title, *History of the Inductive Sciences*, might suggest the exclusion of those subjects not discussed, such as 'Ethnology and Glossology, Political Economy, Psychology', from 'the rank of Inductive Sciences'.[8] He stressed that this was not what he wished to imply and returned to this issue in the *Philosophy*. At several points in both the Preface and Introduction to this work, Whewell explained that his concern was with the philosophy of knowledge and not simply with the philosophy of the physical sciences. While subjects concerned with the material world were the ones most commonly termed sciences, they were to be studied for 'lessons' of wider significance.

The views respecting the nature and progress of knowledge . . . though derived from those portions of human knowledge which are more peculiarly and technically termed

[5] Quoted ibid. i. 112.
[6] Ibid. 145. For endorsement of this, see Maxwell (1876), 206.
[7] C. Darwin (1974), 61.
[8] Whewell (1857), i. 15.

Sciences, will by no means be confined, in their bearing, to the domain of such Sciences as deal with the material world, nor even to the whole range of Sciences now existing. On the contrary, we shall be led to believe that the nature of truth is in all subjects the same, and that its discovery involves, in all cases, the like conditions.[9]

Whewell contended that man's knowledge on a wide range of subjects was progressing towards the 'substantial character' signified by the term *Science*. He was confident that a study of these developments would reveal 'certain universal characters which belong to truth'; this would hold when the survey was extended beyond the realm of inert matter to include the sphere of human thoughts and actions: 'we shall still find', he affirmed, 'some general analogies which belong to the essence of truth, and run through the whole intellectual universe. Hence we have reason to trust that a just Philosophy of the Sciences may throw light upon the nature and extent of our knowledge in every department of human speculation.'[10]

I

It is useful to recall the context in which Whewell formulated his historical and philosophical account of science. For it was during the early 1830s, the seminal period for Whewell's work, that the meaning of 'science' was itself the subject of serious debate. From this time, the definition of the term began to narrow, losing the earlier general sense of systematic or demonstrative knowledge to acquire the more specialized, modern reference to the subject-matter of the physical or natural sciences. Writing later in the century John T. Merz speculated that this process was linked with the foundation of the British Association for the Advancement of Science—a suggestion which has since been accepted and elaborated.[11] However, this shift in meaning did not go uncontested; indeed it was strongly resisted by writers from two important intellectual formations—the Tractarians and the Romantics.

The British Association attempted to explain the meaning of science to the public and to affirm its broad social and educational value. But the meeting at Oxford in 1832 provided a focus for the Tractarians to attack the image of science that the Association promoted. Their critique was not a simple condemnation of science as a threat to the authority of Scripture, but a rejection of the claim, which they detected in the rhetoric of the Association, that physical science could be virtually equated with knowledge and serve as the basis of a wide cultural authority.[12] The *Oxford University Magazine* stressed this issue when it observed that: 'The term *science* equally with that of *art*, has of late been curtailed of its fair proportions . . . Science, in their new

[9] Id. (1847), i. 3. [10] Ibid.
[11] Merz (1896–1904), i. 89–91; Morrell and Thackray (1981), ch. 1.
[12] Ibid. 229–36.

vocabulary is restricted to the investigation of material substances.'[13] This point was central to the extensive critique of the British Association by J. W. Bowden in the *British Critic* during J. H. Newman's editorship. Lamenting the intrusion of the Association in the sensitive issue of the admission of Dissenters to university and its tendency to establish science as a form of secular religion, Bowden located the central problem in the intellectual pretensions of its terminology: 'Physical science, the science, in other words, of matter and material things, now arrogates in effect the name "science" exclusively to itself.'[14] The result was a neglect of the mental and moral sciences, metaphysics, and, of course, theology, the queen of the sciences. Coleridge's censure of the Association, although stemming from a different perspective, was similar. He insisted that the term 'science' encompassed more than physical knowledge and castigated men of science who appropriated the title of philosopher: 'I have met with several genuine Philologists, Philonoists, Physiophilists, keen hunters after Knowledge and Science . . . Truth and Wisdom are higher names than these—and *revering* Davy, I am half angry with him . . . for prostituting the name of Philosopher . . . to every Fellow who has made a lucky experiment.'[15]

It was in this context that Whewell coined the word 'scientist' to describe 'students of the knowledge of the material world'.[16] Unlike some of his other neologisms—anode, cathode, miocene, eocene—which date from this period, this term did not achieve early circulation,[17] partly because men of science did not wish to renounce their status as natural philosophers and, from the perspective of theological critics, perhaps because it seemed to reinforce the identification of science with natural knowledge. But it would be entirely wrong to see Whewell as endorsing any restriction of the term 'science' to the sphere of physical phenomena. He believed that the subjects presently known as 'sciences' represented the 'most certain and stable' examples of human knowledge, but that the intellectual processes which they involved could be extended to other areas. However, he considered that current associations between physical science, empiricist epistemology, and the principle of utility were producing an image of science hostile to moral and metaphysical enquiry. Whewell was aware of this problem before the intervention of the Tractarians at the time of the Oxford meeting, because his friend and correspondent H. J. Rose, who was an editor of the *British Magazine*, had warned against the dangers of an excessive celebration of the mathematical and experimental sciences. In a sermon of 1826 Rose contended that there was a close connection between the popularity of 'Experimental Philosophy' and the current appeal of pragmatic and utilitarian attitudes to intellectual enquiry: present reward had become the

[13] Bowden (1834), 401. See also Anon. (1847), 3–4.

[14] Bowden (1839), 8. See also Morrell and Thackray (1981), 241–2.

[15] Ibid. 20. Coleridge's remark is quoted in Levere (1981), 73.

[16] Whewell (1834*a*), 59.

[17] For Whewell's discussion of terminology with Faraday and Lyell, see Todhunter (1876), ii. 109–14, 178–83. see also Ross (1962), 65–85.

object, and public opinion the arbiter, of the *value* of knowledge. This encouraged an emphasis on results rather than on the process and discipline by which knowledge was achieved; it led to a neglect of those subjects such as philosophy, literature, history, and divinity which cultivated both the moral and intellectual faculties.[18] A concentration on the physical sciences, especially within the framework of a utilitarian philosophy, would produce a 'partial and often dangerous cultivation of the faculties'.[19]

Whewell answered this criticism by distinguishing between science and utilitarianism, by underlining the moral effects of scientific enquiry, and by stressing the metaphysical dimensions of scientific knowledge. He was puzzled by the equation of science and utility and reminded Rose that 'all of the science which we learn here [at Cambridge] is perfectly devoid of all practical use . . . '[20] However, he quickly urged that the pursuit of physical science—for which God had clearly bestowed rational faculties—could engender mental attitudes suitable to a moral and religious being:

> I am persuaded that there is not in the nature of science anything unfavourable to religious feelings, and if I were not so persuaded I should be much puzzled to account for our being invested, as we so amply are, with the faculties that lead us to the discovery of scientific truth. It would be strange if our Creator should be found to be urging us on in a career which tended to a forgetfulness of Him.[21]

But furthermore, Whewell noticed that Rose's image of scientific progress as the accumulation of discoveries and 'the constant rejection of present belief in favour of new views',[22] assumed a mistaken philosophy of science. In amending this view Whewell insisted that 'it is by no means correct to consider advances in science as rejection of what was known for something new. The novelty, if the philosophy have been duly inductive, *includes* old truths and shews them from a new point of view'.[23] This point came to occupy a key place in his mature philosophy of science, but even at this early stage of his career such a statement related to his speculations about the metaphysical aspects of science. Some years later Whewell endorsed the position (advocated by Coleridge and his own friend William Rowan Hamilton, the Irish mathematician) in which sound knowledge of nature depended on principles provided by the mind. He did not fully accept Hamilton's definition of science as 'strict, pure and independent; deduced by valid reasonings from . . . intuitive principles';[24] but

[18] Rose (1826), 12, 17.

[19] Id. (1834), 48. For more on Whewell's response to rationalist and utilitarian approaches to science, see the chapter by G. N. Cantor in this volume.

[20] Whewell to Rose, 19 Nov. 1826, Todhunter (1876), ii. 75. Whewell reinforced his conception of science as purely speculative by distinguishing it from art (or technology). Science did not lead directly to practical applications; in fact, 'Art is the parent, not the progeny, of Science.' Whewell (1857), i. 253; also (1847), ii. 106–12.

[21] Whewell to Rose, 19 Nov. 1826, Todhunter (1876), ii. 76. For Whewell on the limits of natural theology, see the chapter by J. H. Brooke in this volume.

[22] Rose (1826), 12.

[23] Whewell to Rose, 12 Dec. 1826, Todhunter (1876), ii. 78–9.

[24] Quoted in Hankins (1980), 386. For the notion of a connection between a 'metaphysical' and

by the time Coleridge barred the title of 'philosopher', Whewell agreed that physical science depended upon mental activities which could properly be described as metaphysical, and therefore, that the term 'science' must encompass more than the study of the material world.

A closely allied feature of Whewell's early metascientific interests was also apparent by the time he responded to Coleridge by proposing the term 'scientist'. For although this word was contrasted with 'artist', Whewell intended it to underline the common enterprise in which astronomers, chemists, geologists, and botanists were engaged. Significantly, his first published reference to the new term occurred in his review of Mary Somerville's book *On the Connexion of the Physical Sciences*, a work which sought to illustrate the analogies between various branches of 'physics' and the possibility of general laws embracing them.[25] Whewell was sympathetic to this search for relationships between the disciplines, because like William Vernon Harcourt, the first secretary of the British Association, he was anxious about the specialization manifested in the various sectional divisions of the new scientific forum. Following Harcourt's reference to 'the commonwealth of science', he anticipated the prospect of science disintegrating, like 'a great empire falling to pieces', and hoped that the ideas expressed by Somerville might arrest this process of 'separation and dismemberment'.[26] Thus, although the word he coined came to be identified with specialist expertise, Whewell's interests in the early 1830s were very much concerned with the wider relationships—epistemological and methodological—that bound the various sciences together. This chapter attempts to outline Whewell's thoughts on these issues from his early speculation with colleagues to his major writings on science. It considers the tension in his work between the notion of a general philosophy of knowledge delineating the common features and analogies of all the sciences (including moral science), and his stress upon the distinctiveness of different disciplines resting upon peculiar fundamental ideas. By appreciating these two dimensions it is possible to show firstly how he deployed the idea of the unity of science in order to legitimate subjects such as political economy and moral philosophy as 'sciences'; and secondly, how he was compelled to reject interpretations of this ideal which threatened the distinctiveness of concepts (such as final cause) crucial to his natural theology.

an 'objective' science, see Hamilton to Whewell, 25 May 1833, Whewell Papers, Trinity College Library, Cambridge University, Add. MS a. 205[102]. Coleridge had expounded the role of metaphysical ideas in science in *A Preliminary Treatise on Method* (1817); this formed the general introduction to the *Encyclopaedia Metropolitana* which appeared between 1817 and 1845. On the intellectual context in which Whewell's epistemology emerged, see Yeo (1979). For an analysis of key concepts in his philosophy, see Fisch (1985).

[25] Somerville (1834), Preface and p. 413. See also Patterson (1983), especially ch. 7.

[26] Harcourt (1831), 28. Whewell (1834*a*), 59. Interestingly, this image of fragmentation was deployed by critics of institutional science such as Bowden, and also by Carlyle, who celebrated the notion of the solitary genius capable of grasping general laws without the need for detailed experimentation. See Bowden (1839), 17, 28–9; T. Carlyle (1829), 439–58. The threat to the unity

II

The wide scope of Whewell's programme was apparent in his early correspondence with his close friend, the political economist Richard Jones. He regarded Jones as an ally in a campaign to analyse the foundations and preach the merits of what he variously called 'the true faith', 'the Reforming Philosophy', 'my beloved Induction'.[27] These discussions, dating back to the early 1820s, document the genesis of Whewell's trajectory towards the *History* and *Philosophy*; but as he himself stressed in the prefaces of these works, the final parameters of this project were always conceived to extend beyond the physical sciences. Writing to Jones in 1831 Whewell revived Rose's earlier suggestion of establishing a review which would take 'a line of moral philosophy, political economy, and science' and thereby form 'a school which would be considerable in influence of the best kind'.[28] He regarded the scope of this project as an appropriate realization of the plans for the inductive philosophy initiated by Bacon—a method of investigation applicable not only to material but social and moral subjects.[29]

Jones's work on political economy therefore occupied a crucial place in Whewell's speculations about the general features of scientific knowledge. For since it dealt with moral and social phenomena—the interests, passions and behaviour of men—it was a test case for the extrapolation of inductive philosophy. Whewell was hopeful that political economy could be brought within the sphere of inductive sciences and assured Jones in 1822 that the subject involved 'no peculiar principles of observation or deduction'.[30] There was much at stake here, for Whewell and Jones saw themselves as combating David Ricardo's political economy—a strongly deductive system which threatened, in their estimate, to give a misleading image of the true character of scientific knowledge. Whewell was not rejecting a role for deductive thinking but argued that Ricardo's axioms rested upon an insecure foundation. Jones's work on rents and wages in different societies was intended to undermine the generalizations of 'the "downward road" people' by providing a detailed set of observations and a classification of data from which legitimate inductions could be made.[31] Whewell and Jones located one of the key errors of their opponents in Richard

of science from specialization was often met by the notion of a general scientific method. See Yeo (1981) and (1986).

[27] See Todhunter (1876), ii. 115, 172, 267.

[28] Whewell to Jones, 24 Apr. 1831, Todhunter (1876), ii. 118.

[29] Whewell was probably influenced by the Scottish version of Bacon's programme in the works of Dugald Stewart. See Yeo (1985).

[30] Whewell to Jones, 16 Aug. 1822, Todhunter (1876), ii. 49.

[31] Whewell to Jones, 23 July 1831, Todhunter (1876), ii. 125. Whewell applied deductive reasoning to the postulates of Ricardo's doctrines but stressed that 'Mathematical calculations (cannot) supply the place of moral reasoning' and regarded the basic principles of his political economy as inadequately grounded. See Whewell (1830*a*), i, and (1831*g*), 1–3, 43–4.

Whately's opinion that 'principles of action are known by consciousness and do not require detailed observation'.[32] By promoting Jones's researches as an example of proper induction, Whewell hoped to use the dispute on political economy as a means of establishing correct views about the nature of moral science and its analogy with physical science.

Whewell's confidence about this task appears to have fluctuated. In 1827 he told John Herschel, 'I am quite satisfied both of the truth of Jones's general views, and also that they possess the great property of true doctrines, that they are capable of being proved at any step of their induction between the most general and the most particular.'[33] Nevertheless he later reminded Jones that 'the analogy between physical and political or economical science is yet to be shewn', and revealed his doubts about 'the identity of the scientific method (that is, the method of making a science) in physical and moral sciences'.[34] Yet much of Whewell's uncertainty may have related to the possibility of explaining the inductive approach and exemplifying its application to moral subjects, rather than to the strength of the analogy with the study of the material world. In 1831 he explained to Jones that 'it is not that I do not think induction applicable and the examination of its rules important to the notional sciences, but I do not yet see my way to anything which I could propound concerning generals'.[35] Whewell believed that the principles of induction could not be 'either taught or learnt without many examples'; and that even with the history of the successful physical sciences before them, people were still ignorant about the true nature of induction.[36] Here again Whately's views were seen to be dangerous because his work on *Logic* of 1831 did not grasp the process of scientific thinking. Jones remarked that Whately entertained the 'strange notion that the inference made by the mind when its inductive processes are completed is an inference which belongs to *deductive* reasoning and may be called *logical* according to his own strict definition of logic.' As a remedy he suggested that one of them should publish 'some popular views of inductive reasoning' that encompassed examples outside natural philosophy.[37] Both men welcomed the appearance of Herschel's *Discourse* as a partial answer, but Whewell pleaded with Jones to produce his books on rents (and the second volume on wages), which would supply 'illustrations and examples of the ascending method applied to moral sciences.'[38] When it finally did appear Whewell regarded its careful survey of the tenure of land in different countries at different times as an effective answer to the prevailing doctrines of political economy. Thus Jones's conclusion that rent could not be described by one general definition underlined the prematurity of Ricardo's axioms and his

[32] Whewell to Jones, 15 July 1831, Todhunter (1876), ii. 123.
[33] Whewell to Herschel, 23 Nov. 1827, ibid. 86.
[34] Whewell to Jones, 15 July 1831 and 23 July 1831, ibid. 123, 126.
[35] Whewell to Jones, Feb. 1831, ibid. 116.
[36] Whewell to Jones, ibid. 115–16.
[37] Jones to Whewell, 24 Feb. 1831 (postmarked), Whewell Papers, Add. MS c. 52.
[38] Whewell to Jones, 23 July 1831, in Todhunter (1876), ii. 125.

failure to learn from either Bacon or the history of the physical sciences: the deductive character of Ricardo's system was not a mark of its scientific character but rather a symptom of its failure to appreciate the process by which sciences were established.[39]

But apart from the vindication provided by Jones's work, Whewell became more confident about the common features of the physical and moral sciences as he became convinced that he had grasped the true nature of inductive thinking. Discussing their reading of the Saint-Simonians in 1832, he agreed with Jones that there were

> several right notions about the character of science, one in which they have hit on the same word which I have used for nearly the same thing—the *conceptions* which must exist in the mind in order to get by induction a law from a collection of facts; and the impossibility of inducting or even of collecting without this. However, I do not think they have done anything to diminish the necessity of developing and verifying this notion, and I am very sure that when this is done we shall have a much clearer view of the nature of general truth and the way of getting at it—including political and perhaps something of moral truth.[40]

By June Whewell had developed a plan for books on the history and philosophy of science and a third, entitled 'Prospects of Inductive Science' which would deal with the extension of inductive processes to 'philology, art, politics and morals'.[41] Whewell's confidence in the inclusion of these subjects within the sphere of inductive science may have been reinforced by his conviction about the role of 'Ideas' in the attainment of scientific truths. Since these 'mental bonds of connexion, ideal relations . . . sciential conditions' were supplied by the mind, the dichotomy between physical and moral sciences could not be based on traditional contrasts between objective and subjective forms of knowledge.[42] Thus by August 1834 when he explained to Jones that different physical sciences depended upon various 'ideal relations' such as space, time, cause, or life, Whewell promised to 'try to discover the nature of the analogy

[39] See Whewell (1859), pp. ix–xl; see p. xii for the view that Jones's approach was 'more likely to lead to valuable truth, because it was in conformity with that general process of induction, by which all the most substantial truths which man possesses (except only mathematical truth) have been obtained'. Jones made use of Herschel's *Preliminary Discourse on the Study of Natural Philosophy* (1830), quoting a passage referring to the different rates of progress usually occurring in experimental versus observational sciences. The message here was that Ricardo's political economy was prematurely attempting the progress of an experimental subject. Jones (1831), appendix; also page before ch. 1 which gives Whewell's view on definitions. See also Whewell (1833*a*), 271–2.

[40] Whewell to Jones, 19 Feb. 1832, in Todhunter (1876), ii. 141. See Whewell to Jones, 5 Aug. 1834, Whewell Papers, Add. MS c. 51[174], for an indication of his new view of induction. See also Fisch's chapter. This shift is important because although Whewell's search for a general philosophy of knowledge dated from about the 1820s, it stood on different grounds by the 1830s.

[41] Whewell to Jones, 27 July 1834, ibid. i. 90. See also the draft entitled 'The Philosophy of the Progressive Sciences', Whewell Papers, R18 17[8] pp. 3–5 for the intention of using a study of the 'exacter sciences' as a basis for approaching moral subjects.

[42] Whewell to Jones, 21 Aug. 1834, Todhunter (1876), ii. 186.

which exists between these sciences and our knowledge respecting morals, taste, politics, language, and generally, all hyperphysical knowledge'.[43]

We have now seen how Whewell appealed to notions of a general inductive method, or to analogies in the history and construction of various sciences, in order to defend the scientific status of political economy. But this was not simply a general profession about the identity of this subject as a science: it was a more specific defence of a preferred form of enquiry against doctrines, usually named as those of Malthus and Ricardo, which had allegedly transgressed the analogies disclosed by the progress of physical science. Malthus, although producing some sound data on population, had leapt with insufficient evidence to an 'anticipatory thesis' that questioned the wisdom and benevolence of God's social design.[44] Ricardo's system also rested upon inadequate data because it ignored the need for comparative instances; and from this weak base it sought premature deductions.[45] Taken together, the doctrines of these two writers threatened to associate the name of science not only with inappropriate modes of procedure, but with improper moral and religious attitudes. Jones summarized this concern: such views 'having been first insisted on with a dogmatical air of *scientific* superiority', were then asserted as 'an apparent inconsistency between the permanence of human happiness, and the natural action of the laws established by Providence.'[46] When Whewell turned to moral philosophy the ideological investment was equally high because his attempts to establish relationships between moral and material knowledge were informed by conflicts between opposing ethical theories and rival claims to scientific status. In this field we can see how he deployed comparisons from the natural sciences in order to legitimate firstly the notion of moral science as a subject capable of producing knowledge rather than mere opinion; and secondly, to break the pretensions of utilitarian ethics to a privileged affinity with the status of physical science. It will then be possible to see the tension between this reference to the unity of physical and moral science and Whewell's emphasis upon the distinctive ideas in forming different branches of science.

43 Whewell to Jones, 21 Aug. 1834, Todhunter (1876), ii. 186.

44 Whewell to Jones, Feb. 1831, ibid. 116. See also Jones (1831), pp. xi–xix.

45 Ibid., pp. xix–xxiii, appendix; see Whewell (1831*g*), 2–3, 42–3, for the claim that Ricardo's system was not at the stage of having 'a few certain principles' and should therefore be reasoning 'not from principles but to them'. The debate on political economy in the 1830s involving Whately, Scrope, and Senior often raised the question of the nature of scientific method and its relation to this subject. Mill thought that the status of political economy highlighted the problem of the terms 'physical', 'moral', and 'science'. See Mill (1836). Recent work has also pointed to the transposition of concepts between physical and social sciences in this period. See Rudwick (1974), and (1979), 67–73. A later section of this chapter will suggest that Whewell's work contained strong warnings about such transpositions.

46 Jones (1831), pp. xiii, and Jones in Whewell (1859), 3; Whewell, unpublished sermon of 1827, cited in Todhunter (1876), i. 330–1. For further comment on Whewell's political economy, see the chapters by Becher and Ruse in this volume.

III

Whewell's interest in moral philosophy developed well before he took the Knightbridge Chair in this subject in 1838. From the early 1820s his correspondence with H. J. Rose and Julius Hare on epistemological issues often led to questions about the moral implications of philosophical positions and the appropriate form of ethical theory. The central preoccupation here was the need to counter what they perceived as the deleterious effects of Locke's epistemology and Paley's utilitarian ethics, two of the central areas of reading on the Cambridge curriculum.[47] This campaign reached a peak in the 1830s, but as early as 1822 Hare told Whewell that 'Plato is worth ten thousand Aristotles and 100,000 . . . Lockes' and continued to urge the value of Coleridge's work, with its synthesis of Platonic and German thought, as the most suitable foundation for a Christian moral philosophy.[48] Both saw a return to the intuitional school of ethics represented by Joseph Butler as a necessary antidote to the morality of consequences which students were reading in Paley's *Principles of Moral and Political Philosophy*. In its denial of a substantive moral faculty and in its reliance upon notions of utility and expediency, Paley's position appeared dangerously close to the views of the Benthamites. In his *Discourse on the Studies of University of Cambridge*, Adam Sedgwick joined the attack on this position by indicating its relativist implications: virtue and vice would no longer have 'any fixed relations to the moral condition of man but [would] change with the fluctuations of opinion'.[49] Whewell and Hare believed that both epistemology and moral philosophy had to assert the spiritual nature of man and the ability of the mind to reach secure knowledge of ethical principles. Consequently, they both welcomed an undergraduate essay of 1833 on *The Analogy of Mathematical and Moral Certainty* by Thomas Birks (a future incumbent of the Knightbridge Chair) which attempted to overcome the contrast between the security of natural knowledge and the vagueness of moral philosophy. With large debts to Coleridge, Birks argued that knowledge in both fields rested upon intuitive laws of thought and information furnished by the senses; there was therefore no reason 'in the nature of its first principles why the evidence of morals should be placed on a lower footing, or regarded as less purely demonstrative than that of geometry'.[50] From this position the way was opened for the contention that a concern with intuitive ethical principles, in spite of the opinions of the utilitarian school, was consistent with the character of scientific knowledge.

[47] For an analysis of Whewell's thought within this context, see Yeo (1979), 498–503.

[48] J. Hare to Whewell, 13 May 1822, Whewell Papers, Add. MS 77^{127}. For the campaign against Locke, see Aarsleff (1971), 398–9; Schneewind (1974), 371–87; Garland (1980), ch. 4.

[49] Sedgwick (1834), 63, 85. For Whewell's reservations about Butler as an alternative source of ethics and natural theology, see respectively, the chapter by Brooke in this volume and Whewell to Rose, 24 June 1818, in Todhunter (1876), ii. 26.

[50] Birks (1834), 12, 16.

When Whewell took up the Professorship of Moral Philosophy he had published the *History* and was completing the *Philosophy*. He excused his preoccupation with the latter by arguing that although it seemed to concern a field remote from the duties of his Chair, its findings might 'be really found to afford a basis for all philosophy, even that of morals, and, in some measure, an example of the mode in which philosophical truth may be sought'.[51] Thus, although the third major book which he described to Jones was never written, Whewell saw his endeavours in moral philosophy as stemming from this early intention to analyse the '*hyperphysical* sciences', 'keeping in view the analogy of the true type of the progress of science'.[52] Like Birks, he began by confronting the general scepticism about such an analogy, admitting that although speaking of 'Moral and Political *Sciences* . . . we never use the word without a misgiving'. He acknowledged the force of the common antithesis between Physics as the investigation of the material world and Ethics as the study of men's moral actions; but Whewell argued that this concern with men's actions did not establish 'any essential difference between the nature of truth, or the mode of seeking it, in morals and in physics', for human action was involved in all departments of scientific knowledge.[53] Just as cases of moral action necessarily preceded systematic thought about them, so also in physical science, action, either in the sense of practical art or deliberate experiment, preceded theory. He also challenged the view that moral knowledge was 'vague and precarious', while physical knowledge was 'exact and certain', suggesting that the latter was not usually perfect and the former not entirely obscure. Here he made the significant move of drawing attention to the differences within the full range of natural sciences:

> when we look at such sciences as meteorology, and physiology, and geology, we see that the indefiniteness and obscurity of the subjects contemplated, cannot be a reason of imperative authority to exclude morality from the list of sciences, in the most genuine sense of the term; that is, bodies of exact, systematic, progressive knowledge.[54]

Thus although moral philosophy dealt with complex human passions, physiology grappled with the principles of life and geology with the distant revolutions in the 'mysterious history' of the earth. The crucial point for Whewell was that all these subjects manifested a progressive development consisting in the gradual clarification of those ideas on which each subject depended. The lesson was that there is 'a line of progress which is plainly applicable to moral as well as to physical knowledge'. He believed that this was compatible with the recognition

[51] Whewell (1841*d*), 30–1. This comment occurs in the lecture of 1841; see also the conclusion of the first lecture of 1838, 28: 'Inquiries into the nature of truth, the means and method of its discovery, and the philosophy of science, . . . cannot fail to influence . . . our studies with regard to moral truth, moral science and the true philosophy of human life.'

[52] Whewell to Jones, 21 Aug. 1834, in Todhunter (1876), ii. 187.

[53] Whewell (1841*d*), 34, 41.

[54] Ibid. 37, 38–41.

that such progress went in different directions: in physical sciences facts were *idealized*; in moral sciences ideas were *realized*.[55]

But having drawn attention to the shared imperfections of the moral and physical sciences, Whewell pursued another line of argument when he published his major works on *Elements of Morality* in 1845 and *Lectures on Systematic Morality* of 1846. In the first of these he claimed that the basic propositions of a moral system could be considered as analogous to 'the *Axioms* in Geometry'.[56] In taking this position, Whewell, like Birks, was attempting to maintain the possibility of a system of 'Moral Truths, definitely expressed, and arranged according to their rational connection' against the view that such notions were solely derived from the effects of pleasure and pain and justified by utility.[57] But writers sympathetic to his aims were anxious about this move from more general analogies between moral and physical science to the specific comparison of morality and geometry. Thus while Hare welcomed the sermons of 1838, saying that 'the comparison between our scientific and our moral knowledge is locally appropriate and very satisfactory', his friend and relative, Frederic Myers, vigorously rejected any similarity between the elements of morality and 'the elements of Euclid'. And although Whewell subsequently acknowledged that this comparison could be misleading, he continued to defend the value and instructiveness of analogies between moral and physical knowledge, and indeed, between moral and geometrical reasoning.[58]

Whewell suggested that the resistance to such comparisons derived in part from erroneous views about the nature of knowledge. Morality was often said to depend on facts revealed by consciousness and physical science on facts gained by observation. While accepting the need for this distinction, Whewell opposed any extreme antithesis:

> For in mechanics too, those who treat of the elements of the science, have been in the habit of referring to our consciousness. They have usually taught that we derive the conception of *force* from the muscular effect which we are conscious of making. And even in geometry, on what evidence do axioms depend? Is it not on the impossibility which we are *conscious* of, when we try to conceive the opposite of them?[59]

Furthermore, he rejected the assumption behind Myers's opposition to the analogy of morality and geometry: namely, that 'geometry depends upon *arbitrary* definitions and nothing else'. Whewell had previously criticized this position as he found it in the works of Dugald Stewart, and told Myers that 'we must in geometry form our conceptions before we can define them; and that in the process of forming them we establish in our minds axioms which are the

[55] Ibid. 45; Whewell (1860), 315–19. This point was originally made in his 'Second Memoir on the Fundamental Antithesis of Philosophy' (1848*b*).

[56] Whewell (1845*b*). For a detailed discussion of this attempt, see Schneewind (1968).

[57] Whewell (1846*a*), 25. Also Whewell to Herschel, 3 July 1846, in Todhunter (1876), ii. 337–8.

[58] Hare to Whewell, 4 Jan. 1838, Whewell Papers, Add. MS 77[171]; F. Myers to Whewell, 31 July 1845, in Stair-Douglas (1882), 323; Whewell (1846*a*), 44.

[59] Ibid. 53.

first principles of our reasoning'.[60] This also applied to other demonstrative sciences such as mechanics. In fact, Whewell regarded the analogy between mechanics and morality as even more useful to his case because it offered better illustration than geometry of the progressive character of both scientific and moral knowledge. He explained to Myers that 'although we now have axioms, defined conceptions and vigorous reasonings [in mechanics], we can point to persons, and to whole ages and nations, who did not assent to those axioms because the mechanical ideas of their minds were not sufficiently unfolded'.[61] He contended that this process could also be seen in the growth of moral knowledge, and that this answered the objections of those, like Paley, who proposed to test the existence of moral axioms by asking whether they were evident to all men, including savages. But for Whewell, such axioms, in morality, geometry, and mechanics, required a certain 'culture' of the mind: they became evident to all men only as their conception of the terms of the science were 'rendered clear and definite'.[62] In this way Whewell attempted to incorporate morality within the general philosophy of knowledge elaborated in his study of the inductive sciences; that is, one stressing the role of both reflection and experience within a historical process through which fundamental concepts were gradually clarified. And more specifically, he sought to show that an intuitional theory of ethics, or 'high morality', as he called it, conformed more closely than utilitarian ethics with the general methods and processes of the sciences.[63]

IV

We have now seen how Whewell sought to defend the scientific status of political economy and moral philosophy by appealing to methods, epistemology, and modes of progress common to all the sciences. Thus he argued that political economy, although involving the behaviour of men, could be brought within the scope of the inductive method employed in the physical sciences. Morality, while appearing distant from Physics in its close connection with consciousness, nevertheless exemplified the interaction of ideas and experience which had produced the intuitively warranted axioms of the most successful scientific disciplines. Whewell also affirmed the validity of specific approaches or theories in political economy and morality by references to the departure of rival doctrines from the lessons provided by the history of the physical sciences.

These appeals to certain general features of method and epistemology allowed Whewell to construct analogies between the study of politico-moral

[60] Whewell to Myers, 6 Sept. 1845, in Stair-Douglas (1882), 327.
[61] Ibid.
[62] Ibid. 328; Whewell (1846*a*), 34–42.
[63] Id. (1862*a*), 1–3, 126, 132.

phenomena and physical sciences, but they existed alongside comments which indicated *differences* within the sciences. Thus his legitimation of both political economy and moral philosophy exposed various levels and stages of scientific knowledge. Against the charge that the complexity of moral subjects rendered them inappropriate for scientific treatment, Whewell pointed to the indefinite character of enquiries in meteorology and physiology. This enabled him to invoke the notion of stages of development, thereby defending Jones's inductive method as appropriate to the state of political economy, while attacking Ricardo's deductive system as premature. Whewell thus openly acknowledged and exploited these differences, while simultaneously arguing for the inclusion of moral and social subjects within the category of 'science'. But this was not accomplished without difficulty, because there was a tension between Whewell's effort to affirm the relationships between morality and natural science and his need to defend a particular view of moral ideas. In attacking utilitarianism Whewell was rejecting the 'low' morality of consequences in favour of the 'high' morality of principles—principles which supposedly derived from man's moral nature and could be said to be 'necessary, universal, and eternal'.[64] Now it is here that the precise location of the analogy with physical science was significant. When Myers suggested a comparison of morality with 'the lower orders of the Inductive Sciences' rather than with geometry, Whewell remarked that he 'did not know to what classifications of high and low order of sciences you refer'.[65] Yet it could be argued that his preferred ethical theory did indeed require the analogy to be drawn with geometry, not with one of the experimental or observational sciences. The relevant distinction here comes from Whewell's own educational writings, in which 'permanent' subjects such as geometry, Latin, and Greek were set apart from the 'progressive' disciplines such as chemistry, geology, and philology.[66] The former could be taught, authoritatively, as bodies of secure and unchanging laws, rules, and axioms; the latter displayed a history of development which, at certain points, involved dramatic reversals in the central concepts and theories. For this reason Whewell considered the 'progressive' physical sciences inferior to geometry as the basis of the undergraduate curriculum. But equally, his conception of morality as a body of 'permanent truths'[67] was not easily aligned with the mode of progress which he had delineated in the history of the natural sciences, namely, one in which the emergence of new fundamental ideas could radically alter the theories of a discipline. That is, the attempt to legitimate systematic morality by emphasizing its correspondence with the epistemology of science threatened to undermine Whewell's commitment to the immutable and universal character of moral truths. While being able to manage the progressive realization of permanent moral truths by different people—on the analogy with geometry—and although he accepted the possible 'development of new truths, out of those

[64] Ibid. 10. [65] Stair-Douglas (1882), 328.
[66] Whewell (1837), 5–10, and (1845), 5–6, 18.
[67] Id. (1846*a*), 62.

already familiar', it is not clear that he could entertain the prospect of a Copernican revolution in moral philosophy.[68]

Another difficulty in Whewell's attempt to bring moral sciences within a general philosophy of knowledge derived from the need to ensure the special character of moral ideas. Whewell confronted this issue directly in his *Lectures on Systematic Morality*, assuring his students that the pursuit of analogies across the sciences did not imply that there were no differences between moral and material knowledge. While maintaining that there were 'conditions of knowledge' sufficiently general to comprehend all truths, Whewell stressed that moral conceptions must 'be regarded in a moral light, defined by moral lines, treated by moral arguments; not defined and reasoned about in a mathematical manner'.[69] This tension surfaced most clearly in the case of morality because any tendency to assimilate it too closely to the physical sciences would weaken the specific view that Whewell was affirming, namely, the integrity of moral ideas stemming from the nature of men rather than from the consequences of actions in the social world. But this tension was not restricted to morality; rather it pervaded Whewell's entire philosophy of knowledge.

V

It is possible to see Whewell's interest in political economy and moral philosophy as an illustration of his concern with the relationships between various fields of knowledge. In debates about both these areas, Whewell drew upon his study of scientific knowledge in order to affirm the possibility of common methodological and epistemological principles which could be extended to the region of politico-moral subjects, thus constituting them as moral sciences. But in arguing this case he appealed to the differences within the sciences and, particularly in the case of the debate on morality, was anxious to retain the distinctiveness and integrity of the ideas involved. By now returning to Whewell's writings on the natural sciences, it is possible to see how he combined references to the common features of intellectual progress with an emphasis on the specific characters of different disciplines. Whewell believed that both facets could be integrated without relinquishing the ideal of the unity of science and knowledge. However, we can gain a contemporary perspective

[68] Whewell (1846*a*), 38–41, 63. For another statement of this tension, see Schneewind (1977), 111–17. The implication of Whewell's views on the progressive nature of morality is unclear. When speaking of the Fundamental Ideas of the physical sciences, such as Cause, Substance, Kind, Final Cause, he claimed that the progress of science did not 'bring into existence new elements of truth in the human mind, but it may bring them into view'. Whewell (1860), 372–3. That is, the Fundamental Ideas did not change but their emergence and the interpretations of conceptions derived from them were historical in character. It is not clear whether Whewell was prepared to conceive moral science, on the analogy with material science, as a domain which could expect to see the discovery of new Fundamental Ideas and radically different constructions of them.

[69] Whewell (1846*a*), 44, 48–9.

on his work by considering the writings of Baden Powell—another metascientific commentator—who regarded Whewell's position as a threat to the progress of scientific enquiry and the true unity of the sciences.

Whewell's confidence in the potential relevance of his *History* and *Philosophy* to other forms of enquiry rested upon the analogies and relationships delineated amongst the natural sciences. As man's knowledge in these fields developed, 'certain universal characters' or 'general laws' were revealed.[70] Having embarked on '*a connected and systematic survey of the whole range of Physical Science*', Whewell reported that it was possible to consider the various sciences as 'all members of one series, and as governed by rules which are the same for all'.[71] Referring to the acknowledged advances in sciences from Astronomy to Physiology, he recalled Bacon's aspirations and asked whether these developments revealed 'some common process, some common principle'; whether 'the organ by which discoveries have been made has had something uniform in its structure and working'. If this was so, he suggested, it may be possible to extract from the history of scientific progress 'the elements of a more effectual and substantial Method of Discovery', which could be applied to new areas of enquiry.[72] But Whewell also warned against the danger of forcing analogies between the nature of physiological discoveries and those of mechanics; the comprehensiveness of his survey, which included many of the 'intermediate sciences', revealed the pitfalls of this tendency.[73]

Similarly, while celebrating the success of mathematical physics and the virtues of mathematical training, he stressed that this form of reasoning could not be applied to all subjects. Furthermore, he qualified the concept of a common method capable of simple transfer across various fields: although extremely tempting, an *Art of Discovery* was not strictly possible, partly because historical study suggested that 'different sciences may be expected to advance by different modes of procedure, according to their present condition'.[74] There were various stages in the development of science—chemistry, for example, was in its infancy compared with the maturity of astronomy—and different modes of enquiry were thus appropriate. Most crucial in this connection in Whewell's view was the grasp of the peculiar Fundamental Idea which constituted the conceptual framework of a particular discipline. It was this extension of Kant's teaching on a priori categories, beyond the Ideas of space and time, that epitomized the controversial character of Whewell's work, not only because of its idealist epistemology, but, as we shall see from the reaction of Powell, because its emphasis on the special metaphysical foundation of different disciplines threatened the notion of the unity of science and the apologetic ends it served.

[70] Id. (1847), i. 3.

[71] Ibid. 8.

[72] Ibid., p. vi.

[73] Ibid. 9.

[74] Ibid., p. viii. For Whewell's scheme of historical stages—prelude, inductive epoch, sequel—and its inapplicability to many of the sciences, see Todhunter (1876), i. 107, 135–8 and the chapter by Cantor in this volume.

Whewell's philosophical and historical enquiries led to the conclusion that 'Each science has for its basis a different class of Ideas; and the steps which constitute the progress of one science can never be made by employing the Ideas of another kind of science.'[75] Thus although he was willing to explore links between various branches of science[76] (for example, he sought a fruitful alliance between mineralogy and chemistry), Whewell was deeply suspicious of attempts to explain substantially new phenomena in terms borrowed from a previously established science. He advised that discussions of 'attraction' in chemistry should resist the tendency to employ concepts from mechanics, as if the forces in question were of the same kind. The chemical phenomena to be explained (changes in colour, transparency, taste, etc.) manifestly indicated the need for a new concept which recognized chemical combination as a 'peculiar principle, a special relation of the elements, not rightly expressed in mechanical terms'.[77] Similarly, an adequate account of Living Things demanded a 'peculiar idea', not conceptions derived from physics or chemistry. Whewell admitted that this need for novelty went against the proclivities of human nature to rely on old ideas; but the lesson of his *History* was that progress in the understanding of mechanical, chemical, and vital phenomena depended upon the appreciation that each involved 'new characters'. To advance to a 'Science of Cause' in a new area, the challenge was to break away from familiar ideas and to form 'distinctly new conceptions'.[78] Whewell acknowledged that in sciences such as chemistry and physiology the appropriate Fundamental Idea had not been sufficiently clarified—the idea of 'Polarity' was too uncertain to form part of an undergraduate curriculum and there was no consensus on the idea of 'Life'—but the need for distinctive conceptions in these fields was a precondition of their future progress.[79]

But in Whewell's view, these differences between disciplines did not mean that the less exact subjects fell outside the sphere of scientific knowledge. Thus, when discussing the issue of classification in Natural History, he replied to the contention that its inexactness rendered it unscientific. Such a judgement arose from the fact that 'the mathematical and mathematico-physical sciences have, in a great degree, determined men's views of the general nature and form of scientific truth; while Natural History has not yet had time or opportunity to exert its due influence upon the current habits of philosophizing'.[80] Moreover, he claimed that the apparent inconsistency of classifications and definitions in

[75] Whewell (1847), ii. 18–19.

[76] For his views on the close connections between certain disciplines, see Whewell to Herschel, 15 Feb. 1829, and Whewell to J. D. Forbes, 16 Dec. 1831, in Todhunter (1876), ii. 98, 136. For the problem posed by the relationships between chemical and electrical action for Whewell's taxonomy of the sciences, see the chapter by Cantor in this volume. See also Knight (1970), 53–4.

[77] Whewell (1847), i. 387–8.

[78] Ibid. ii. 99–100. Whewell's criticisms of Newton's *vera causae* principle can be understood in this context. See ibid. ii. 279–86. For Newton's confidence in the future explanatory power of his approach, see McMullin (1985), 36.

[79] Whewell (1847), i. 545, and ii. 366.

[80] Ibid. i. 494.

Natural History was also found in other sciences, and that the progress which this subject had made towards exact distinctions could provide lessons for 'the best modes of pursuing truth of all kinds'. Thus the recognition of differences between disciplines did not entirely banish the notion of a general unity: it was legitimate to conceive '*definite* differences' among the sciences as included 'within a finite boundary of similarity'.[81] There could then be a movement from 'the Philosophy of the Sciences' to the 'Philosophy of Science'.[82] A general philosophy of knowledge was possible because the sciences, although having distinct conceptual foundations, all involved the dialectic of Facts of Ideas, the Fundamental Antithesis of Philosophy. It was the presence of this process in the establishment of all real knowledge which convinced Whewell that his findings were 'suited to throw light upon Moral and Political Philosophy, no less than upon Physical'.[83]

The previous discussion has shown that Whewell's pursuit of relationships between the branches of science coexisted with a stress upon the distinctiveness of disciplines constituted by special Fundamental Ideas. This brought his work into conflict with an interpretation of the unity of science which had considerable power in contemporary scientific and theological debates. In the writings of Charles Babbage, William Carpenter, Robert Chambers, and Baden Powell this notion asserted the possibility of a uniformity of method across all the sciences, leading to the discovery of uniform, general laws explaining all natural phenomena. Babbage anticipated such a result in his *Ninth Bridgewater Treatise*:

> All analogy leads us to infer, and new discoveries continually direct our expectation to the idea, that the most extensive laws to which we have hitherto attained, converge to some few simple and general principles, by which the whole of the material universe is sustained, and from which its infinitely varied phenomena emerge as the necessary consequences.[84]

Carpenter drew upon the methodological norms entailed by this ideal when he complained that Whewell's treatment of physiology suggested that it should be conducted on different principles from those recommended for the physical sciences.[85] But the conflict can be best focused through a discussion of Powell's concerns about the implications of Whewell's philosophy.

VI

Like Whewell, Powell was a mathematically trained don and cleric who adopted the task of commenting on science and its cultural significance.

[81] Whewell (1841*d*), 32. [82] Ibid.; id. (1847), i. 3, 6. [83] Ibid. i.x.

[84] Babbage (1838), 32. See also Chambers (1846), 3–5, 147–9.

[85] Carpenter (1838) noted that Whewell excluded final causes from '*physical enquiry*', but made them central to physiology (338–9). For the debate about the unity of method across the sciences, see Yeo (1986), 273–82.

Although not always naming the target of his criticism, some of the key features of Powell's work on the philosophy of science and natural theology were arguably directed against his Cambridge counterpart.[86] Throughout his writings, Powell deployed the notion of the unity of science at both methodological and substantive levels. Firstly, he maintained that inductive reasoning rested upon '*a conviction of the universal and permanent uniformity of nature*'; it had no reference to a distinction between past or present, and scientific explanations must assume the uniform and universal action of presently known causes, which could then be legitimately extrapolated beyond actually observed phenomena. Powell claimed that the unity of science consisted in the adherence to this principle of uniformity and, predictably, rejected the suggestion that geology, in dealing with unwitnessed past events, should be in any way distinguished from other inductive sciences. He therefore strongly objected to Whewell's contention that such palaetiological sciences, in searching for historical causes, could admit a *qualitatively* different order of past forces that lay beyond the scope of inductive investigation.[87] Powell believed that this position was similar to earlier theological attacks upon the status of geology—it excluded this discipline from the 'type of unity which binds together the whole range of inductive sciences'.[88]

But Powell's conception of the unity of science went beyond a commitment to a principle of methodological uniformity; it embraced the vision of a union of all the sciences under common, general laws. In his view, the inductive principle pointed to 'the great archetype of *Unity*';[89] and the tendency of scientific thought since the ancients was away from isolated and disconnected branches towards 'harmony, simplicity, and unity of character'. Progress towards this ideal was the direction of history: 'All the first great modern advances were directed towards *combining* and *uniting* branches hitherto distinct, and tended to evince a *unity* of idea and principle.'[90] He cited the work of Faraday on the relationships between chemical and electrical action. It was a mistake to insist on an essential distinction between '*molecular* forces and those acting on matter in larger *masses*': this was merely a temporary distinction, like that between chemical and mechanical action, which would eventually be dissolved.[91] Similarly, he argued, it was important to recognize that distinctions between the various scientific disciplines for the purpose of classification were

[86] Powell (1834), 1, for the reference to his attempt at a 'connected view of the principles and results of each department of science'. For a contrast of the approaches of Whewell and Powell to the tasks of natural theology, see Young (1980), 76–88. Corsi (1988) which offers a thorough study of Powell, was not published when this chapter was completed.

[87] Powell (1849), 35.

[88] Id. (1855), 52–9. See also id. (1838), 60–1, for the defence of geology.

[89] Id. (1849), 35, and (1855), 98.

[90] Ibid. 41, 63.

[91] Ibid. 49–50. See Cantor in this volume for an example of the way in which Whewell could deploy the taxonomic distinctions Powell was rejecting. Thus in replying to the claim that the emission theory of light explained certain natural phenomena, Whewell argued that these fell outside the *physical* theory of optics which, he asserted, was 'as a blank' before the emergence of the undulatory theory. See Whewell (1857), ii. 319–20, 478.

only practical and provisional. In a clear reference to Whewell's philosophy, Powell asserted that:

It is a reversal of the order of inductive advance to endeavour to isolate each department of science, and to place it on a separate base, by a theory which would assign to each branch certain real differences of principle and peculiar fundamental ideas essentially characterising it. If such a distinction could be made out, it could be but a temporary and provisional ground of classification, in time to be superseded by a reduction to a higher common principle.[92]

There was therefore disagreement over the lessons provided by history and the manner in which future progress in science would be achieved. Powell attacked the proposition that various subjects were associated with peculiar ideas and conceived scientific progress as the explanation of previously mysterious phenomena in terms of '*mechanical* laws'. There was no 'mysterious connection' of mechanics, as such, with the idea of causation; rather this subject displayed the simplicity which could be expected when other phenomena were reduced to 'simple and intelligible causes of force and motion'. When this was done for new areas, he explained, 'we bring those branches of science into the domain of exact science, and break down the line of demarcation which hitherto seemed to separate them'.[93] On the other hand, Whewell claimed that new truths were usually reached by 'the application of new ideas, not by the modification of old ones'. Researchers had to escape the powerful tendency to employ familiar causes in order to seek 'distinctly new conceptions' for 'newly-studied phenomena'.[94]

Like several of Whewell's other scientific contemporaries, (for example, Jones, Herschel, and de Morgan) Powell rejected, or at least strongly qualified, the idealist aspect of his epistemology.[95] But furthermore, he saw the general thrust of Whewell's history and philosophy as a threat to his own conception of the unity of science. The reason for this is not difficult to find, for in important ways, one of Whewell's central points was antithetical to Powell's position. Thus, despite Whewell's references to the unity of knowledge, and his conviction that consilience of inductions produced a convergence of theory towards simplicity and unity, his work stressed the particularity and integrity of different disciplines.[96] George Sarton probably had this last feature in mind when he remarked that Whewell's work 'is not a history of science as we understand it today, but a juxtaposition of various special histories, which is

[92] Powell (1855), 44–5; and p. vi for mention of his disagreements with Whewell.

[93] Ibid. 45–6. He also criticized those, such as Whewell, who regarded the sciences of 'statics and dynamics' as 'based on inherently distinct principles', ibid. 46, and Powell (1849), 35. See also Powell (1850) for their epistemological dispute.

[94] Whewell (1847), ii. 100–5.

[95] Like Herschel and Jones, Powell preferred to explain apparently intuitive conceptions in terms of 'successive and gradual *abstraction*', Powell (1849), 35. See also Jones to Herschel, 1 June 1841, Royal Society Library, Herschel Papers, H 10 370; and Herschel (1841).

[96] Whewell's inductive tables showed the 'constant advance towards unity, consistency, and simplicity', but this was within different branches of science, such as astronomy and optics. Whewell (1847), ii. 78.

something very different; it represents a lower stage of integration'.[97] But from Whewell's perspective, a recognition of the distinct epistemological foundations of different branches of science was essential for an understanding of the manner in which scientific discoveries had been achieved. Whereas Powell predicted the eventual inclusion of fields such as mechanics and chemistry under some general law, Whewell stressed their connection with separate, appropriate conceptions. As noted earlier, Whewell had spoken of science as a 'great empire falling to pieces' under the pressure of specialization, but he did not believe this unity could be restored by ignoring the conceptual distinctiveness of its components. 'The Mechanical, the Secondary Mechanical, the Chemical, the Classificatory, the Biological Sciences', together formed 'so many great Provinces in the Kingdom of Knowledge, each in a great measure possessing its own peculiar fundamental principles'.[98] In keeping with this political image, Whewell emphasized the intellectual sovereignty of those discrete fields, and did not anticipate the kind of reduction to comprehensive, general laws which Powell envisaged. He recognized that the empire of science could no longer be held together by great men equally competent in several disciplines, but maintained that it could be comprehended by the philosopher of knowledge who saw the unity of epistemological processes in different fields of enquiry. Drawing upon another favourite metaphor, the architectural, he conceived scientific knowledge as a 'lofty palace . . . where hundreds of different apartments belong to a common plan'.[99]

These conflicting epistemological positions had clear consequences in the debates about the status of physiology (or biology) and the species question. It was here that an appeal to the unity of science was crucial to Powell's defence of the life sciences and their ability to explain organic phenomena in terms of natural causes. From this perspective he was bound to resist the contention, central to Whewell's philosophy, that 'the *sciences of organization and life* . . . involve altogether a *new class and order of ideas* of so peculiar a kind that they must stand out as entirely exceptional cases to the general *unity of the sciences*'.[100] Powell rejected the existence of a special teleological principle as essential to the understanding of organized life, arguing that notions of design and final cause were experiential inferences and did not point to qualitative differences between physical and organic phenomena. In any case, he believed that the wider principle of 'unity of composition' subsumed the relationships to which teleology referred, and was better able to reveal 'the proper place of physiology among the sciences . . . in its relations to that *unity* of principle

[97] Sarton (1936), 63. For an earlier, more positive view, see Anon. (1875–89). For the place of this issue in Whewell's historical perspective, see Yeo (1987).

[98] Whewell (1847), ii. 19.

[99] Ibid. i. 15; also 708 on the 'lofty temple of Truth' and the 'Great Architect'.

[100] Powell (1855), 63. He entered a protest against the word 'Biology', arguing that it should denote the history of life, not the principle of vitality. He preferred 'Zoonomy—science of the *laws* of life'. See Powell (1859), 172.

which pervades them all'.[101] Here Powell made a characteristic conflation of the levels to which the notion of unity applied: thus morphological analogies in anatomy were aligned with general laws of physics and chemistry to erase any doubt that

> there really exists . . . [a] connexion of *some kind* between the manifestations of life and the simplest mechanical or chemical laws and that this connection and dependence is but one component portion of the vast chain of physical causation whose essential strength lies in its universal continuity, which extends, without interruption, through the entire world of order, and in which a real disruption of one link would be the destruction of the whole.[102]

Powell was thus prepared to embrace human intellectual and psychological phenomena as part of man's 'animal' nature and therefore part of the physical order of things.[103]

Here again, Whewell's appeal to the ideal of the unity of science produced different results. Like Powell, he regarded it as a sanction for the possibility of incorporating areas of human behaviour such as language, law, government, and psychology within the sphere of inductive knowledge: his major work strongly suggested that the investigation of such subjects 'must be governed by the same general rules to which all science is subject'.[104] But such a licence for scientific research was linked with a stress on the integrity of the fundamental ideas constituting these new domains. Just as chemistry could not be reduced to, or subsumed by, mechanics, so physiology could not be understood without reference to final cause, the peculiar idea appropriate to the phenomena of organic life. By guarding the conceptual boundaries between disciplines Whewell thereby protected the values and beliefs associated with the idea of man's unique place in a purposeful universe.

But, as suggested previously, this philosophical—and strategic—emphasis on distinctions did not lead him to relinquish the ideal of the unity of science. In fact, whereas Powell separated the physical and moral spheres of creation, claiming that scientific and philosophical conclusions about the former could not impinge upon the spiritual domain, Whewell pursued a wider unity of knowledge. He contended

> We have seen that biology leads us to psychology, if we choose to follow the path; and thus the passage from the material to the immaterial has already unfolded itself at one point; and we now perceive that there are several large provinces of speculation which concern subjects belonging to man's immaterial nature, and which are governed by the same laws as sciences altogether physical.[105]

Whewell was able to adopt this position, later elaborated in his *Philosophy of Discovery*, because his conception of the unity of science (and knowledge)

101 Powell (1855), 74.
102 Ibid. 67.
103 Ibid. 74–8, and Powell (1859), 270–3, 470.
104 Whewell (1847), i. 657.
105 Ibid. 642; also ii. 115–16 and 640–2.

rested upon epistemological analogies, not substantive natural laws. There was a uniformity in the way knowledge of nature was achieved, not a unity of the laws of nature. Furthermore, his emphasis upon the special intellectual foundations of different sciences and their various stages of progress towards axiomatic conceptions, allowed him to embrace moral ideas within a general philosophy of knowledge without endangering their integrity. The present number of necessary truths was small, for the attainment of ideal conceptions was limited to a few sciences. 'Most facts', said Whewell, 'are not *idealized*'; and this meant that we should not be surprised to find that moral ideas were often '*unrealized*'.[106] But all scientific knowledge depended upon a consonance between metaphysical ideas in the mind and the nature of the physical world. The recognition of such fundamental ideas indicated an affinity between the human and divine Mind, and since this was a progressive process, the emergence of new fundamental ideas strengthened 'the *proofs* of a Divine origin'.[107] By stressing distinctive ideal conceptions as the basis of the sciences, Whewell opposed Comte's view (shared by Powell) in which progress of knowledge consisted in the emancipation of science from metaphysical thought. For Whewell, this was a necessary part of the inductive movement,[108] and it was precisely the continuing contribution of ideas from the human mind which demonstrated, more convincingly than Powell's limited physico-theology,[109] the unity of the moral and physical spheres of God's creation.

VII

When referring to Whewell's idealist epistemology in his work on *The English Utilitarians*, Leslie Stephen announced that: 'I shall not attempt to explain a theory which seems to be radically incoherent, and which made no converts.'[110]

[106] Whewell (1860), 386, also 356–7, 372, 378.

[107] Ibid. 372–3.

[108] Whewell (1847), i. ix; ibid. ii. 378–9. This key feature of Whewell's work was appreciated by several philosophers and theologians who sought a counterweight to the positivist view of science. 'Dr Whewell . . . has boldly grappled with the metaphysical conceptions which lie at the basis of science, overturned the sensationalism which too often has attached itself to the physical inquirer . . . ' Morell (1847), ii. 246. Morell believed that his position was the basis for a stronger natural theology. 'Writers, for example, like M'Culloch and Whewell, who have applied the highest scientific knowledge to maintain the validity of natural religious conceptions, are, philosophically speaking, most evidently idealistic in their tendency.' Ibid. 604. James McCosh, John Tulloch, and James Martineau also appreciated Whewell's answer to positivism and to the theological nescience of Henry Mansel's Bampton Lectures of 1858. See Yeo (1977), ch. 7; id. (1979), 512–16.

[109] For a criticism of Powell's natural theology as one which regarded the 'moral and spiritual world as a region not only distinct from, but disconnected with, the physical', see Anon. (1855). This was probably written by James Martineau, the Unitarian theologian, who welcomed Whewell's idealist philosophy of science (but not his moral philosophy).

[110] Stephen (1900), iii. 142–3. But in his entry on Whewell in the *DNB* Stephen quoted Henry Sidgwick's view that Whewell 'did more than anyone to introduce some interest in philosophy' at Cambridge. Stephen (1885–90), and Sidgwick (1876).

On this last point, Stephen was essentially correct; and it is only recently that Whewell's philosophy, generally disregarded by the nineteenth-century scientific community, has attracted serious attention. Yet it is still largely the case that Whewell's views on the method and epistemology of the natural sciences—often approached from the parameters of modern debates—have been the prime focus of scholarship, while his wider project and its context have been neglected. This chapter therefore suggests that a study of Whewell's philosophy of knowledge and the tensions within it allows a greater appreciation of his epistemology and its implications for nineteenth-century debates about the relationships between the sciences.

8. *The History and Geography of the Intellectual World: Whewell's Politics of Language*

by SIMON SCHAFFER

VICTORIAN metaphors, like Victorian values, are back in style. A transformation in our reading of the natural sciences, especially Darwinism, is promised. The language of the philosophers has been deconstructed. William Whewell seems a suitable case for treatment. He was a major philosopher of language, a 'verbarian Attorney General'.[1] Hence the attraction of taking his verbal imagery seriously. Whewell defined those who were allowed to make new scientific terms and also defined the terms they could use. Metaphor was to be shunned. Yet he advocated this plan through some of the most powerful tropes of Victorian science. Whewell's verbal figures were notable even for his own audience. In 1841 John Herschel observed that the style of Whewell's *History* and *Philosophy of the Inductive Sciences* was marked by 'a great assemblage and variety of metaphorical allusion, much greater indeed than we should like to see adopted by an author less capable of curbing the exuberance of a lively fancy into an entire subordination to his reason'.[2] Readers were as impressed by Whewell's 'fancy' as his 'reason'. Herschel argued in the same review that 'half the labour of the modern inductive philosopher' was to make language 'a perfect daguerrotype' of nature. 'Common language is a mass of metaphor, grounded not on philosophical resemblances, but on loose, fanciful and often most mistaken analogies.'[3] Unless Whewell helped make scientific language coldly mimetic he would, in Herschel's view, lose his status as an inductive philosopher.

This chapter examines the connection between Whewell's status and his work on language. Amongst early Victorian savants Whewell's central metaphors were commonplaces, notably his geographical imagery: sciences progressed through a strenuous ride across a dramatic landscape, and the realm of élite scientific enquiry was pictured as an isolated and bright habitation surrounded by outlying darkness, where error and illusion reigned. It was unoriginal to use

[1] Beer (1985); Young (1985); Benjamin, Cantor, and Christie (1987); Levine (1987); Jordanova (1986). For the 'Attorney General' see Coleridge (1830), 24.

[2] Herschel (1841), 238; id. (1857), 256.

[3] Herschel (1841), 190; id. (1857), 166–7.

combative and topographic imagery in the scientific culture of the 1830s: Martin Rudwick has demonstrated the presence of metaphors of battle and competition, sports field and lawcourt, during the Devonian controversy in which Whewell participated. It was equally unoriginal to suppose that better language would cure corruption and sophistry. Bentham's *Handbook of Political Fallacies* (1824) was a notable contemporary example: 'it seemed not too early in the nineteenth century to take up the subject on the ground of morality, and to invite common honesty for the first time to mount the bench and take her seat as judge'.[4] But Whewell put his metaphors to an idiosyncratic, challenging use. He rejected Benthamite 'honesty' and 'utility'; he fought 'Steam Intellect' and Radicalism with different linguistic tools. These tools involved the construction of a map of the geography of the sciences and the promulgation of rules of travel.

This analysis of Whewell's linguistic enterprise begins with an outline of his images of landscape and journeying to show how he pictured the embattled realm of scientific enquiry. His celebrated argument for the singularity of this inhabited world is connected with his advocacy of an élitist account of scientific organization. This outline is followed by a description of Whewell's mature theory of good language developed during the 1830s and of the political context in which this theory emerged. The culture of Cambridge philology at this period is shown to be closely linked with Whewell's own search for a vocation as priest or natural philosopher within the academy. A comparison with earlier attempts by S. T. Coleridge in this field illuminates the specific character of Whewell's endeavours at changing the pattern of language in the world of learning. The balance of the chapter is devoted to Whewell's activities as linguistic reformer and innovator. He saw threats to morals and philosophy in the bad language of the populist press and Radical politics. He aimed at reform in public culture and specific sciences alike. This was also an ethical and critical aim: Whewell's work on the history and terminology of architecture is discussed in order to bring out his self-appointed role as cultural critic. Then the final sections of the chapter consider three such efforts: Whewell's attempt at a linguistic purge in mineralogy; his public fights in reviews of political economy, inductive philosophy, and popular science; and, last, his negotiations with Faraday over the terminology of electromagnetism. It can no longer surprise historians that the rhetorical and the figurative were deeply intertwined in such a set of scientific projects. The point is not to demonstrate the mere existence of such verbal features, but to see how they worked, to describe Whewell's linguistic tool-kit and the philosophical and political uses to which these tools were put.

[4] Rudwick (1985), 435–7; Bentham (1824), 5.

'The High Priori Pegasus'

Whewell compared his philosophical introduction to the *History of the Inductive Sciences* with 'the geographical sketch of a country, with which the historian of its fortunes often begins his narration'. His 'Metaphysics' was the 'geography of an Empire'; the sketch was 'a slight outline of the Geography of that Intellectual World of which we have here to study the History'.[5] What kind of world was Whewell mapping? How did one travel through it? The progress of knowledge was a mountain climb to Pisgah, whence the past desert and the future 'land of promise' were both visible.[6] Whewell's promised land was impressive: Alexander von Humboldt compared it with his own view of the Amazonian forest from the Andes, while Herschel read it as an imperialist epic: 'the occupation and settling of the country under the dominion of the conquerors, quelling the insurrectionary movements of ignorance and prejudice under the new regime, and partitioning out the land in provinces and domains'.[7] For Herschel and his colleagues, Whewell provided an energetic journey on a 'high priori Pegasus'. The aerial ride was exciting and unfamiliar.[8]

Consider the imagery of exercise and horsemanship. A 'bold, bad rider', Whewell rode good horses into the ground: at Lord Fitzwilliam's he 'learned for the first time that the powers of a horse are not inexhaustible'. Like that of William Wordsworth, Whewell's correspondence bristles with astonishing itineraries. He stormed through geological Britain, architectural France, Alpine Italy, and mineralogical Germany.[9] Telling Sedgwick of the delights of Rhineland rail travel, he dreamt of 'gliding along on the back of a dinotherium—if they ever condescended to take passengers in the time of the pre-adamite lake'. And, notoriously, the Master of Trinity died in 1866 following a fall from a horse during a ride through the Gog Magog Hills.[10] His real-life travels shade into his intellectual travails. To his peers and juniors, Whewell seemed a towering figure of embodied energy.[11] Roderick Murchison, prominent fox-hunter and geologist, described the Trinity man as 'the best bottomed hack-hunter we have ever had in our stable, for all work comes alike to him, no day is too busy and no fence too high'. Murchison was well placed to perceive Whewell as a noble steed: Jim Secord has persuasively traced the filiations between field sports and field science in Murchison's geology, contrasting this with the mathematical interests of the Cambridge group, including Whewell, Sedgwick,

[5] Whewell (1857), i. 15.

[6] Ibid. 4.

[7] Humboldt to Whewell, Aug. 1840, Todhunter (1876), i. 148; Herschel to Whewell, 22 July 1837, Royal Society Library, Herschel Papers, H21 224; Herschel (1857), 159.

[8] Ibid. 227.

[9] W. Clark (1866), 551–2, for horsemanship. For travel see e.g. Whewell to Richard Jones, summer 1829, Todhunter (1876), ii. 99–105 (a journey along the Rhine to Switzerland and the Alps). For Wordsworth's influence on Whewell, see Wordsworth (1978–88), iv. 251 n. 3.

[10] Whewell to Sedgwick, 3 Sept. 1849, Stair-Douglas (1882), 366, and for his death see ibid. 553.

[11] Murchison, in Morrell and Thackray (1984), 328; Secord (1986), 62–3.

and Herschel.[12] Whewell often represented the role of the mathematical sciences, and of Cambridge's hegemony, in similarly gymnastic terms. As self-appointed arch-enemy of the Cambridge Network, David Brewster saw this point: Whewell was 'an individual whose speed neither time nor tide can stay—whose movements are neither equable nor straightforward, who comes into perpetual collision with those who are struggling in the same cause'.[13]

Key issues between Whewell and his enemies—including the role of mathematics in pedagogy and the right way of conducting geological enquiry—were represented as choices of means of travel. In the geological case, Whewell was well aware of the strenuous work-rate demanded of field workers.[14] This gave point to his critique of Lyell's actualism in the early 1830s. What Whewell decided to call 'uniformitarianism' seemed to him an inappropriate means through which to traverse the landscape of the geological past. Uniformitarians gaily abandoned the help which history offered them: 'Mr Lyell throws away all such crutches; he walks alone in the path of his speculations . . . He requires time, no doubt; he will not be hurried in his proceedings. But if we will allow him a free stage in the wide circuit of eternity, he will ask no other favour.' 'Catastrophism' promised a more fulfilling, less lonely and time-consuming means of locomotion.[15]

Similar imagery recurred when Whewell confronted Scottish and utilitarian critics of liberal education. In his 1836 review of Whewell's *Thoughts on the Study of Mathematics*, the Edinburgh professor William Hamilton insisted that mathematical reasoning should not be given the role of master discipline which Whewell assigned it. Hamilton denied that 'we shall come best-trained to the hunting field of probability by assiduous locomotion on the railroad of demonstration', and compared the 'gymnastic of the mind' promised by rigorous mathematical demonstration with 'the routine of the treadmill in the gymnastic of the body'. The Scotsman detailed the disastrous morals, deadened minds, and servile habits prompted by Cambridge mathematics and penal labour alike. Both routines disqualified their practitioners from worldly life.[16] As Williams shows in this volume, Whewell's defence involved the reformulation of the place of mathematics and philosophy in the formation of university students. This included a renewed contrast between geometry and analysis in Whewell's critique of the mathematical strategy of his collegues such as Babbage, Herschel, and Peacock. Hamilton's charge that mathematical logic was mechanical and formalist was transformed by Whewell into a distinction between the gymnastics of geometry and the machinery of analysis. The

[12] Herschel (1841), 238; id. (1857), 256.

[13] Brewster (1842), 139.

[14] Whewell slammed John MacCulloch for his fraudulent claims for expenses during the Scottish geological survey and his reports of journeys far surpassing 'the locomotiveness of ordinary mortals': Whewell quipped that MacCulloch must have been 'in actual possession of the seven league boots', Whewell (1832), 105 n.; D. Cumming (1985).

[15] Whewell (1831), 185.

[16] W. Hamilton (1836), 414, 428–9.

references to Babbage's celebrated analytical engine and to the political connotations of machinofacture were marked:

In the one case, that of geometrical reasoning, we tread the ground ourselves, at every step feeling ourselves firm, and directing our steps to the end aimed at. In the other case, that of analytical calculation, we are carried along as in a railroad carriage, entering it at one station and coming out of it at another, without having any choice in our progress in the intermediate space . . . It is plain that the latter is not a mode of exercising our own locomotive powers; and in the same manner analytical processes are not a mode of exercising our reasoning powers . . . It may be the best way for *men of business* to travel, but it cannot fitly be made a part of the gymnastics of education.[17]

Education as gymnastics was a Grecian commonplace. Whewell gave it a special meaning. He told his Lancaster teacher in 1817 that 'the mind seems to be a mad jockey to the body'. The aim was to guide and constrain the intellectual powers so that they acted as a balanced and progressive guide to the material powers of creation. Whewell's vocation was the search for an appropriate role for the intellectual élite. The resonance with Coleridgean concepts of clerisy was deliberate.[18] In his *Constitution of Church and State* (1830), Coleridge aimed at a balance between what he baptized the 'permanent' and 'progressive' forces in the social order, guaranteeing advance without the costs of disruption, and treating the claims of precedent and reform with equal attention. The 'Nationalty' sustained a 'clerisy' to embody its intellectual accomplishment. Whewell explicitly shared these views of 'the National Church and the National Constitution not as mere formulae but as living things'.[19] He applied Coleridge's terminology to a discrimination between the 'permanent' and 'progressive' features of the curriculum: the former to be safely taught in the universities, the latter to be the objects of enquiry by élite savants.

Whewell emphasized the élitist singularity of creative endeavour. The origin of life on Earth, of humanity, and most recently of language, were not events subject to natural causation. The natural world was a uniquely privileged stage on which the dramas of creation, incarnation, redemption, and intellectual advance could all successively be performed.[20] His essay *Of the Plurality of Worlds* (1853) argued for the uniqueness of this planet, 'the one *fertile flower*' in

[17] For Whewell's mathematics see Becher (1981). The citation is from Whewell (1845), 40–1. Whewell and Babbage clashed over the 'analytical engine' in 1837. See Whewell to Babbage, 30 May 1837, Whewell Papers, Trinity College Library, Cambridge University, 226 c. 80 149[16].

[18] Whewell to Morland, 27 Mar. 1817, Stair-Douglas (1882), 26; Rothblatt (1981), 34. For the clerisy see Knights (1978).

[19] Coleridge (1830); Whewell to Marshall 27 Dec. 1842, Stair-Douglas (1882), 282–3. Debating Catholic relief with William Wordsworth in 1829, Whewell wrote that while 'persons officially connected with our academic bodies as well as others who are elevated by the civil and religious institutions of the country should consider themselves as most particularly bound to resist the busy and impatient love of change to keep steady and secure the maxims and institutions which have descended from other times', yet 'human institutions must accommodate themselves to the ever-changing relations and forms of society which the tendency of ages brings into existence', Wordsworth (1978–88), ii. 48–9 n. 5.

[20] Hilton (1988), 308.

the universe. Since 'the material world must be put in an inferior place, compared with the world of mind', it followed that humanity was provided with 'one school of moral discipline, one theatre of moral action, one arena of moral contests for the highest prizes'. All else was a world of superfluous 'lumps' tenanted, if at all, by 'corals and madrepores, fishes and creeping things'.[21] The reference to the habitable world as a theatre picked up the claims of Robert Chambers's anonymous *Vestiges of the Natural History of Creation* (1844), which argued that each globe might be 'a theatre of organic being, or in the way of becoming so'. Whewell felt this challenged the divide between moral, intellectual humanity and base, unthinking matter. He rejected its 'System of Order' in which 'men grew out of apes, . . . language is the necessary development of the jabbering of such creatures, and reason the product of their conflicting appetites'.[22]

Whewell's stress on the privileged theatre and the moral elevation of humanity was not lost upon his audience: 'they say in Cambridge that Dr Whewell's book . . . reasons to this end: the planets were created for this world; this world for man; man for England; England for Cambridge; and Cambridge for Dr Whewell'. Whewell allegedly 'treated planets, stars and nebulae with a graduated scorn exactly proportional to their distance from the Cambridge Observatory'.[23] Connections would be made between the intellectual geographies of the stellar universe, the Earth, the polity of science, the nation, the university, and the college. Whewell joked to Mary Somerville about the comparison between reform in star maps and political reform in Britain. Endorsing the 'imperial theme' of early Victorian geology, he told the Geological Society that the British geological sequence had been providentially arranged as an '*Index Series* of European formations'.[24] Astronomical and geological élitism went with political chauvinism and its defence. In his texts on university education Whewell developed the conservative and visionary theme of England's mission. In this unique island a Coleridgean balance prevented stifling autocracy or revolutionary excess:

> Has England not taught the nations how to live? By an admirable combination of active and original intellect with unequalled practical sagacity and force of character, England has constantly impelled the progress of thought and of institutions in Europe, while, at the same time, she has held back from the extravagances and atrocities to which the progressive impulse has urged more unbalanced nations . . . England had a Reformation of Religion without the abolition of her Church, she had a Revolution of dynasty without the destruction of her loyalty; she has had a Reform of her Parliament, we trust, without any fatal wrench to her Constitution.[25]

[21] Whewell (1853), 224, 244, 256, 243; Crowe (1986), 288–92.

[22] Whewell (1846), 12, 8; Chambers (1844), 161; on Whewell versus Chambers see Brooke (1977); Raub (1988).

[23] Crowe (1986), 314; W. Clark (1866), 548.

[24] Whewell to Somerville, 29 Aug. 1838, Stair-Douglas (1882), 194; Whewell (1839), 80; Herbert (1985), 487; Secord (1982).

[25] Whewell (1837), 138–9.

The threat to this secure island was manifest both in the 'critical spirit' of extreme German idealism and in the 'extravagantly democratic' regimes in France. In 1837 Whewell argued that 'an education which invokes the critical spirit', such as those of Germany and France, would inevitably breed 'revolution after revolution'. By 1848 he saw his prophecy confirmed. 'What are we to say of the Revolution?', Whewell asked J. D. Forbes after his trip to Paris in winter 1847–8. Were the troubles to spread, through Chartism and its ilk, 'we have the world thrown back into the condition in which it was when we were born, and have been grinding on for so many years to no purpose'.[26] Whewell and his allies 'ground on' to prevent this return to the anarchic past and to defend the virtues of the embattled realm of the present. The strenuous exertions of the citizens of the intellectual world would encompass uncharted territory beyond their walls.

The Thraldom and Illusion of Words

A good example of dangerous error in an uncharted zone of knowledge was in the field of mesmerism and spiritualism, an important battlefield between the clerisy and the demos.[27] In June 1853 Faraday reported in the press on his experiments designed to support the claims of the physiologist William Carpenter that 'table turning' was the result of 'a *quasi* involuntary muscular action'. Faraday put his equipment on show at a Regent Street instrument maker's shop. He offered the platform of the Royal Institution to Whewell with the aim of debating the 'mental condition of the public body'.[28] The previous year Faraday had invited Whewell to witness experiments performed on muscle contraction at the Royal Institution by the great German physiologist Emil Dubois Reymond. Comparing mesmerism and electrophysiology, Faraday told Whewell that 'the social phenomena presented by table turning etc. fully shew that for those who esteem themselves among the fully educated there is still an education of the mind required'.[29] Whewell's lecture, 'On the Influence of the History of Science upon Intellectual Education', was delivered in April 1854. His theme was that contemporary liberal education was based on the excellent classical models of the deductive sciences of geometry and jurisprudence, which had been the original intellectual response to the evils of facile sophism. Yet there was now need for training in inductive sciences. Only thus could 'the

[26] Ibid. 52, 47–8; Whewell to Forbes, Mar. 1848, Todhunter (1876), ii. 344.

[27] Barrow (1986), 74, 146–59. In 1854 T. H. Huxley compared table-turning and the public appeal of *Vestiges*, see Huxley (1854), 439 in Yeo (1984), 14.

[28] Faraday (1971), 690–2; Oppenheim (1985), 242–3, 328. For Maxwell on Faraday and spiritualism in 1853, see Campbell and Garnett (1884), 135, 341–3; 'the next phase of the dark sciences is that in which the phraseology of physics is exchanged for that of psychology. The verb to will acquires a new and popular sense, so that everyone is able to will a thing without bequeathing it.'

[29] Faraday to Whewell, 4 May 1852, Whewell Papers, O15 49[32]; Faraday (1971), 716.

mind escape from the thraldom and illusion of words', learn to distrust the search for mere definitions, and avoid the dangers of 'a number of scientific words current among us, which are applied with the most fantastical and wanton vagueness'. Chief amongst these were the vocabularies of electricity and magnetism, the diseases of which Whewell saw as the source of popular superstition in 'animal magnetism'.[30]

Let us examine the resources Whewell was using in such public statements, resources developed as a result of his immense labours of the 1830s. During that decade he tried to make scientific language fit for public consumption. He provided Lyell, Faraday, and Jones with terminology for their sciences, seeking to direct geology, galvanism, and political economy away from danger and towards the right moral ends. In the *History* and *Philosophy* he reflected aphoristically on the place of terminology in the history and structure of excellent knowledge.[31] Language was a political weapon for Whewell's contemporaries. Like them, he connected moral conduct and the construction of an acceptable tongue.[32] Coleridge had argued in *On the Constitution of Church and State* that 'there is found even among our educated men a vagueness in the use of words, which presents, indeed, no obstacle to the intercourse of the market, but is absolutely incompatible with the attainment or communication of distinct and precise conceptions'. Purified language was 'a new organ of thought for the mind', and this was, for Coleridge as for Whewell, as much a spiritual as a material necessity. Whewell argued that language was, at its best, an ethereal food for the mind, both 'instrument' and 'nutriment of thought'.[33]

A well-formed language carried the sciences' immortal progress. Whewell's metaphors for such a linguistic system referred to treasure, 'pearls' which did not hang without a string, treasure-ships 'laden with this, their precious freight'. The continuity of knowledge was guaranteed by the permanence of past work embodied in right language: 'this is the fortune, not only of the great and rich in the intellectual world . . . who have accumulated treasures of their own, but the humblest inquirer, while he puts his reasonings into words'.[34] This imagery gave historical philology a key role in Whewell's moral vision for the sciences. It taught respect for traditional tongues, but it also highlighted the dynamical transformation of language. When explicating the move from galvanism to chemistry in his *History*, a move over which Whewell agonized in his drafts, he informed his readers that galvanism, based on the fundamental idea of polarity, was only comprehensible through an analysis of chemistry, its better half. Chemical terms 'open to us a new drama', they were terms 'which

[30] Whewell (1854), 21, 29, 32–4.

[31] Whewell to Rose, 22 Sept. 1822, Stair-Douglas (1882), 79; for studies of language see Stoll (1929), 69–74; Schagrin (1973); Belsey (1974).

[32] Aarsleff (1982), 120–45; O. Smith (1984).

[33] Coleridge (1830), 167; id. (1840), 62–3; for Coleridge's linguistic theory here see Corrigan (1980); McKusick (1986), 119–31; for Whewell's linguistics and mechanics see Whewell (1847), i. 270–2.

[34] Id. (1858), 73; id. (1858*a*), i. 55; id. (1847), i. 272.

we cannot translate into any ordinary language'. Tracing the heroic work of chemical discoverers was the sole means of interpreting this novel language. Terminology dictated the pattern of exposition of the *History*; it warranted its continuity; finally, it was the reason for the catastrophic breaks between the sciences which Whewell repeatedly presented. 'The meaning of words is to be sought in the progress of thought; the history of science is our dictionary; the steps of scientific induction are our definitions.'[35]

The task of a language was to help *permanent* knowledge emerge, as 'the truths at first ascertained become so familiar and manageable that they can be employed as elementary facts in the formation of higher inductions'.[36] Linguistic reform would often be needed as the ideal component of a science became clarified and theories became familiarized in common parlance. This principle caused the permanent to aid the progressive. A further question needed to be resolved: who was qualified to introduce new terminology? In his argument for the connection between the historical development of language and that of society, Coleridge had answered this question using his concept of 'desynonymization'. He argued that the innovative philosopher of genius, especially in the sciences, should strive to increase vocabulary by demonstrating that public usage was applying alleged synonyms in different senses, and by coining new terms to match these desirable proliferations of special senses. The historical progress of learning matched this linguistic technique: 'the whole process of society as far as it is human society depends upon . . . the progress of desynonymizing, that is, the feeling that there is a necessity for two distinct subjects which have hitherto been comprehended in one'. Innovation was the crucial task of a linguistically skilled élite of philosophical discoverers. In similar fashion, Whewell's politics of science dictated that 'the only persons who succeed in making great alterations in the language of science are not those who make names arbitrarily, and as an exercise of ingenuity, but those who have much new knowledge to communicate'.[37] Whewell concluded that while common usage should be privileged and radical innovation shunned, yet the élite of scientific discoverers were the only legitimate reformers. Their work succeeded because through reform new knowledge became embodied in what was permanent in scientific parlance. Whewell used these views about the cultural responsibilities of the élite throughout his career. Without their custody, the intellectual world would revert to Babel.[38]

Whewell produced an intriguing set of arguments about emergence and language which were designed to be mutually supportive. His interpretation of geological dynamics implied a great break between originating and continuing causes of progress. His interpretation of philology then confirmed that this was the course in linguistic development. He insisted that the history of human

[35] Whewell Papers, Add. MS a. 217[4] on polarity; Whewell (1857), iii. 90.
[36] Id. (1847), ii. 526.
[37] See P. Hamilton (1983), 69–70, 75–6; McKusick (1986), 94–5; cf. Whewell (1847), ii. 518.
[38] Id. (1851), 12.

knowledge was part of this enterprise, and so must be subject to the same palaetiological principles: 'the science of the causes of change includes the productions of Man as well as of Nature'. History of science was a palaetiological discipline. Scientific terms were different solely because they could be deliberately designed. These terms must be designed in accordance with Whewell's theory of emergence. Then they would embody theories, moral, linguistic, geological, or cosmological, which themselves described emergence in the world's past. Finally, the philosopher would be able to tell an excellent story of the emergence of knowledge by a palaetiological analysis of the history of these terms. This was why 'history of science' could and should be read as a dictionary, explicitly the historical dictionary which Whewell and his fellow etymologists were seeking to design.[39]

Permanence and Progress: In Search of Vocation

This design emerged as a result of Whewell's 'induction project', a quest for a fit position in the intellectual world of early Victorian Britain: 'the reform of our Philosophy is the work to which I have the strongest vocation, and which I cannot give up if I were to try'.[40] The geography of the intellectual world which Whewell proposed helped him find his own place on the map somewhere between materialist science and idealist philosophy. In March 1834 he told William Rowan Hamilton that 'I have persuaded myself that I have got the mud to subside, and have been trying to distinguish how much of the stuff comes from the clear spring of intellect, and how much is taken up from the base mud of the material world.' Hamilton, devout Coleridgean Tory, was impressed by Whewell's endeavours. He told Coleridge that while Cambridge was dominated by 'men of professed science . . . too apt to absorb the mind', Whewell stood out as possessed of 'more of the philosophical spirit'. The feeling was mutual. Whewell deliberately excused Hamilton ('a superb analyst and noble fellow') from the charge he levelled in his Bridgewater Treatise (1833) against the atheism of deductive mathematicians.[41] He sent Hamilton his paper 'Of the Nature of the Truth of the Laws of Motion', an effort 'to render the principles of intellectual necessity and empirical reality what belongs to them—to each its due'. The paper was a key moment in Whewell's new aim at drawing and defending boundaries in the sciences.[42]

His exchanges with his old friend, the Coleridgean divine and Germanist historian Julius Hare, show how the aim was developed within the academy. Whewell had engineered Hare's appointment as classics lecturer at Trinity in

[39] Whewell (1857), iii. 401–2; Aarsleff (1967), 219.

[40] Whewell to Herschel, 9 Apr. 1846, Todhunter (1876), ii. 234.

[41] Graves (1882–9), i. 554, 559; Whewell to Jones, 2 Feb. 1833, Todhunter (1876), ii. 154; Whewell (1834), 339.

[42] Graves (1882–9), ii. 40–1; Whewell to Hamilton, 27 Mar. 1834, Todhunter (1876), ii. 178; for the essay on the laws of motion, see Whewell (1847), ii. 592–3.

1822. He hoped that Hare would begin 'infusing good principles of taste and scholarship'. Hare replied with a vigorous defence of Plato against Locke and the Benthamites and an agreeable faith that 'philosophy is the true centre of an university education, or at least that philosophy and religion are its two foci'. Whewell and Hare collaborated in studies of German literature and philosophy. Hare and his colleague Connop Thirlwall prepared a translation of Schleiermacher's *Essay on the Gospel of St Luke* (1825) and Niebuhr's *History of Rome* (1828). Whewell participated in the work of the Etymological Society and its organ the *Philological Museum* which Hare and Thirlwall founded in 1831.[43] He wrote on the formation of adjectives from verbs with Coleridge as an authority, and provided evidence of his scepticism of the received nominalism of English etymologists such as Horne Tooke, who argued that linguistic development followed the demands of utility alone. For Hare, 'philology, in its highest sense, ought to be only another name for Philosophy, its aim should be to seek after wisdom in the whole series of its historical manifestations'.[44]

Whewell was now aligned with a broad front against radicals and utilitarians. In controversies surrounding Thirlwall and the emancipation of Dissent, Trinity was racked by the surrounding crisis. Spokesmen like Whewell, Thirlwall, and William Wordsworth, made explicit connections between moral philosophy and political epistemology.[45] In autumn 1834, after Coleridge's death, Whewell debated with Hare the wisdom of the proposal to establish a memorial Prize on 'some question, speculative or historical, of metaphysical philosophy', especially connected with the religion of 'our National Church'. Any work which made divines realize Coleridge's value as a religious thinker would be of 'great benefit to themselves', and he recruited Jones to the cause.[46] Whewell's worry was that this version of philosophy had not yet passed from its progressive, disturbed stage to that of a permanent system. Only a discipline like that of 'Schleiermacher and the best of the Germans' could comprehend such a progressive field.[47] The attack on Locke, Condillac, Horne Tooke, and their followers was especially strong in Trinity. It was alleged that the 'sensationalists' had all held that languages were systems of arbitrary signs developed out of unmediated experience.[48] Whewell and Thirlwall shared the view that Athenian sophists were the predecessors of late eighteenth- and early nineteenth-century materialists. In the work of Thirlwall's student John Donaldson, the *New Cratylus* (1839), the analogy was linked with the Niebuhrian argument that language was the repository of cultural identity and the bearer of ancient

[43] Distad (1979), 41; McFarland (1964); Preyer (1981); Thirlwall (1881), 65–6; Aarsleff (1967), 220; Schneewind (1977), 95–101.

[44] Whewell (1832*a*); O. Smith (1984), 122–3. For the Etymological Society see Aarsleff (1967), 216–21, 228; Whewell (1852), 133–4.

[45] Thirlwall (1881), 108–12; Wordsworth (1978–88), ii. 709–10; Whewell (1834*g*).

[46] Hare to Whewell 12 Oct. 1834, Whewell Papers, O18 H2[11–12]; Todhunter (1876), I. 90; Whewell to Hare and to Jones, autumn 1834, ibid. ii. 195–9; Distad (1979), 154–5; Harding (1985), 96–7.

[47] Whewell to Hare, 19 Oct. 1834, Todhunter (1876), ii. 196.

[48] Preyer (1981); Burrow (1967), 186–7; Aarsleff (1982), 35.

wisdom. Commentary on Plato's *Cratylus* was a standard means of expressing hostility to models of language based on extreme empiricism; identification of sophists and utilitarians was a common manœuvre.[49] Whewell used the history and philosophy of the inductive sciences in the war of idealist philology against what he described to Sedgwick as the 'ultra-Lockian' school.[50]

Once the *Philosophy* was completed, Whewell asked Hare whether he should take a living in Yorkshire and leave Cambridge, a departure he had also contemplated at the beginning of his philosophical work in 1831.[51] The Sussex clergyman responded vigorously that Whewell should remain as moral instructor within the academy: 'all are not meant to be *pastors*: some are to be apostles, some *doctors*': it was unlikely that Whewell 'could unravel the whole network of your thoughts, to weave them anew, whether you could descend like Apollo to become a shepherd'. In any case, 'it would be an evil thing for England' if he did.[52] The role of doctor of the Church which Hare offered to Whewell was designed to make sense of the volumes he had published and the task that lay ahead. Whewell's acquisition of power in late 1841 changed this task, but the moral function of his analysis of the sciences remained uppermost. This also distanced him from the 'gentlemen of science'. Murchison expostulated to Sedgwick that while Whewell 'had risen on the *backs of the men of science*' he was now turning upon them 'as a *high priest* who abjured their ways'.[53] This was a palpable hit. Whewell's 'priestcraft' alienated him from the intellectual world of the managers and specialists. This was a concomitant of Whewell's vaunted omniscience and of the position he won through his 'induction project' and his acquisition of collegiate and varsity power.

'The Emergency which has Arrived'

The 'power' of the philosopher was defined by the fundamental antithesis of philosophy, permanently enshrined in 1840. The protagonists of the sciences colligated facts under conceptions, but philosophers had the task of considering facts and theory as distinct, and then mapping the successive clarification of the fundamental ideas which characterized each science. Whewell coined the term 'scientist' in 1833 in response to Coleridge's argument that '*philosophers* was felt to be too wide and lofty a term' for delegates at the meetings of the British

[49] Burrow (1967), 192–3; Distad (1979), 111–12; Turner (1981), 278–9; McKusick (1986), 4–8, 34–5. For Whewell's commentary on Plato's anti-sophist dialogues directed at a wider audience (1860), see Todhunter (1876), i, 279.

[50] Whewell (1847), i. p. iv; Schneewind (1977), 95.

[51] Whewell to Hare, 13 Dec. 1840, Stair-Douglas (1882), 207; Whewell to Jones, 13 Nov. 1833, Todhunter (1876), ii. 172.

[52] Hare to Whewell, 17 Dec. 1840, Whewell Papers O18 H2[23]. Cf. Whewell's response: 'If you had called me a persevering farmer of systems, or had said that in architecture, as in some other matters, by trying to catch the principle of the system, I had sometimes been able to judge right of details, I should have recognized some details of myself.' Stair-Douglas (1882), 217.

[53] Clark and Hughes (1890), ii. 71; Morrell and Thackray (1981), 428–9.

Association. 'The inconveniences of this division of the soil of science into infinitely small allotments have been often felt and complained of,' Whewell wrote. In 1840 he affirmed the need for such an all-embracing term.[54] There was a division of labour between cultivators and philosophical landscape planners, a familiar distinction discussed both by Coleridge and by the Scottish metaphysicians led by Dugald Stewart.[55] Whewell's 'facts in the history of science' were exemplifications of the process of idealization, subject to higher philosophical management. This might offend his scientific readers, but he said he had aimed at 'general, or at least metaphysical, readers'.[56]

The contrast between these readerships was significant. Whewell was closely concerned with the various audiences for esoteric and popular expositions of correct and improving views of science and culture. The 1830s was the key decade in the emergence of the mass Victorian reading public. The war of the 'unstamped press' raged from 1831 until 1836, while during 1832 the first major popular 'informative' periodicals, such as the *Penny Magazine*, backed by Brougham's 'Useful Knowledge' campaign, and Chambers's *Journal*, were widely distributed.[57] Artisan radicalism and utilitarian propaganda seemed in danger of flooding from the print-shops. As Adrian Desmond has shown, this was seen as a real threat by the academic savants. Whewell's own publisher John Parker also distributed the pious *Library of Christian Knowledge* and its *Saturday Magazine*.[58] Whewell, however, could not easily ally with the scientific élite, who were bound to be hostile or, at best, unconcerned with his analysis of fundamental ideas: 'I do not think that any practical cultivators of special sciences will feel any deference for a person who has presumed to speculate about them all.' Deference was what Whewell needed. He would not win it from those of the vulgar whom he described to his brother-in-law Frederic Myers when discussing the readership of *Vestiges*. They lacked 'the apprehension of the difference of the nature of truth and falsehood in science and philosophy, and this is a support which no argument can remove'. As for the élite savants, 'I do not see what we have to do, except seek another audience.'[59] The 'people' of the 1840s who flocked to the materialists' banner were not those of the early 1830s at whom Whewell aimed in his lengthy review of John Herschel's *Preliminary Discourse*: 'I intended it to be as good an attempt as I could make to get *the people* into a right way of thinking about induction.' Nor were they the 'non-reading men who have a very limited taste for metaphysics', his description to

[54] Whewell (1834), 59–60; id. (1847), ii. 560.

[55] Coleridge (1818), i. 446–7; id. (1830), 68–70; Levere (1981), 71–6; Corsi (1987).

[56] Whewell to Lubbock, 7 July 1840, Todhunter (1876), ii. 284.

[57] Altick (1957), 274, 332; Webb (1955), 77–80; Hollis (1970). For an excellent bibliography see L. James (1976), 347–73.

[58] Desmond (1987); Webb (1955), 77–8.

[59] Whewell to Lord Northampton, 5 Oct. 1840, Todhunter (1876), ii. 293, cf. Whewell to Murchison 18 Sept. 1840, ibid. 286–7; Whewell to Myers, 16 Mar. 1845, Stair-Douglas (1882), 317–18; Yeo (1984), 26.

Hamilton of the intended student market for his *Introduction to Dynamics* (1832).[60] The meaning of 'induction' and 'metaphysics' changed in the interim—so did Whewell's expectations of and attitude to his possible audiences. A crisis had now arrived. Whewell put the point this way in the introduction to his *Philosophy*:

> It appears to be absolutely necessary, in order to check the prevalence of grave and pernicious error, that the doctrines which are taught concerning the foundations of human knowledge and the powers of the human mind, should be from time to time revised and corrected or extended. Erroneous and partial views are promulgated and accepted; one portion of the truth is insisted upon to the undue exclusion of another; or principles true in themselves are exaggerated till they produce on men's minds the effect of falsehood. When evils of this kind have grown to a serious height, a *Reform* is requisite. The faults of the existing system must be remedied by correcting what is wrong, and supplying what is wanting. In such cases, all the merits and excellencies of the labours of the preceding times do not supersede the necessity of putting forth new views suited to the emergency which has arrived.[61]

Whewell reckoned 'erroneous and partial views' dominated public discourse. With a better philosophy, language could be a means for the restoration of health to the body of knowledge. The 'English school of geology' was praised just because of its greater resemblance to 'one of the schools of philosophers of the ancient world, where conversation and action were the means of communicating knowledge, than a sect of these modern times, in which the press is supposed to be the source of all information, and popular circulation the criterion of all progress'.[62] Benthamite reviewing and 'Useful Knowledge' polluted language and intellectual progress. 'You begin with the attempt to *popularize* science but you will only effect its *plebification*. It is folly to think of making all, or the many, philosophers', exploded Coleridge.[63] The clerisy's failure was evident. It seemed necessary to affirm the spiritual mastery of what was permanent in knowledge.[64] For example, in October 1838, Whewell welcomed J. S. Mill's retraction of what he saw as the evil of utilitarianism, but commented that the 'destructive and outrageous' aims of his journal, the *London Review*, had not changed in like measure. 'If John Mill be a Coleridgian or a Conservative in any intelligible sense, he must be the most infatuated of men to imagine that he can do any good by promulgating his opinions under such auspices.' Whewell's venom against the 'savage exclusive party spirit' of Whigs in government and Radicals in opposition was part of his work to establish his credentials as a reforming moralist and philosopher.[65]

[60] Whewell to Jones, Feb. 1831, and Whewell to Hamilton, 17 May 1833, Todhunter (1876), ii. 115, 163.
[61] Whewell (1847), i. 7.
[62] Id. (1831), 180.
[63] Coleridge (1830), 68–70; id. (1818), i. 447; Nesbitt (1934).
[64] Hamilton (1983), 200–1; Knights (1978), 20; Newsome (1974), 115–18.
[65] Whewell to Hare, 15 Oct. 1838 and 22 Sept. 1840, Todhunter (1876), ii. 270–2, 285.

By 'emergency' and 'reformation' Whewell meant a falling away from the Idea expressed in a social or moral structure which demanded a reaffirmation of truths handed down as a legacy from the past. The role of the guardians of Church and State (and, by implication, philosophy) was to actualize and sustain this Idea in its purified form. In the succession of sciences 'the relics and ruins of the earlier states are preserved, mutilated and dead, in the products of later times'. Whewell denied this was a mere simile. 'The analogical figures by which we are tempted to express this relation are philosophically true. It is more than a mere fanciful description to say that in *languages, customs, forms of society, political institutions*, we see a number of formations superimposed upon one another.'[66] So insights into the history and philosophy of the sciences could generate better conduct in public affairs. They promised a reformation of *taste*. Whewell told Jones in 1834 that after his history and philosophy of the sciences he would 'write reviews of the progress of morals, taste and the rest, much as any other critic would, only keeping in view the analogy of the true type of the progress of science'.[67] As usual, his metaphors illuminate this point. A passage from the *Philosophy* connected the fight with sensationalism and problems of aesthetic judgement:

> What train of ideas of beauty and symmetry and stability, in the mind of the statuary or the architect, has produced those great works which mankind look upon as among their most valuable possessions;—the Apollo of the Belvedere, the Parthenon, the Cathedral of Cologne? When this is what we want to know, how are we helped by learning that the Apollo is of Parian marble, or the Cathedral of basaltic stone? We must know much more than this, in order to acquire any insight into the principles of statuary or of architecture. In like manner, in order that we may make any progress in the philosophy of knowledge, which is our purpose, we must endeavour to learn something further respecting ideas than that they are transformed sensations, even if they were this.[68]

This made explicit the connection between his philosophy of science and his work with the engineer Robert Willis and the architect Thomas Rickman on Gothic churches. In his *Philosophy*, Rickman's taxonomy for the Gothic was offered as the linguistic embodiment of an exemplary system.[69] Whewell's argument was that in the emergence of High Gothic a disconnected vocabulary of merely utilitarian features, 'breaking up of the connexion of parts', was, by a mental act of idealization, transformed into 'reformation, selection of the good, rejection of the mere customary'.[70] The argument implied a new taxonomy of styles, and promoted a re-evaluation of the ethical basis of Christendom. The political audience for Whewell's aesthetics is evidenced by those who found his architectural work appealing. These included the evangelical

[66] Whewell to Marshall, 27 Dec. 1842, Stair-Douglas (1882), 281–3; Whewell (1857), iii. 399.

[67] Whewell to Jones, 21 Aug. 1834, Todhunter (1876), ii. 186–7; Whewell Papers, R18 17[11].

[68] Whewell (1847), i. 35.

[69] Stair-Douglas (1882), 136; Cannon (1964), 178; Whewell (1847), ii. 531 n. Kenelm Digby recalled that in France in 1823 Whewell drew a table of the '*epochs* of Gothic architecture', an early use of the term by Whewell, Stair-Douglas (1882), 38.

[70] Whewell (1835), 222.

High-Churchman William Sewell, Whewell's opposite number in the Moral Philosophy chair at Oxford. In 1834 Sewell argued that 'a nation of Newtons could no more produce a gentleman than a nation of infidels could create a Christian'. In the same spirit, the great Oxford Tractarian John Henry Newman in 1841 attacked claims of Brougham and the apostles of 'steam intellect' that 'mighty spirits' such as Newton might teach moral lessons.[71] It is striking, therefore, that Sewell's 1841 review of Whewell and Rickman enthusiastically endorsed the connection between philosophy of science, the idiom of religious architecture, and moral idealism. Sewell summarized Whewell's view that the pointed arch 'lay for ages like every other simple fact in nature, each of which to common men means nothing, and to the eye of genius alone contains a multitude of applications and deductions'. Sewell understood the Cambridge philosopher's argument that 'a philosophical inquirer into the history of science would inquire not by whom or at what time an apple was first seen to drop to the ground . . . but under what circumstances these simple facts dropped like seeds into a suitable soil . . . and brought forth the theory of gravitation'.[72]

A Christian spirit was the essential idea through which the Gothic type was realized. This brought Sewell, and by implication Whewell, worryingly close to the Catholicizing architectural reformer A. W. Pugin, reviewed in the same 1841 essay. Pugin's *Contrasts* (1836), reissued in 1841, argued for an idealist account of the 'master-mind' which designed the Gothic as an embodiment of Christian faith, now corrupted by utilitarianism and paganism. These readings, and episodes such as Hugh Rose's early dealings with the Oxford Movement, brought Whewell into dangerous polemics in the Church.[73] Dangerous, too, was the appropriation of Whewell's epistemology by the Cambridge Camden Society, a visionary group led from Trinity which aimed to found an 'ecclesiology', or restored Church architecture, 'like Astronomy and Geology, an Inductive Science'.[74] 'Ecclesiology', established in 1839, became a science of enthusiastic fact-gathering, 'taking a Church', and working to make design conform closely to the immanent spiritual Idea of religious architecture. John Neale, co-founder of the Society, used Whewell's combination of philosophy, philology, and aesthetics to win believers back to the faith. William Wordsworth, for one, assumed Whewell was behind the Camden Society's manœuvres. Whewell's close friend and patron, Bishop Blomfield of London, resigned from the Society because of his suspicion of its links with Tractarianism. In 1845 the Camden Society stormily collapsed, following members' efforts to erect a Romanizing

[71] Corsi (1988), 121, 136, 123, and Hilton (1988), 172, for Sewell; Newman (1841), 13. For a Tractarian view of Whewell and his colleagues, see Hurrell Froude to John Christie, 15 Apr. 1833, Guiney (1904), 103: 'they know everything, examine everything and dogmatize about everything . . . Whewell's book, and Sedgwick's Lectures, and Thirlwall's research, and Hare's taste, pop upon one at every turn'.

[72] Sewell (1841), 114.

[73] Pugin (1841), 2–5; Stanton (1968); on church politics see Hilton (1988), 26–9; Brent (1987), 148–9.

[74] Chadwick (1971), 212–14; White (1962), 49.

stone altar in the Round church, and legal action against Trinity's senior tutor.[75] The reaction of men such as Sewell and Neale to Whewell's doctrines indicated the uses to which his critique of 'sensationalism', Benthamism, and popular error might be put in political theology and public culture.

Similar dilemmas arose in the central area of Whewell's concern, the rapidly expanding realm of scientific journalism. When discussing the scandalous British response to Thomas Young's early papers on undulatory optics, Whewell was confident that the notorious impact of Brougham's comments in the *Edinburgh Review* testified to a want of élite responsibility. The public sheets were untrustworthy guides to judgement: 'there was in England no visible body of men, fitted by their knowledge and character to pronounce judgement on such a question or to give the proper impulse and bias to public opinion'. The villains were 'the writers of *Reviews* alone, self-constituted and secret tribunals'. The Scottish reviewers treated Young just as Mill and the London reviewers behaved to contemporary philosophy and religion. Whewell was sceptical, as were most conservatives, of a figure such as Brougham, 'Lord Facing-both-ways', who told his Glasgow University audience in 1825 that 'real knowledge never promoted either turbulence or unbelief', yet boasted to his colleague John Allen two years later that in his science lectures, 'I stept aside from Cavendish's discovery to praise the Whig families and to lecture the enemy, and from Napier's logarithms to lecture of feudal nobility. I don't conceal that my object is at least partly political.'[76]

So was Whewell's. But the puzzle was to work out how specialist men of science could manage the forum of public debate. Whewell was made aware of the compromise needed to fight publicly against the author of *Vestiges*, or Darwin, or Macaulay. In October 1848 Macaulay lectured at Edinburgh Literary Association on the value of superficial knowledge, given the enormous progress of modern over ancient science. Whewell counselled his ally J. G. Forbes to preserve silence, until Macaulay spoke to 'eminent men'. Then Whewell changed his tune and demanded a swift and damning response which would not 'call up a plea for quick knowledge'.[77] He had turned down an invitation to review *Vestiges* in the *Edinburgh*, was deeply disturbed by the violent review Sedgwick contributed to the journal, and told Forbes that there was 'no handle' to correct the views of *Vestiges* readers. What Whewell felt he could offer readers of *Vestiges*, and later of Darwin, was a general argument on palaetiology: 'the impressiveness of these views to my mind arises from their being not the propositions of a system of doctrines, but the corollaries of the history and philosophy of science'.[78] Whewell sought a means of addressing

[75] Ibid. 33; Wordsworth (1978–88), iv. 626; for the Camden Society's fate, see Chadwick (1971), 221.

[76] Whewell (1857), ii. 346, on Brougham and Young; for 'Facing-both-ways' see Peacock (1963), 656, 811; for Brougham at Glasgow see Stewart (1985), 183, 203.

[77] Whewell to Forbes, 14 and 24 Oct. 1848, Todhunter (1876), ii. 346–50; Whewell (1849).

[78] Brooke (1977*a*), 139–40; cf. remarks on Darwinism in Whewell to Brown, 26 Oct. 1863, Todhunter (1876), ii. 434.

the scientific leadership rather than the general public, informing the leadership of what the precedent of their past conduct suggested for right action now against the enemies of truth.

Whewell's interventions in actual scientific work are marked by this aim. In autumn 1818, commenting to John Herschel upon recent developments in experimental optics, Whewell implied that because of his neophilia Herschel might miss the philosophical teleology taught in optics. In the 1830s, debating the wave theory with Hamilton, Whewell made the same point. Any new fact 'at variance with the undulatory theory' would generate 'a new property or a contradiction'. The moral meaning of wave theory was not, therefore, dependent on empirical data with which its protagonists were obsessed.[79] The same feature was visible in other sciences. J. S. Mill's failing was his use of recent science, such as Leibig's physiological chemistry. 'The most profound and sagacious physiologists and chemists cannot yet tell which of these will stand as real discoveries . . . how then should Mill do it, to whom the whole subject is new?' And here Herschel was at fault, because he had encouraged neophilia amongst his readers through discussions of 'the process of discovery from a single example'. Whewell defended the precedented and the permanent against the over-enthusiasm of the innovator. Reflecting on events at the British Association in 1857, where the dynamical theory of heat had been canvassed by Humphrey Lloyd, William Thomson, and their allies, Whewell reacted bitterly that such a model of heat was 'different enough from any theory which sound science has ever established'. He bridled at the public role which exponents of novelty abrogated to themselves. 'Presidents of the British Association and Edinburgh Reviewers ought not lightly to adopt such fancies; for theirs is a judicial position, and they give ordinary readers a wrong notion of the progress of science.'[80] The problem was to develop a technique by which these judges could instruct their readers in the 'right notion of the progress of science'. Whewell believed he had access to such a technique through linguistic reform. The rest of this chapter concerns itself with the way Whewell used it.

'Mineralogical Supernaturalismus'

To detail the ways Whewell chose to use his 'historical dictionary', consider three kinds of position which he developed from the mid-1820s. First, he experimented in linguistic reform in his efforts to win the Cambridge chair of mineralogy, a subject he maintained was a scene of 'anarchy in which every man did what seemed right in his own eyes'. Second, he wielded his developing interpretation to reform public discourse. The cases considered here are

[79] For Whewell and Herschel's optics see Todhunter (1876), ii. 28–9; cf. Whewell to Hamilton, 17 May 1833, Whewell Papers O15 47[95]; Cantor (1975).

[80] For comments on Mill, see Todhunter (1876), ii. 313; Herschel's 'single example' is in Herschel (1830), 159–64. For response to heat theory, see Todhunter (1876), ii. 411.

Whewell's responses to three different contemporary attempts in this field: Herschel's *Preliminary Discourse* (1830); encyclopaedia articles on logic, and lectures on political economy, by the Oxford professor Richard Whately; and the immensely popular *Connexion of the Physical Sciences* (1834) by Mary Somerville. Last, so Whewell told Herschel in September 1831 during a discussion about terminology in optics, he was 'solicitous that my friends should not use bad language'.[81] An example is galvanism, which was his finest case of excellence in progressive science. Whewell found in Faraday an apparently enthusiastic devotee. When he edited Whewell's 'Employment of Notation in Chemistry' for the Royal Institution's *Journal* in 1831, Faraday found 'it a most unfortunate thing that men who as experimentalists & philosophers are the most fitted to advance the general cause of science . . . should by the promulgation of their own theoretical views under the form of nomenclature . . . actually retard its progress'. Whewell congratulated Faraday as one of those who 'look forwards . . . to the general bearing and future prospects of the subject; and it is an additional advantage when they humour philologists so far as to avoid gross incongruities of language'.[82] His experience was not always so happy, whether in the university disciplines, the public sphere of the reviews, encyclopaedias and lecture rooms, or the private one of the laboratories of Mohs, Berzelius, and Faraday.

Whewell's work on mineralogical nomenclature raised the thorny issues of the relation between his role as philologist, self-appointed representative of the wranglers, and experimentalists in fields such as chemistry and galvanism. He began his reformist efforts in the last years of the incumbency of Cambridge's first mineralogy professor, Edward Clarke, and during the 1820s campaigned for the chair on the resignation of Clarke's successor, John Henslow. He published papers on the mathematics of crystallography in Cambridge and in the *Philosophical Transactions*. In summer 1825, when Henslow's departure became imminent, Whewell went to Germany to study with Werner's successor at Freiberg, Friedrich Mohs. He pursued the great mineralogist to Vienna to learn of the dispute between the Wernerians, who used a geometrical scheme to classify species by their outward form, and the chemists, led by J. J. Berzelius, for whom mineralogical taxonomy should be subordinated to considerations of chemical composition. Eilhard Mitscherlich, a student of philology and medicine, had won the patronage of Berzelius and, in 1825, the chemistry chair at Berlin through his promulgation of the principle of isomorphism, that crystal form was independent of the chemical constitution but related to atomic proportion. The debates of the early 1820s revealed mineralogical taxonomy as a site of major controversy. Whewell saw it as 'sectarian', the contest between mineralogists and chemists in need of treatment.[83]

[81] For mineralogical anarchy see Whewell (1857), i. 517–18; for bad language see Todhunter (1876), ii. 132.

[82] Faraday (1971), 90, 266.

[83] For Cambridge mineralology see Becher (1986), 61–3; for crystallography see Deas (1959), 135–8; Burke (1966), 120–3, 162–5; Whewell (1828*a*), pp. v–vi.

This intervention promised major pay-offs. Clarke was an influential anti-Wernerian, commending chemical analyses of mineral types to Sedgwick, who was geology professor from 1818.[84] However, the later 1820s and early 1830s saw the decisive shift of Cambridge geology to its mathematical mode, abandoning chemical mineralogy for 'mountain geometry'.[85] Whewell became part of a high-level interdisciplinary move, linking geometry, chemistry, mineralogy, and geology. Second, as he told Jones in autumn 1825, this was a major opportunity to teach true epistemology: 'do not imagine I am doing all for the material sciences with my mineralogy . . . my science shall without ceasing to be good and true mineralogy, be also a most profitable example of that higher philosophy of yours which legislates for sciences'. His German trip highlighted ways in which Teutonic scholarship could be reprocessed by Cambridge inductivism. In 1825 Hugh Rose preached a series of fierce sermons in Cambridge against the materialism of German theology. Whewell deliberately compared his mineralogical pilgrimage with Rose's attack. He told Rose that the Germans 'have got a natural history system of my science . . . and a very dangerous heresy it is considered to be'. Unlike Rose, however, here as elsewhere, Whewell was more favourably inclined to this German view: 'I find my mineralogical *supernaturalismus* giving way.'[86]

Whewell's rejection of Rose's 'supernaturalism' exactly matched his proposals for taxonomic reform. These were designed to show, for the first time, how the linguistic basis of a new science could secure its moral and historical probity and permanence. Where better than in an anarchic German discipline? Chemists relied too much on their experience with blow-pipes, mineralogists were obsessed by goniometry. What was needed was the overview which mathematical discipline could provide, making clear the fundamental ideas at the root of devoted empiricism.[87] His immediate recourse was to consult Herschel on chemical terminology, following this with an explanatory essay and chart in May 1828 on his successful assumption of Henslow's chair that spring.[88] Whewell told Herschel that a reform both of chemistry and mineralogy was needed. Chemical symbolism was to be whipped into line with good algebra; mineralogy would take on the form of a refined geometry. Only then could crystal forms be properly analysed with respect to chemical and spatial relations. Whewell asked whether Herschel believed the Berzelius scheme was too embedded to be shifted. 'It is very monstrous that the people who have invented notations have not been sensible of the nature of mathematical symmetry.' Herschel was sceptical of the need for such a dramatic imposition of wrangler culture on mineralogy and chemistry. Whewell responded that

[84] Oldroyd (1972); Secord (1986), 64; C. Smith (1985). For Whewell's late comments on Clarke and his development of these mineralogical ideas into a full-blown fantasy about meteorites see Todhunter (1876), i. 378–9.

[85] Secord (1986), 66–7.

[86] Whewell to Rose, 15 Aug. 1825, Todhunter (1876), ii. 60; Corsi (1988), 28–30.

[87] Whewell to Rose, 19 Nov. 1826, Todhunter (1876), ii. 76–7; Whewell (1832*b*), 359.

[88] Id. (1828*a*), p. i.

'names are not to be imposed by *definitions*': they must embody fundamental mathematical ideas of gradation and symmetry, connection of form and law-like regularity.[89] For Whewell, this was proof positive that while local expertise must be respected and mere neologism avoided at all costs, there were nevertheless intellectual principles, apparent to the mind if well trained in clear and distinct apprehension of fundamental ideas, which could guide the classificatory and kindred sciences.[90] This was part of his riposte to theological hostility to such scientific work.

He pursued this insight about the connection between intellectualist mathematicization and pragmatic reform of critical structure throughout his tenure of the chair. In 1831 he sent Faraday a key paper on chemical nomenclature and in 1832 he gave a valedictory survey of mineralogy at the Oxford meeting of the British Association. Pessimism about his reform emerged at this moment and stayed with him throughout his life. In his 1831 paper he attacked Berzelius's nomenclature as arbitrary, forgettable, biased, and impractical. Such chemical forms, unenlightened by knowledge of the fundamental idea of the science, 'are merely unconnected records of inferences which are in some degree arbitrary . . . the analysis itself, the fundamental and certain fact from which the inferences are made, is not recorded in the symbol'. By breaking algebraic rules and neglecting the right combination of experimental knowledge and ideal interpretation, Berzelius and his followers still erred from the path of a progressive science.[91] The same disease afflicted the followers of Mohs. Even though 'crystallography is essentially a mathematical science', the mineralogists 'violated' mathematical principles in their terminology and practice. Optics and geology were 'sciences in which striking advances in theory, new and widening views, and the bright promise of future progress attract the attention of all', yet 'such, at least in our own country, mineralogy of late has not been'.[92] Neither in his *History* nor his *Philosophy* did Whewell ever retract this condemnation of a science which he abandoned to his loyal successor William Miller on his resignation from the chair in 1832.

Whewell's farewell to mineralogy in Oxford in summer 1832 contained a wide-ranging condemnation of the national mentality, a highly charged comment on empiricist hostility to fundamental principles, given just a fortnight after the Reform Act became law. His address also intervened in the Oxonian debates on the place of the natural sciences in the university curriculum.[93] Whewell compared the failures of mineralogy with the 'apathy or contempt with respect to method' which 'appears to be a part of the intellectual character of this country':

[89] Whewell to Herschel, 20 Sept. and 28 Oct. 1828, Whewell Papers O15 47[126, 123].

[90] Whewell (1828*a*), p. x.

[91] Whewell (1831*c*), 438–9, 453; Todhunter (1876), i. 307–10.

[92] Whewell (1832*b*), 329, 335, 323.

[93] For the milieu in Oxford see Morrell and Thackray (1981), 390; Corsi (1988), 115–17.

Men here appear to feel no interest with regard to rules and systems till they are so complete, so clearly developed as to principle, their apparent difficulties so far explained, that the general rule will bear a strict application in each particular instance. They are disposed to despise the dim glimmerings of dawning principles in cases where, though *a* connexion may be probable or certain, *the* asserted connexion is clearly not exact.[94]

This was a pointed appeal for a sympathetic analysis of discovery, the project Whewell inaugurated at his resignation from his professorship. It also made urgent his efforts at stewardship of the discourse of the public journals and debates.

'Thieves and Robbers': Political Economy and Popular Philosophy

In January 1830 Whewell discussed with Hugh Rose the chances of launching a conservative Cambridge journal against 'the liberal revolutionists': 'I have for my own part a profound horror of a personal share in the responsibility of providing food for any periodical creature—even if it were to be like a household dog to protect you from thieves and robbers.' During the following year, nevertheless, Whewell planned a journal with Jones and leapt into print with reviews of Lyell and Jones in the recently revived conservative Anglican *British Critic*, an organ linked to Rose. The review of Jones's *Essay*, a searing indictment of Ricardian doctrine through an assault on political economists' terminology, was initially intended for Lockhart's more prestigious *Quarterly*, where Whewell did manage to print his analysis of Herschel. 'Review away, my boy,' Herschel exclaimed. This essay took him a fortnight. Its connection with the attack on Ricardo and on the philosophical radicals' plans for Reform was clear. Whewell was disturbed that the text had been printed in the populist series of Lardner's *Cabinet Cyclopaedia*. His review set out an early version of the analysis of the troubles of scientific and popular languages he diagnosed in the review of Jones: knowledge was made up of generalizations from particular experiences, but these generalizations were only expressible as an ordered language which embodied and yet concealed the process of formation.[95]

Herschel gave brief attention to the problem of nomenclature, solely considered as an aspect of taxonomy in botany and mineralogy. Whewell responded that it was much more serious a problem than was realized by Herschel, or by Bacon, who omitted the topic entirely. Herschel erred because he neglected the point that language expressed laws and not merely the names of objects. As Coleridge also argued, its chief function was verbal, not nominative. The moral of this claim was hammered home: 'terms of art, well

[94] Whewell (1832*b*), 363.

[95] Whewell to Rose, 10 Jan. 1830; Whewell to Jones, Feb. and July 1831, Todhunter (1876), ii. 106, 108–9, 117–18, 123; Herschel to Whewell, 15 Feb. 1831, Royal Society, H 21 79; Schweber (1980), 280–5; Corsi (1988), 17–18; Jones to Whewell, 24 Feb. 1831, Whewell Papers, c. 52[30].

chosen, contain a concentrated essence of past labours'. As Whewell began work on his 'induction', he already argued for language as a bearer of historical continuity, directing his polemic against sensationalists and Benthamites in explicit terms.[96] The problem of dealing with popular speech was clarified: it was necessary to guard against 'the popular love of the striking and the wonderful', which neglected the *linguistic* acts 'of those who have earned the fame of true philosophers' and lavished praise on the 'sudden and unprepared inspirations' of their heroes. Fighting popular illusion meant defending the philosophical élite and its labours. But it also meant working hard to introduce a better terminology into popular circulation. Without this, the plebs of science would lapse into delusion.[97]

The responses of Whewell and Jones to the moral, economic, and epistemological doctrines of Oxford Noetics, such as Whately, were examples of such work. Whewell gave papers in 1829 and 1831 at the Cambridge Philosophical Society in which principles of the 'higher department of the science of Political Economy which is concerned with the moral and social principles of man's actions and relations' could be cast in deductive, axiomatic form, and thus shown to be simply fallacious. Whewell argued that the arbitrary assumption of 'visionary and unphilosophical' principles led the Ricardians to 'rush forwards and seize upon . . . a general law' which failed to match the facts. General equilibrium theory violated the inductive principles he and Jones believed they had developed as the criterion of good science.[98] This methodology rested on such deep insights into proper linguistic and moral behaviour.[99] Whewell decided to clarify these 'principles' in the public sheets. He told Jones in February that his *Essay* was one of the very few examples of the 'principles of induction' to use against the 'absurdities' of the Oxford school. The following June he affirmed that 'though it is a filthy business having to do with these reviews' he did, nevertheless, 'enjoy the prospect of a row'.[100]

Whewell used the platform of the *Philological Museum* to link his idealism with its linguistic base.[101] In spring 1833 he published a brief paper 'On the Use of Definitions', directed at the relation between the evils of Richard Whately's political economic doctrine and his bad language. Whewell cited Whately's Bampton Lectures delivered at the University Church in Oxford in 1822 on *The Use and Abuse of Party-Feeling in Matters of Religion* and his

96 Whewell (1831*a*), 391, 404.

97 Ibid. 391.

98 Id. (1829), 3–4; id. (1831*b*), 7; de Marchi and Sturges (1973).

99 In his *Preliminary Discourse* Herschel had suggested that 'political economy, at least, is found to have sound principles founded in the moral and physical nature of man', making it approach the condition of 'an experimental science', Herschel (1830), 73. In private, when Whewell sent him a copy of Jones's *Essay* in February 1831, he described political economy as 'a subject that seemed to have been raked into being only to bring up noisome conclusions and bewildering paradoxes'. Herschel agreed that Jones had gone far to establish the subject 'on broad inductive principles'. Todhunter (1876), ii. 115; Herschel to Whewell for Jones, 17 Feb. 1831, Royal Society, H21 80; Checkland (1951); Rashid (1979); Corsi (1987).

100 Whewell to Jones, Feb. and July 1831, Todhunter (1876), ii. 115–16, 123.

101 Checkland (1951), 57; Hilton (1988), 53.

Introductory Lectures on Political Economy (1831). New terms could only be introduced to embody established knowledge, only discoverers of such knowledge could legitimately neologize, and such language must depend on the theoretical achievements of a properly 'progressive' science. These strictures applied to political economy because its subject matter was coextensive with that of philology, which had profited from its emulation of the behaviour of the best and most progressive natural sciences. Whately and the Noetics seemed to claim that arbitrary definitions would settle dispute in contested fields: Whewell replied that good language followed upon the closure of such argument and must be based on the permanent result of closure. Whewell used what was to become his standard gallery of cases of good language: Herschel's optics, Lavoisier's chemistry, Howard's meteorology, and Cuvier's taxonomy.[102] He also pointed out the scandals of those sciences which lacked knowledge enough to embody adequate terms, of which his standard case at this period was mineralogy, and of those which lapsed into 'logomachies' about terms without respect for custom, such as recent disputes in the Geological Society.[103] He used his favourite metaphor, that of the lamp. Two years earlier he praised catastrophists who accepted the 'manifestation of a creative power' in the transition from one population to another: 'geology has thus lighted a new lamp along the path of natural theology'. Noetics and Ricardians achieved the opposite. They presented themselves as a 'venerable corporation of The Lamplighters' but 'there is in these lamps no drop of oil . . . the way is just as dark as ever, and the only consequence is that, in addition to other lumber, we stumble over the lamps themselves'.[104]

The paper was a remarkable development of previous views and a signal of the path his researches in induction had taken. In summer 1822, he had explained to Jones that the term 'metaphysics', 'those fatal four syllables', etymologically referred to the philosophy of mind, and that therefore they could be applied to mathematics or philology. No 'metaphysics of political economy' was possible: he proposed that Jones, Malthus, and their anti-Ricardian allies be baptized 'the *ethical* school'. Efforts to promote an inductivist 'statistical society' within the British Association and in London were energetic but abortive: Whewell remained sceptical of their chances of success through lack of 'zealous theorists'.[105] He also feared the dangers of making such theory part of the common stock of popular debate. The mid-1830s was notoriously a conjuncture of major efforts at popular synthesis of existing knowledge, and Whewell's depression was accompanied by keen concern for the effects of this stream of publications, encyclopaedias, lectures, and reviews.[106] Chief among these was Mary Somerville's *Connexion of the Sciences*. The book, dedicated to

[102] For Whewell on Whateley see Whewell (1833*a*), 263, 266–9; cf. id. (1847), ii. 519–20.

[103] Id. (1833*a*), 269.

[104] Id. (1831), 194; id. (1833*a*), 272.

[105] Whewell to Jones, 16 Aug. 1822, Todhunter (1876), ii. 48–9 for 'metaphysics'; for statistics see Goldman (1983); Whewell to Quetelet, 2 Oct. 1835, Todhunter (1876), ii. 228–9. For later work see Hollander (1983).

[106] Yeo (1984) discusses popularization of science.

the Queen, was a bold effort by the translator of Laplace to summarize a mass of new scientific material. Whewell helped edit the book during autumn 1833 and then reviewed it in the *Quarterly*, praising its author as 'a person of real science'. He cited the work as a good case of the correction of public error, as for example in popular fears of cometary collisions with the Earth.[107] He took the opportunity to develop two key features of his analysis of popular and élite scientific language: what was the best means of displaying the whole of science to a public? What was the principle which connected the sciences together?

Whewell responded by contrasting the behaviour of popular lecturers who showed 'models, machines, diagrams', with authors who traced the language of science through the ages. Lecturers were compared to Swift's 'philosophers of Laputa, with their *real* vocabulary', those sages who believed they had captured the essences of things by carrying them in sacks upon their backs. Somerville, and, by implication, Whewell himself, aimed for a broader, linguistic vision of science and its past rather than the shows and wonders of the contemporary experimental demonstrations. The right metaphor was that of a Turneresque painting, in which the whole was grasped though detail was lost:

> The office of language is to produce a picture in the mind; and it may easily happen in this instance, as it happens in the pictures of some of our un-Pindaric artists, that we are struck by the profound thought and unity displayed in the colouring, while there is hardly a single object outlined with any tolerable fidelity and distinctness. The long-drawn vista, the level sunbeams, the shining ocean, spreading among ships and palaces, woods and mountains, may make the painting offer to the eye a noble expanse magnificently occupied; while even in the foreground, we cannot distinguish whether it is a broken column or a sleeping shepherd which lies on the earth, and at a little distance we may mistake the flowing sleeve of a wood nymph for an arm of the sea. In like manner, language may be so employed that it shall present to us science as an extensive and splendid prospect in which we see the relative positions and bearings of many parts, though we do not obtain from it precise notions of optical phenomena, or molecular actions.[108]

Whewell rehearsed his own dilemma in the presentation of all science as a choice between showman and poet. The underlying binding principle of the sciences was linguistic, 'in the same way in which a kindred language provides the common stock and relationship of nations'. This language was mathematical. When Somerville, via Lady Murchison, raised this problem with him in autumn 1841, he answered brutally that 'mathematical language is the very plainest English which can be used to those that can understand it'.[109] No doubt this captured the principle he developed in the mid-1830s on the relation between the formation of mathematics and the fundamental ideas which bound sciences as coherent bodies of knowledge. The lesson he started

[107] Whewell (1834*a*), 58; Patterson (1983), 142–3, 130.
[108] Whewell (1834*a*), 55.
[109] Ibid. 60; Whewell to Mrs Murchison, 19 Nov. 1841, Todhunter (1876), ii. 302–3.

to teach in his 1834 review, was that a *via media* must be sought between the histrionics of performers such as Faraday and the vague sketches of progress which Somerville provided. Significantly, it was her accounts of Faraday's work on polarity which he cited *in extenso* when debating the connection between the sciences of heat, galvanism, and mathematics.[110] Galvanism provided Whewell with his best opportunity for directing a progressive science through the creation of its terms.

From Tubal Cain to Faraday

Faraday first consulted Whewell on terminology in early 1834, having been directed thence by their mutual friend Willis. The relationship has been read in terms of Faraday's search for 'some new names to express my facts in electrical science without involving more theory than I could help'.[111] Certainly he sought to avoid using notions of terms such as 'poles' which implied an action-at-a-distance or atomistic model; he also tried to distance himself from Berzelieus's fluid theory of electrochemistry and Ampère's electrodynamics, to the latter of which Whewell remained loyal. On the other hand, Faraday was an indispensable resource for Whewell's programme; his work was 'the greatest event which has ever happened in the history of chemistry', an 'epoch' through which Whewell was living. In 1831 Whewell already diagnosed this science as the best case of rapid progress. Ampère's development of the Oersted effect between 1820 and 1822 answered 'nearly to the entire history of astronomy from the first contemplation of the moving skies by Chaldean shepherds to the demonstration of universal gravitation by Newton'. Herschel echoed Whewell in 1846. He told Murchison that Oersted's discovery 'affords the second instance which science has witnessed of the complete *concilience* [*sic*] into one of two separate and previously independent sciences'. The career of electromagnetism would thus be the proving-ground for several rival theories of excellent science.[112] During the years after Faraday's development of electromagnetic induction in summer and autumn 1831, he rapidly acquired heroic status: Whewell heard Faraday's triumphant introduction of the new terms at the Royal Society in 1834, while the Duke of Somerset wrote to Charles Babbage in 1835 that 'the story of Faraday is just one that is sure to make a great noise. There is something romantic and quite affecting in such a conjunction of poverty and passion for science . . . he comes out as the Hero of chemistry.' Whewell wanted to make sure that this romantic story had a moral which suited his purposes. He described his induction project to Herschel as 'a history of all the physical sciences . . . from Tubal Cain to Faraday'.[113] The

[110] Whewell (1834*a*), 62–3.
[111] Faraday (1971), 264; Williams (1971), 258–9: Ross (1961).
[112] Faraday (1971), 294–5, 297; Whewell (1831*a*), 395; Gooding (1985), 233.
[113] Somerset to Babbage, Berman (1978), 174; Whewell to Herschel, Todhunter (1876), ii. 248–9.

mythic role of the Royal Institution professor was an essential element in Whewell's dealings with him.

The moral Whewell wanted the Faraday story to teach was simple: each progressive science advanced through clarification of its appropriate fundamental idea. The fundamental idea for Faraday's science must be *polarity*. In *The Friend* Coleridge argued that 'the greatest improvements in the *method*, the nearest approaches to a *system* of electricity' would come from 'those philosophers who . . . rejecting the imagination of any material substrate, simple or compound, contemplate in the phenomena of electricity the operation of a law which reigns through all nature, the law of POLARITY'. His claim was that the progressive clarification of this vision distinguished advancing electricity from static magnetism. Whewell shared these views: he argued that in Faraday's strategy 'all the superfluous and precarious parts gradually drop off from the hypothesis which we devise in order to represent polar phenomena; and the abstract notion of polarity—of equal and opposite powers called into existence by a common condition—remains unencumbered by extraneous machinery'.[114] Whewell read Faraday as a complete immaterialist. It was mere convention to imagine some such 'machinery' or material substrate: 'the idea of *Polarity* becomes most pure and genuine, when we entirely reject the conception of *Poles*'.[115] This reading needed considerable interpretative work.

In May 1834, for example, Whewell discussed the new terms 'anode' and 'cathode' for the surfaces of the electrolyte in contact with the electrodes. Faraday had considered the names 'Voltode and Galvanode', to which Whewell immediately protested. They needed a term which did not imply the transfer of some matter.[116] But Whewell wanted it to imply his own account of polarity: 'the terms which you mention . . . show that you are come to the conviction that the essential thing is to express a *difference* and nothing more'. Whewell saw polarity as 'a contrast of properties corresponding to a contrast of positions'. He would insist that Faraday's isolation of the fundamental idea of polarity was not primarily experimental: an '*a priori* conviction that co-existent properties must be connected'.[117] In 1834 he instructed Faraday in these terms: 'it is very desirable in this case to express an *opposition*, as well as a difference'. Faraday was by no means so committed to this interpretation. In December 1835, after finishing the sections on galvanism in his *History* and reading the decisive seventh series of Faraday's *Experimental Researches*, Whewell suggested that Faraday should ascertain the balance of polarized forces within the battery to link galvanism with chemistry: 'the more I think of it crystalline forces are *polar* in their own way, and must, it would seem, be connected with your polar chemical forces'.[118] During this period Faraday was arguing with William Sturgeon and his colleagues on the problem of chemical action in the galvanic

[114] Coleridge (1818), 478–9; Whewell (1847), i. 349. Whewell's view of Faraday's debt to Humphry Davy was crucial here: see ibid. 356, 363–4.

[115] Ibid. 356.

[116] Faraday (1971), 271.

[117] Ibid; Whewell (1847), i. 347, 366, 433.

[118] Ibid. 270–1, 299.

cell. He resisted Whewell's suggestion and began to abandon the interpretation based on polarity.[119]

Evidence of this is provided in Faraday's unhappiness with the term 'electric capacity' in 1837. He wished to refer it to 'the more or less ready polarity *or other condition* assumed by the particles between the inducing surfaces'.[120] His language of 'electrotonic state' and his appeal to magnetic 'curves' were not endorsed by Whewell.[121] In 1846, as David Wilson shows in his chapter, Faraday famously challenged the fundamental tenet of Whewell physics: the luminiferous ether. Even 'polarity' was ultimately abandoned. Commenting on John Tyndall's use of Whewell's term, Faraday exclaimed that 'that phrase *polarity* in its present undefined state is a great mystifier'.[122] In successive editions of his *History*, Whewell sought vainly to make sense of the way electromagnetism had developed. 'The fermentation at present is perhaps too great to allow us to see clearly the truth which lies at the bottom.' He was sceptical of the hasty rejection of Coulomb's account of action-at-a-distance between electric and magnetic poles. In his *Philosophy* he was forced to apologize for the brevity of his treatment of the fundamental idea of polarity. In the third edition of the *History* (1857) he proffered his own '*polar* relations' as one among many competing models, including those of Faraday and Ampère, for the causal explanation of electromagnetic induction.[123]

The events of 1845–8, when Faraday achieved crucial experimental results on the interaction between crystallography, optics, and electromagnetism, were decisive for Whewell's views of the role polarity must play as the regulative idea for this science and its relations with chemistry. Faraday released news of his magneto-optical work to Whewell privately in November 1845: 'I am sure that when *you* have *this discovery you will understand and appreciate it*.' However, Faraday would use this work precisely to destroy the polar interpretation of the field.[124] Whewell complained to Forbes of 'the overcharged importance of Faraday's view of his recent discoveries'. Clarification of the fundamental idea needed to interpret the magneto-crystalline experiments required 'a clear mathematical head', a theoretical skill which Whewell and Forbes believed that Faraday conspicuously lacked. Whewell asked Herschel whether he might help: 'have you seen or repeated Faraday's experiments; and can you get any theoretical hold of them?' Herschel himself was engaged in a fraught exchange with Faraday about the interpretation of these experiments and on Faraday's reading of them.[125] Conflict with Faraday seemed inevitable.

[119] Faraday (1839–55), i. 259–79; Morus (1988); Gooding (1982); L. Williams (1971), 437.

[120] Faraday (1839–55), i. 393–4; id. (1971), 315, my italics.

[121] Ibid. 532; Whewell to Forbes, 10 Mar. 1855, Todhunter (1876), ii. 404; Whewell (1857), iii. 88.

[122] Faraday (1846); id. (1839–55), iii. 360–2; id. (1971), 805. See F. James (1985), 149.

[123] Whewell (1847), i. 374; id. (1857), iii. 88, 531; L. Williams (1971), 372.

[124] Faraday (1971), 466; F. James (1985), 144–6; Faraday (1839–55), iii. 1–26, on the rotation of the plane of polarized light.

[125] Whewell to Forbes, 5 May 1846, Todhunter (1876), ii. 334–5; Gooding (1985).

An opportunity for resolution of this trouble emerged in autumn 1848. Faraday informed Whewell of his exciting experiments on what he proposed calling the 'magneto crystallic force'. These were trials which seemed initially to satisfy Whewell's demands of the 1830s, revealing effects on crystals detectable in strong magnetic fields involving the orientation of the crystal along its 'magnecrystallic axis'. Faraday found 'no traces of polarity though the setting force be strong', a problem which led the two men to develop the term 'paramagnetism'. Whewell came to see that Faraday held that 'bismuth &c have no real polarity but only a seeming polarity arising from each particle having a tendency to go from strong to weak magnetic spaces'.[126] Whewell found ways of saving his fundamental idea. In the third edition of the *History* he interpreted these trials with bismuth crystals as a good case of 'co-existent polarities'.[127] When Faraday told him he had abandoned the term 'polarity' in favour of 'axiality', Whewell replied that 'axiality' might emerge as a term for 'some cases which we have been accustomed to call *polarization*'. Fortunately, an ideal opportunity emerged for clarification. In the same letter, Faraday invited Whewell to present a lecture in London. Whewell responded enthusiastically that he would speak on 'The Idea of Polarity'.[128]

The lecture provides our final example of Whewell's efforts to place Faraday's work in the history and geography of the intellectual world. The doctrine he presented to the Royal Institution audience was to contrast the family of sciences embraced by 'polarity' with his 'pattern science' of celestial mechanics. The 'mechanical sciences' progressed from individual entities with varying qualities to a single mechanical property, namely gravitation. But in galvanism, chemistry, and magnetism, apparent simplicity, the phenomena of attraction and repulsion, revealed itself as highly individuated qualitative complexity. The advance to complexity needed Whewell's best rhetoric: 'New ideas in science gradually and slowly become clearer and clearer . . . Polarity is a new idea which is not yet clear.' Whewell traced it in magnetism, since Gilbert; in galvanism; and in Faraday's work. He brought in the polarization of light, condemning those leaders in the field who misled their public into the faith that 'polarization' was 'a mere confession of ignorance'.[129]

Whewell teased apart wave theory and polarization, noting Malus's effort to clarify the idea through the notion of 'particles with poles'. Whewell sought to place this fundamental idea in Faraday's most recent work. In a large chart displayed at the meeting he listed a series of polarities Faraday had identified matched with the corresponding 'antithesis': induction and the start of current; diamagnetism and the repulsion of bismuth; magnetic-crystalline action and the direction of the axis. The entry corresponding to the antithesis associated

[126] Faraday (1971), 590–1; id. (1850); id. (1839–55), iii. 87; Gooding (1981), 253–63, on bismuth.

[127] Faraday (1971), 528; Whewell (1857), iii. 535.

[128] Faraday (1971), 530–1.

[129] The lecture is drafted in Whewell Papers, Add. MS a 217[5] fo. 2; cf. Whewell (1847), i. 272–3.

with the 'electrotonic state' was filled in with a query mark. Above all, Whewell cleverly placed Faraday's work in its connection with other sciences, with electrochemistry and optics, with crystallography and the wave theory of heat. This was, then, a bold attempt to show that the challenge which Faraday's recent work seemed to pose to the fundamental idea did not diminish its value as an organizing principle whose meaning was in the course of emergence. Whewell's lecture and his draft chart are testimony to his attempt to interpret a new science. He summarized the convergence of 'Voltaic Phenomena' and 'Chemical Phenomena': 'These two kinds of action co-exist, their polarities correspond and the numbers which express their quantities are the same: (Faraday) electric polarity and chemical polarity are the same'.[130] This was a visionary conclusion. Whewell never printed such a chart in his *Philosophy*: he found no map on which to place the work of the Royal Institution's laboratory nor a language with which to describe it.

Conclusion: Whewell's Critical Path

Using such maps we have traced the ways in which Whewell tried to direct past and current sciences, to chart their progress, and render their excellences permanent. This topography was designed to give Whewell his own place as philologist and 'high priest' as part of the politics of Victorian culture. Such cultural politics has often been examined through the world of the periodicals and reviews.[131] Whewell's notions of correct usage are thrown into sharp relief when placed in this public setting. His role has been notoriously difficult to categorize: alongside more familiar types, such as philosopher, historian, naturalist, and priest, let us now place that of 'critic'. The advantage of adding this label is that it captures rather well the peculiar position which Whewell adopted towards the sciences of his time. Numerous examples of his self-presentation as critic of contemporary culture have been cited. In December 1828 Whewell discussed with William Wordsworth the views on classical pronunciation developed by the celebrated aesthetician Uvedale Price. Whewell was struck that Wordsworth shied away from such philological matters, citing Sedgwick's remark that 'a man may be too good a mineralogist for it to be possible for him to be a geologist'. 'In the same way', Whewell supposed, 'a person may be too good a writer of verses to be a good critic of versification, or rather I should say a good anatomist of verse'.[132]

The images fit Whewell well. He used his critical stance to full effect to promulgate his philosophy and politics. Furthermore, this stance points our

[130] The chart is sketched in Whewell Papers, Add. MS a. 217[4].

[131] Excellent accounts of the role of the Victorian periodical in the politics of culture include Young (1985), 134; Yeo (1984); Shattock and Wolff (1982).

[132] Wordsworth (1978–88), i. 681 n. 3. For Price's aesthetics in the formation of Romantic sensibility, see Watson (1970).

attention towards unjustly neglected literary and critical resources which Whewell used to make his imagery of the sciences' progress. For example, one area which deserves further study is that of Romantic poetics, at least as crucial a resource as the more commonly cited relationship with Kantianism. An excellent case is his exchange with Herschel in December 1842 over the latter's translation of Schiller's great philosophical poem, *Der Spaziergang* (1795), to be published in Whewell's collection of hexameters five years later.[133] Schiller's poem is the archetype of the Romantic *Bildungsreise*, an astonishing meditation on the development of human society from primitive bliss through civilization to the horrors of Revolution, matched with an equally striking evocation of a long mountain climb from meadowland to rocky height.[134] Whewell's modifications of Herschel's effort all point to the profound message Whewell read, in striking contrast to Herschel's scepticism of Schiller's literary and philosophical virtues. Whewell described himself as 'a great admirer of Schiller's philosophical spirit as shewn in his poems'; he argued that *Der Spaziergang* 'conveys in the most poignant and finished form an image of the progress of society: rural life, classes, cities, government, patriotism, commerce, art, science, unbelief, corruption, revolution, refuge in the repose of nature, which is the poet's final solace'. Schiller charted a path which Whewell also took in his own work on the history and epistemology of human creativity.[135] It seems important to understand this path in order to revivify our comprehension of the authority of Romantic and conservative intellectuals in the formation of nineteenth-century sciences. No doubt our own 'division of the soil of science', in Whewell's phrase, has made it very hard to recapture this sense of an integration between languages of nature and culture.[136] For Whewell they were intimately linked. The point of this chapter has been to show this connection and explore its political and philosophical meanings.

[133] Todhunter (1876), i. 286. Herschel's text is in Herschel (1857), 685–97.

[134] Abrams (1973), 453–7.

[135] Herschel to Whewell, Royal Society H22 286, 6 Nov. 1846; Whewell to Herschel, 3 and 23 Nov., and 27 Dec. 1842, Whewell Papers O15 47[132, 135–6]. Cf. Whewell (1831*a*), 386: 'the progress of inductive philosophy' is like 'the wanderings of the alpine traveller'. 'The august forms which surround him and the pure atmosphere into which he rises make his toil a sport and his perseverance a delight.'

[136] The reference to 'division' is in Whewell's discussion of the term 'scientist', cited above: see (1834*a*), 59–60.

9. *Convergence: Metaphysical Pleasure Versus Physical Constraint*

by DAVID B. WILSON

WHEWELL insisted that physics and metaphysics must be combined. Not merely an inferior stage of understanding that mankind progressed through on the way to true physics, 'the metaphysical is a necessary part of the inductive movement'.[1] Successful physicists were characterized 'not by having *no* metaphysics in their heads, but by having *good* metaphysics while their adversaries had bad; and by binding their metaphysics to their physics, instead of keeping the two asunder'.[2]

Physics and metaphysics were among Whewell's earliest studies. In addition to pure mathematics, Whewell's Senate House examination in 1816 covered mechanics, astronomy, optics, and hydrodynamics. At about the same time, he was studying metaphysics, probably to meet requirements for his fellowship at Trinity in 1817. Concentrating on one aspect of the interaction of these early interests of Whewell's from the 1820s into the 1840s, this chapter examines the role of metaphysics in his deliberations on the possible convergence, or unity, of the physical sciences.

These were decades of important developments in the physical sciences. They witnessed not only the introduction of the undulatory theory of light but also the labours of such as Oersted, Ampère, Faraday, and Forbes in heat, electricity, magnetism, and chemistry. P. M. Harman's book, *Energy, Force, and Matter*, clearly shows how 'unity' could be a kind of umbrella term for what were different early nineteenth-century approaches to physical theory.[3] For Laplace and his school, physical phenomena were caused by the motion of microscopic material particles interconnected by action-at-a-distance forces. Unity meant successfully finding the same type of physical explanation in each separate area of physics. Fourier emphasized mathematical formalism over underlying physical causation. Unity of physics meant successfully mathematicizing each of its branches, especially if one could demonstrate that different

I am grateful to the Master and Fellows of Trinity College for permission to quote from the Whewell Papers, to T. D. Hobbs for his help with the Whewell Papers, and to N. H. Robinson, former Librarian of the Royal Society, for his assistance.

[1] Whewell (1847), i. p. ix. [2] Ibid. [3] Harman (1982), 12–33.

branches possessed analogous mathematical structures. Fresnel hoped to reduce all physical phenomena to the activity of one universal fluid ether.

Of these French options, Fresnel's was closest to the approach of those Cambridge graduates, William Thomson and James Clerk Maxwell, in their mid-century reformulations of heat, electricity, and magnetism. Thomson's 1847 paper, 'On a Mechanical Representation of Electric, Magnetic, and Galvanic Forces', would have 'represented' electricity, magnetism, and light as motions or strains in one elastic–solid medium, and he spent much of his career searching for a 'great comprehensive theory' of physical phenomena.[4] Maxwell's researches pursued Thomson's unifying physical insights.[5] In 1874, Maxwell distinguished between different sorts of unity in a review which contrasted Mary Somerville's *Connections of the Physical Sciences* with W. R. Grove's *Correlation of Physical Forces*. Going through eight editions from 1834 to 1849, Somerville's book at least demonstrated through its large sales 'that there already existed a widespread desire to be able to form some notion of physical science as a whole'.[6] In Somerville's work, however, the chief unity seemed to be provided by the book's binding. She showed merely that research in one science could depend on knowledge of others and that the physical sciences all involved mathematical analysis. 'The unity shadowed forth in Mrs Somerville's book is therefore a unity of the method of science, not a unity of the processes of nature.'[7] By contrast, Grove's book, published in 1847 and reaching its sixth edition in 1874, argued that 'of the various forms of energy existing in nature, any one may be transformed into any other, the one form appearing as the other disappears'.[8] Unlike Somerville's, Maxwell appeared to suggest, Grove's book was one of those which were important in the history of scientific ideas because they produced 'a concentration of thought in that part of the field of science which at that particular season ought to be cultivated'.[9]

The question addressed by this chapter might thus be put: To what extent would Whewell's views have contributed to a unified view of physics in that early Victorian season? The tripos examinations taken by Thomson and Maxwell in 1845 and 1854, respectively, displayed no concern for unifying physical conceptions and embodied little more of the subjects of heat, electricity, and magnetism than did Whewell's tripos of 1816. Given Whewell's general prominence in Cambridge and his other influences on younger Cambridge scientists, it is possible that his own considerable writings on these non-tripos subjects, including whatever he thought about their possible unity, would have seemed especially important to mid-century Cambridge undergraduates

[4] Thomson (1882–1911), i. 76–80.

[5] For Thomson's influence on Maxwell, see the chapters by Harman, Knudsen, and Siegel in Harman (1985). Thomson and Maxwell also used Fourier's sort of approach to good effect.

[6] Maxwell (1927), ii. 401.

[7] Ibid. 402.

[8] Ibid.

[9] Ibid. 401.

who were interested in those areas of physics.[10] However, though the existence of Thomson's and Maxwell's work accentuates the importance of examining Whewell's ideas, one ought not, of course, simply to point Whewell's thinking towards Thomson's and Maxwell's. Only by examining Whewell's thought in its own context can we imagine how it might have influenced later thought.

During the period from the 1820s into the 1840s, the idea of unity did occupy a place in Whewell's thinking. As the centre of a small group of Cambridge men interested in heat, electricity, and magnetism, Whewell was the most important Cambridge writer on those subjects from 1820 to 1840. Joined by many others, including major Cambridge figures like G. B. Airy and George Green, Whewell also, of course, wrote on astronomy, mechanics, and optics. Whewell's changing views on convergence among all these branches of science had both metaphysical and physical components. Metaphysically, convergent theories were pleasing and, therefore, to be sought. But the search was always constrained by physical considerations—by conceptual plausibility and observational reality. Examining some of Whewell's unpublished manuscripts from the 1820s and his well-known books of the next two decades, this chapter traces his developing thoughts on the convergence of the physical sciences.

1. Metaphysical Pleasure

In 1818, Whewell recommended the *Encyclopaedia Britannica* article on 'Metaphysics' to a friend preparing for an examination on the subject.[11] Evidently quite familiar to Whewell, the article is an appropriate starting point for exploring his ongoing consideration of the subject. If we cannot pretend that Whewell necessarily accepted the article's arguments in all their detail, we can at least read it to discover the overall scope and general direction of the subject as the young Whewell encountered it. Indeed, at that level, the article's basic viewpoint appeared in unpublished and published writings of Whewell's from the 1820s into the 1840s. In a particular way, the essential similarity of human and Divine intellect provided a metaphysical framework for Whewell's thoughts about the unity of the physical sciences.

'*Mind* or intelligence, and especially the *supreme intelligence*, which is the cause of the universe, and of every thing which it contains,' wrote the *Encyclopaedia Britannica* author, 'is the principal subject of this science.'[12] The article presented an essentially Lockean view of human understanding. Devoid of innate ideas, the human mind relied upon experience of the external world and reflection upon its own activity to construct a body of contingent

[10] For discussions of the early-Victorian mathematical tripos, see Becher (1980) and D. B. Wilson (1985). For discussions of Whewell's influence on Stokes and Maxwell, respectively, see D. B. Wilson (1984) and Harman (1985).

[11] Whewell to H. J. Rose, 7 Sept. 1818, Whewell Papers, Trinity College Library, Cambridge University, R2 99[11].

[12] Anon. (1817), 555.

knowledge. External objects impinged upon sense organs sending impulses through nerves to the material brain where, through a means unknown and perhaps unknowable, they were sensed by the non-material mind. For this process of sensation, 'the Supreme Being, who made us and placed us in this world, has given us such powers of mind as he saw to be suited to our state and rank in his creation'.[13] In addition to contingent knowledge, the human understanding possessed a few demonstrative truths based on 'intuitive evidence'. 'The mathematical sciences, categorical logic, and that part of metaphysics which demonstrates the being of God, are therefore the only branches of human knowledge which admit of strict demonstration.'[14] The final part of the article considered mind and Mind, concluding with a section on the being and attributes of God. Our own mind is the only one with which we were '*intimately* acquainted', and its powers—sensation, consciousness, reason, and so on—clearly established it as a non-material entity distinct from body.[15] Animals also appeared to possess minds, indicating the existence of a range of minds 'with different kinds or degrees of power'.[16] The most powerful of these was that of God, whose necessary existence as the cause of us was 'founded on the intuitive knowledge which we have of our own existence, and therefore independent of all theories about the nature and reality of the material world'.[17] To help establish the *unity* of God, this a priori argument was supplemented both by 'the common arguments *a posteriori* drawn from the *order* and *harmony* of the world' and by the Bible.[18] Just as the existence of mind entailed the existence of Mind, the ordinary attributes of mind entailed the grander attributes of Mind. 'We find in ourselves such qualities as *thought* and *intelligence, power* and *freedom*, &c. for which we have the evidence of consciousness as much as for our own existence.'[19] We know we have not had them from eternity so they must have had some cause. 'Now this cause, as it must be *superior* to its *effect*, must have those perfections in a *superior* degree.'[20] Indeed it must be an unlimited degree, and therefore God was, for example, omniscient.

The article inherited and blended aspects of the tradition of Descartes, Locke, and others.[21] It presented metaphysics as a philosophical and especially theological discipline, which agreed with Scripture. It presented a basic metaphysical view of God, man, and nature that Whewell repeated: God created man with a mind essentially similar to, though less powerful than, His own—a mind suitable for experiencing and comprehending God's other creation, nature.

As Whewell expressed the view in 1825, pleasure indicated scientific truth. Just as God took pleasure in creating a simple universe, we took pleasure in

[13] Anon. (1817), 559. [14] Ibid. 604–5. [17] Ibid. 674.
[15] Ibid. 645. [16] Ibid.
[18] Ibid. 678. [19] Ibid. 681. [20] Ibid.
[21] But not Kant whose philosophy, of course, greatly influenced the development of Whewell's philosophy of science.

understanding it. The world was constructed so that one scientific conclusion, if it were the true one, could explain many phenomena. And our emotional satisfaction in contemplating such conclusions was itself, therefore, an indication of the truth of that conclusion. Whewell developed the argument in an unpublished essay, 'Reflexions on God', in which he rejected the notion that the existence of natural laws eliminated 'the necessity for a perpetually sustaining and preserving Providence in the world'.[22] 'The discovery of such laws', Whewell began his attack, 'is undoubtedly an employment which gives a strong and peculiar pleasure to a speculative mind.'[23] Their beauty lay in their generality and simplicity. In such laws, 'we see innumerable facts of the most remote and dissimilar kind arranging themselves under one class, ⟨and⟩ becoming as it were only different ⟨representations⟩ exhibitions of a principle which we represent to ourselves as the most simple & single imaginable'.[24] Contemplating them in all their simplicity arouses thoughts and emotions within us. We do not think they exist only in our minds, not operating in nature itself. 'We feel this irresistible conviction that they are truly & rightly deduced from nature . . .; and if this conviction is accompanied with a glow and an exultation as if we had ⟨with⟩ in it obtained something which indeed deserves the name of knowledge, we cannot but suppose that they have some ⟨other⟩ meaning & origin beyond ourselves.'[25] Indeed, since the laws are neither arbitrary nor capricious, 'it appears that their only meaning & reality consists in their existing in the mind of an intelligent being, and thus becoming the rules according to which he contemplates or acts'.[26] God has written the 'book of nature' in the language of these laws, 'and ⟨they⟩ we ⟨had⟩ can read it so far as our minds are of the same nature with his'.[27] Moreover, the greater the generality of our laws the closer we come to God's thoughts, for 'the laws by which he delights to act we delight to contemplate'.[28]

Themes from the 1825 manuscript ran through Whewell's writings on natural theology, morality, and the philosophy of science in the 1830s and 1840s. His mature views of the philosophy of ethics and the philosophy of science appeared, for the most part, in separate publications during the decade from 1837 to 1847. Before that divergence, however, Whewell showed in his Bridgewater Treatise of 1833 how they were, in reality, unified. Indeed, the convergence of ethical and scientific reasoning into the concept of God amounted to an application in theology of what Whewell later called a 'consilience of inductions' in science.

The primary task of Whewell's Bridgewater treatise was to employ physics and astronomy in supporting the design argument for God's existence. 'My prescribed object is to lead the friends of religion to look with confidence and pleasure on the progress of the physical sciences, by showing how admirably

[22] Whewell, 'Reflexions on God', 13 Nov. to 1 Dec. 1825, Whewell Papers, R18 9[13], p. 1. I have supplied page numbers for this six-page unpaginated manuscript.

[23] Ibid.

[24] Ibid. 1–2. Angle brackets indicate deletions in the manuscript.

[25] Ibid. 3.

[26] Ibid.

[27] Ibid. 4.

[28] Ibid. 6.

every advance in our knowledge of the universe harmonizes with the belief of a most wise and good God.'[29] Whewell accomplished part of his goal, for example, by introducing his views of the luminiferous ether, some of which are discussed in the next section. In going beyond its 'prescribed object', however, the Bridgewater Treatise picked up the thread of Whewell's ideas as it emerged from the 1825 'Reflexions' and stretched towards the later, more specialized *Philosophy of Science* of 1840 and *Elements of Morality* of 1845. First, advanced sciences were simple. 'The sciences which have at present acquired any considerable degree of completeness, are those in which an extensive and varied collection of phenomena, and their proximate causes, have been reduced to a few simple general laws.'[30] Second, man and nature dove-tailed harmoniously showing that they were created for one another by the same Being. 'The creator of the atmosphere and of the material universe is the creator of the human mind, and the author of those wonderful powers of thinking, judging, inferring, discovering, by which we are able to reason concerning the world in which we are placed.'[31] Third, given the interconnections between our physical understanding and our moral sensitivities, we must conclude, as Whewell's chapter title declared: 'The Creator of the Physical World is the Governor of the Moral World.'[32] Fourth, in comprehending nature and ethics, mind was thinking the thoughts of Mind. 'The Divine Mind must be conceived by us as the seat of those laws of nature which we have discovered . . . All the laws which govern the moral world are expressions of the thought and intentions of our Supreme Ruler.'[33] Finally, the existence of our well-founded knowledge of 'harmonious relation' in the physical world—along with the pleasure we derive from contemplating that harmony—practically guaranteed that a similar, not-yet-articulated system must exist in the moral world. 'It would be strange indeed, if, while the mechanical world, the system of inert matter, is so arranged that we cannot contemplate its order without an elevated intellectual pleasure; . . . the rational faculties and moral tendencies of man should belong to no systematic order.'[34] Thus, given Whewell's unifying metaphysical conception of an essential similarity between the human and Divine intellects, pleasure aroused by physical simplicity constituted evidence for the reality of an analogous moral simplicity.

When fully developed, Whewell's ethical theory attacked prevailing utilitarianism. Rather than the goal of happiness dictating right behaviour, innate moral principles dictated right behaviour which in turn would naturally produce happiness. In his *Foundations of Morals* of 1837, Whewell supported his position with Scripture. Romans 2: 14 informed us that there was 'a law written on men's hearts by God. There are in the mind internal powers, a

[29] Whewell (1834), p. vi. The first edition was published in 1833.

[30] Ibid. 367–8. Lest one think that Whewell's reference to laws in this quotation contradicts his view of causation discussed below, it should be noted that he gave as an example of what he had in mind the reduction of phenomena to 'the mechanical properties of the luminiferous ether', ibid. 368.

[31] Ibid. 257–8. [32] Ibid. 254. [33] Ibid. 379–80. [34] Ibid. 290.

natural bearing of parts, a fixed constitution, which are the most important works of our Maker's hand.'[35] Whewell asked rhetorically: 'How can we imagine that the Divine thoughts are not the perfect archetype of all which conscience approves and admires?'[36] Scripture may have revealed the presence of such moral ideas, but precise awareness of them only emerged over the course of history, through reflection upon the demands of conscience. The first volume of Whewell's *Elements* laid out these ideas in detail, concluding that the one fundamental truth was that 'Duty is the way to Happiness'. Unable to be established on 'merely philosophical grounds', however, this truth required religion—the topic taken up at the beginning of the second volume of *Elements*.[37] In the second volume, Whewell endorsed two arguments for God's existence: the a posteriori design argument, which he had discussed at length in his Bridgewater Treatise, and the a priori cosmological argument, a version of which had been supported by the author of the article on 'Metaphysics' in the *Encyclopaedia Britannica*.[38] Natural religion or theology, by proving the existence of a creator God, allowed us to see that 'the Structure of our Minds, as well as of our bodies, is the work of God the Creator'.[39] We could better understand that 'the Supreme Object of human action is Happiness' and that 'Moral Rightness leads to Happiness'.[40] Christian revelation deepened our understanding. In thought and deed, we should strive to emulate basic moral precepts, which were represented to the highest degree in God—benevolence, justice, truth, purity, and wisdom. 'To conform our minds to those Ideas, is to conform our character to the Character of God. To approach to this Character, is to approach to *the Image of God* . . . '[41] Hence, while it fell to reason to identify the innate principles of morality, revelation guaranteed that there were such and that following them generally produced happiness. To the extent that it knew and followed Mind's moral ideas, mind would be happy.

The analogues in science to moral principles like benevolence were Whewell's 'Fundamental Ideas' like space, time, and causation. These and other parts of Whewell's philosophy of science revolved around his central notion of the 'consilience of inductions'.

The historical-philosophical argument of the *Philosophy* maintained that the most powerful scientific theories were produced by consiliences of inductions. Separate inductions from a wide spectrum of diverse phenomena all pointed to the same conclusion. 'The cases in which inductions from classes of facts altogether different have thus *jumped together*, belong only to the best established

35 Whewell (1837*d*), 19.

36 Ibid. 26.

37 Whewell (1845*b*), i, sect. 565.

38 Ibid. ii, sects. 568–9.

39 Ibid., sect. 570.

40 Ibid., sects. 573–4.

41 Ibid., sect. 584. For a discussion of Whewell's views in the context of Victorian moral philosophy, see 'The Cambridge Moralists', in Schneewind (1977). Schneewind writes that, for Whewell, 'In science and in moral matters our power of intuition is directed upon Ideas in the mind of God', thus giving us, with limitations, 'some genuine knowledge of God' and 'a new assurance of "the Divine nature of the human mind" ', ibid. 102.

theories which the history of science contains.'[42] The two best examples were gravitational theory and the modern undulatory theory of light. Whewell divided each of these subjects into formal and physical components, corresponding to the laws of phenomena and their causes. His 'Inductive Tables' for these subjects diagrammed their historical progress, as recognition of laws and causes increasingly joined disparate phenomena into a unifying explanatory scheme. The celestial and the terrestrial converged in the theory of universal gravitation. A wide variety of optical phenomena came together in the undulatory theory. Gravity and transverse vibrations were thus the underlying causes of an enormous range of physical phenomena. 'The Consiliences of our Inductions give rise to a constant Convergence of our Theory towards Simplicity and Unity.'[43]

Whewell's metaphysical belief that emotional pleasure indicated scientific truth helped him to defend his methodology of consilience. The methodology required that knowledge of causes be possible, and this was a requirement that Whewell had to argue for, especially in opposition to Auguste Comte. Two parts of the argument were that necessary truths (i.e. the laws of motion) could be successfully derived from the Fundamental Idea of causation, and second, that scientists had, in fact, generally given causal explanations in their highly successful enterprise. More to my point, however:

> In many subjects the attempt to study the laws of phenomena, independently of any speculations respecting the causes which have produced them, is neither possible for human intelligence nor for human temper. Men cannot contemplate the phenomena without clothing them in terms of some hypothesis, and will not be schooled to suppress the questionings which at every moment rise up within them concerning the causes of the phenomena.[44]

The search for physical causes was thus required not only by the analogy of past science but also 'by the irrepressible speculative powers of the human mind'.[45] This theme continued in the *Philosophy's* concluding sections on the induction of 'ulterior' causes and the Supreme Cause. The search for the (ulterior) causes of already known causes (like gravity) was 'an inevitable result from the intellectual constitution of man'. Finding one cause, 'he still *desires* a further insight' into even deeper causes.[46] If restricted to the realm of science, however, this search for causes failed to produce emotional *satisfaction*. 'The strain of music from the lyre of Science flows on, rich and sweet, full and harmonious, but never reaches a close: no cadence is heard with which the intellectual ear can feel satisfied . . . In the utterance of Science, no cadence is heard with which the human mind can feel satisfied.'[47] As in music, the idea of a 'closing strain' was essential for our 'relish' of science. The Idea of Causation thus led to 'the assumption of a First Cause'.[48] Moreover, since we recognized

[42] Whewell (1840), ii. 230.
[43] Ibid. 239.
[44] Ibid. 269.
[45] Ibid. 270.
[46] Ibid. 580, my italics.
[47] Ibid. 580–1.
[48] Ibid. 581.

personality in ourselves, we must also attribute it to the First Cause, which though 'immeasurably elevated' above lesser causes 'must still include in itself all that is essential to each of them'.[49] According to Whewell: 'It would render our philosophy utterly incoherent and inconsistent with itself, to suppose that Personality, and Ideas, and Will, and Purpose, do not belong to the Supreme Cause from which we derive all that we have and all that we are.'[50]

A Comtean would have had obvious objections. First, he would presumably have denied that it was, in fact, impossible to contemplate phenomena without speculating on their causes. But even if it were impossible or difficult to do so, one should guard against that tendency and, in the end, put such causes to the side, recognizing only established laws as the valid parts of positive scientific knowledge. For a Comtean, any dissatisfaction with mere laws would simply have been understandable frustration resulting from our wanting to know more than we were able. Such frustration would prove the existence neither of a Fundamental Idea of Causation nor of a First Cause. Instead, it would prove the existence of strict boundaries to human knowledge. Whewell's arguments from the pleasure found in causal theories would merely demonstrate how difficult it always was to break with familiar customs. Whewell's reply, I think, would have been the theological metaphysics which went largely unstated in the *Philosophy*. Given God's existence, impossibility, or even difficulty (to concede a point for argument's sake), in doing without causes meant that that was not the way God intended us to comprehend His creation. Pleasure, accordingly, arose when our thoughts emulated God's thoughts, about the key causes around which He had constructed the natural world.

In summary, Whewell presented two, ostensibly independent approaches to mind–Mind similarity. The one, founded on revelation and moral conscience, disclosed innate principles of morality which, when obeyed, led to happiness. The other, relying on the history and philosophy of science, revealed a set of Fundamental Ideas. Whewell presented feelings of satisfaction as consequences of the presence in us of Fundamental Ideas and, therefore, as evidence that Fundamental Ideas did indeed exist. But this argument had force for Whewell, I would suggest, because such feelings functioned as implicit bridges between the two approaches. Approved by the *Encyclopaedia Britannica* article on 'Metaphysics', the Scriptural approach to God certainly established a similarity between mind and Mind. God did, after all, create man in His own image, and in true morality, said Whewell, we approximate the 'Image of Character'. Biblical guarantee of mind–Mind similarity would sanction feelings as indicators of truth: happiness indicated one's conformity to Divine morality; pleasure or satisfaction showed one's comprehension of Divine science. Whewell had employed pleasure in this way in his 1825 manuscript, 'Reflexions on God', which did not present the apparatus of Fundamental Ideas. In the *Philosophy*, I am suggesting, Whewell's earlier notion was retained and embedded in the

[49] Ibid. 583. [50] Ibid.

foundation for the theory of Fundamental Ideas. There, given its place in the development of Whewell's thinking, the notion provided Whewell with genuine support for the Idea of Causation, support independent of and perhaps stronger than that afforded by history and philosophy. Hence, though Whewell held up the philosophy of science as a kind of guide showing how to construct a philosophy of morality, it is also true that the religious views so bound up with his ethics guided him to legitimate arguments for a philosophy of science.

In this way, metaphysical pleasure reinforced the search for physical causes, and—to make the transition from metaphysics to physics—knowledge of causes was inextricably linked through the theory of consilience to the ideal of simplicity. Speculating on future scientific progress, Whewell tied the goal of convergence to the necessity of employing causal theories:

> It is not only allowable, but inevitable to assume, as the vehicle of heat and light, a medium possessing some of the properties of more familiar kinds of matter. But the idea of such a medium, which we possess, and on which we cannot but reason, can be fully developed only by an assiduous study of the cases in which it is applicable. It may be, that as science advances, all our knowledge may converge to one general and single aspect of the universe. We abandon and reject this hope, if we refuse to admit those ideas which must be our stepping-stones in advancing to such a point.[51]

A successful unification of physical processes, as envisaged by Maxwell, would, therefore, have provided Whewell with enormous metaphysical pleasure, indicating mind's comprehension of that which Mind created. Metaphysics must have been a driving force fostering support for unifying physical theories.

2. Physical Constraint

The question is, however, did Whewell actually adopt a unified view of physical phenomena? To answer, this section first considers work by Cambridge men around Whewell in order to help define the context of his early deliberations on physical unity. As the rest of the section discusses, Whewell certainly did deliberate on the issue, and his views changed, especially in response to the experimental researches of James David Forbes and Michael Faraday. The metaphysical desire for convergence was constrained by physical research.

Though Cambridge was noted for its mathematical studies of the Newtonian sciences of astronomy, mechanics, and optics, during the 1820s Whewell and others were also interested in the subjects of heat, electricity, and magnetism. One interested party was James Cumming, tenth wrangler in 1801 and professor of chemistry at Cambridge from 1815 to 1861. Another was Francis

[51] Whewell (1847), ii. 328. This quotation is contained in a section that Whewell added to the second edition to extend his argument against Comte. Another added section on knowledge of causes is: 'Of the Conception of a First Cause', ibid. i. 700–8.

Lunn, an 1818 Cambridge graduate who became a fellow of the Royal Society in 1819 and sponsored Whewell's successful candidacy for fellowship in 1820.[52] Lunn was introduced to readers of the *Encyclopaedia Metropolitana* (to which he contributed several articles in the 1820s) by H. J. Rose, editor of the encyclopaedia and correspondent of Whewell: 'In Cambridge, in a considerable circle of persons qualified to judge in these matters, his talents were justly appreciated, and his acquirements acknowledged to be of the highest order.'[53]

Early nineteenth-century chemistry courses generally contained a fair amount on the subjects of heat and electricity, and so it was not surprising that Cumming was interested in them. In 1823, he published an experimental paper 'On the Development of Electro-Magnetism by Heat'[54] and, four years later, a translation with additional notes of J. F. Demonferrand's *A Manual of Electro-Dynamics*, which had been recommended to him by Whewell. Cumming seconded Demonferrand's endorsement of Ampère's reduction of magnetism to electric currents. On the other hand, though he discussed 'the connexion between electrical and chemical action', he did not specifically support the reduction of chemical affinity to electric force.[55] It was too soon to decide about the relationship between heat and electricity in electric conductivity: 'Till the causes of heat and electricity are known, and of that peculiar constitution of matter, which excites the one, and transmits or propagates the other, our reasoning on this subject must be inconclusive.'[56] He was, however, willing to speculate on the possible unity of electricity and light:

> The immediate cause of electrical attraction and repulsion may be supposed to consist in the vibrations of an ether, similar to, if not identical with that, the existence of which has been imagined, to account for the phenomena of light. If there exist such a fluid, resulting from the reunion of the two electricities, it must be universally diffused, nearly incompressible, and capable of permeating all bodies.[57]

A similar mix of caution and speculation appeared in Lunn's encyclopaedia articles on 'Heat', 'Chemistry', and 'Electricity', the mathematical portion of the last being written by Whewell. Lunn thought that it was not yet established whether electricity and heat consisted of imponderable fluids or were properties of ordinary matter. For 'convenience', he wrote of electricity and caloric as imponderable fluids.[58] He thought that electricity and galvanism 'are but modifications of the same agent' and that they 'are now clearly proved to exert an overruling power in magnetic phenomena'.[59] The heat generated by an electric current may simply have been forced out of the wire by the mechanical

[52] Lunn's is one of the signatures on Whewell's certificate of candidature. Cumming's signature is on Lunn's. Both certificates are in the Royal Society.

[53] Rose (1845), p. xv. Individual articles were written and circulated earlier than 1845. Though Lunn's contributions carry no dates, the plates accompanying them do (see nn. 58 and 61).

[54] J. Cumming (1823).

[55] Demonferrand (1827), 262–6.

[56] Ibid. 291.

[57] Ibid. 248.

[58] Lunn (1824–5, 1826). Plates accompanying 'Electricity' are dated 1824 and 1825, 'Heat' 1826.

[59] Id. (1824–5), 55.

action of the material electric fluid, or it may be that the heat was just the electric fluid itself in a new form.[60] Lunn discussed the similarities of radiant heat and light, concluding: 'The identity of heat and light is however yet far from being established, and there is great room for a well directed series of experiments on this subject.'[61] Whether or not electric action was the cause of chemical affinity 'is a question which the Chemist is deeply interested in resolving'.[62] Lunn explained that heat, light, electricity, and magnetism were being treated in the *Encyclopaedia Metropolitana* as separate divisions, but by no means insisted on their permanent separation: 'To what extent the researches of future times may prove all these agents to be modifications of some universal all-pervading principle, it is not for us to determine.'[63]

Hence, not only did Cumming and Lunn write about the non-Cambridge subjects of heat, electricity, magnetism, and chemistry, they also considered various convergences among them. Moreover, they largely agreed on the latter topic. They both thought that magnetism had been reduced to electricity, but were wary of equating electric force with chemical affinity or of assuming that the same imponderable fluid underlay both electric and thermal phenomena. Lunn showed similar scepticism about heat and light (not discussed by Cumming), while Cumming seemed favourably inclined towards a convergence of electricity and light (not discussed by Lunn). In the book that Whewell persuaded him to write on electricity and magnetism, to render those subjects suitable for the mathematical tripos, Robert Murphy in 1833 pointed out connections between electricity, on the one hand, and chemistry, light, and heat.[64] Experimental researches, Murphy wrote, seemed 'completely to establish an identity between electricity and magnetism'.[65] Shaped within this Cambridge context, Whewell's views on these matters begin to become apparent to us around 1830.

In what may have been a preliminary study for his Bridgewater Treatise, Whewell wrote an extended essay on 'The Undulatory Theory of Light' around 1830. Though the high status of optics was 'only of a few years standing', it had joined mechanics and astronomy as 'the only sciences in which a multiplied and varied ⟨mass⟩ assemblage of phenomena have been traced up to simple and general causes so as to leave us nothing to desire on this head'.[66] Not only did the essay explain how the undulatory theory explained the assemblage of phenomena, it concluded with five ways that the theory reflected and supported 'The Theistical View of the Universe'—'the main

[60] Lunn (1824–5), 99.

[61] Id. (1826), sect. 273.

[62] Id. (1827). The plates accompanying this article are dated 1827.

[63] Ibid. iv. 587. These writings by Lunn and Cumming further support Becher's thesis that much attention was paid to science at Cambridge in the first half of the nineteenth century, including areas of science not represented on the mathematical tripos, Becher (1986).

[64] Murphy (1833), 35–7, 103, 129.

[65] Ibid. 134.

[66] Whewell, 'The Undulatory Theory of Light', Whewell Papers, R18 17^7, p. 53. The page numbers (52–215) are on the manuscript itself. Generally, Whewell has written only on the odd-numbered pages.

purpose and object here, of the preceding speculations'.[67] The five arguments depended heavily on the ether:

> If the existence of an ether, the means of diffusing light through the world, be ⟨proved by⟩ established by proofs of the same kind as those which ⟨support the⟩ shew the existence of gravitation . . . it is then as right and reasonable for the believer in God to take into account the former truth as the latter, ⟨in⟩ when he thinks concerning God's government of the world.[68]

The key words for Whewell's five arguments were *simplicity, scale, full, variety,* and *harmony*. Whewell thought that 'we have in the existence of the luminiferous ether, the same *simplicity* of means which we find in all other parts of nature. What more simple than a single elastic fluid pervading all space', giving rise to all the phenomena of light.[69] Moreover, contrasted to ordinary matter, the ether possessed an extreme rarity and an enormous elasticity. Such a *scale* of quantities in nature could only increase our admiration for God, who 'is not limited in his means by numerical magnitude. Billions of billions are as easily managed by him as dozens by us.'[70] Since the universe must be *full* of this tenuous, elastic ether, we have 'fresh ground to believe that the quantity of contrivances goes incalculably beyond anything we can at present conjecture'.[71] Furthermore, knowledge of the ether with its peculiar properties helped refute the view that concerned Whewell in his 1825 essay, 'Reflexions on God'—that in a material world of necessary properties and laws, God was superfluous. On the contrary, the existence of a *variety* of properties for matter indicated that *choice* was involved, especially when the natural world was considered with respect to man. For example, regarding the differences between sound and light: 'We must allow then that the medium and the mode of vibration in each case are adjusted to the ⟨sense?⟩ instruments of perception so as to make those do their proper office. The machinery of light and of sound are matters of contrivance and adjustment.'[72] Finally, since light obviously was closely 'connected' with heat, electricity, and chemical action, 'there is a remarkable *harmony* prevailing among them . . . The ether must have such properties as enable it to act and be acted upon when such agents are present.'[73]

Given such theologically directed remarks on simplicity, as well as his metaphysical approval of unifying theories, it is hardly surprising that Whewell would also consider the possibility of a unified physical world. He did, but cautiously. Contemplating the universe's fullness, Whewell wrote that the ether 'seems susceptible only of two kinds of motion, a transfer of its parts and a vibration', analogous to the movements of air in wind and sound. Because it seemed 'improbable' that heat, electricity, and magnetism could be reduced to such movements, Whewell concluded that the ether could not be responsible for such phenomena and that the universe must therefore be interlaced by several imponderable agents.[74] Indeed, in his portion of the article on 'Electricity',

[67] Ibid. 193. [68] Ibid. [69] Ibid. 195. [70] Ibid. 201.
[71] Ibid. 203. [72] Ibid. 209. [73] Ibid. 211–13. [74] Ibid. 203–5.

Whewell had taken a realistic view of electric fluids, in contrast to Lunn's strict neutrality. Whewell prefaced his mathematical treatment of electricity with the explanation that various electric phenomena led to 'the supposition of two electric principles or fluids; each of them consisting of particles which repel those of the same fluid, and attract those of the other'. If this general theory, when developed with mathematical exactness, still agreed with the intricacies of electric phenomena (as his part of the article showed to be the case), 'it will be difficult not to believe that the theory truly expresses the laws according to which nature operates'.[75] We know, Whewell wrote in the optics manuscript, that the universe must be 'still fuller' than astronomy and optics by themselves would lead us to conclude, even though simple theories in other physical subjects had not yet been formulated.[76] Though Newton, in concluding his *Principia*, had suggested that 'some one fluid or ether may account for the properties of light, of heat, of cohesion, of electricity, of sensation and of voluntary motion', Whewell could not conceive that this was 'probable'.[77] As he pointed out, however, such a unified, Newtonian viewpoint would not have undermined his theological arguments. In fact, what Whewell's natural theology would have lost in fullness (in a still full universe), it would have gained in the universe's greater simplicity. However, despite the direction of his theological metaphysics, the significance of the ether, and hints of convergence, Whewell stiffly resisted the vision of a physical world unified in ethereal activity.

Whewell's Bridgewater Treatise of 1833 conveyed essentially the same view of these matters as had the optics manuscript. The book was not restricted to optics, of course, but it repeated Whewell's arguments involving such properties as variety and scale that had appeared earlier in the optics manuscript. Newton's first law of motion and his law of gravity were universally true, for example, but neither was necessarily true. Since they therefore must have been *selected* to be as they were, they provided evidence of God's design in the universe.[78] In addition, realizing the enormous scale that exists in the universe—in its distances and the densities of its materials, for instance—kept us from mistakenly 'supposing his [God's] powers to have any limits'.[79] The ether retained its prominence. Still conceived by Whewell as a substance of extreme tenuity and great tension, the ether was a substance whose 'offices are so numerous and so beneficent'.[80] Its functions fitted harmoniously with those of ordinary matter. 'How should the properties of these three great classes of agents, visible objects, air, and light, so harmonize and assist each other, that order and life should be the result? . . . We conceive there can be but one answer; a most wise and good God.'[81]

Though largely unchanged, Whewell's views of the reality of the electric fluid and the possible unifying role of the ether seemed to be modified

[75] Whewell, part 2 of Lunn (1824–5), 140.
[76] Whewell Papers, R18 17[7], pp. 203–5.
[77] Ibid. 210.
[78] Whewell (1834), 222, 233, 238.
[79] Ibid. 276.
[80] Ibid. 134.
[81] Ibid. 80.

somewhat. Whewell declared that 'every one knows that a thunder-cloud is strongly charged with the electric fluid, (if it be a fluid,) and that the stroke of the lightning is an electrical discharge'.[82] Moreover, he thought the connection between magnetism and galvanism was so close 'that they may be said to be almost different aspects of the same agent'.[83] And electricity and galvanism seemed to differ from one another only 'in the same manner in which a fluid at rest differs from a fluid in motion'.[84] Whewell's chapter on 'The Ether' listed the connections of light with heat, electricity, galvanism, magnetism, chemistry, and vegetable life. All of these connections indicated that the ether must possess many contrivances, of most of which we were ignorant. Thus repeating his earlier view of a universe full of contrivances, Whewell now seemed marginally more willing to place the non-optical contrivances in the optical ether, rather than in separate imponderable fluids. He was far from insistent on such a view, however, for though connections existed among them, 'whether heat, electricity, galvanism, magnetism, be fluids; or effects or modifications of fluids; and whether such fluids or *ethers* be the same with the luminiferous ether, or with each other; are questions of which all or most appear to be at present undecided, and it would be presumptuous and premature to take one side or the other'.[85] Hence, though still quite sceptical about the unifying role of the ether, Whewell in 1833 appeared to accept magnetism and electricity as manifestations of the same agent—presumably the (probably) real electric fluid—and to be willing at least to consider the possibility that these and other phenomena were unified in the (clearly) real luminiferous ether.

By 1835, Whewell was even more confident of his views regarding electricity and magnetism. In the first place, he now unequivocally proclaimed the reality of the electric and magnetic fluids. In a report to the British Association, Whewell explained that Newton's first rule of philosophizing called for the acceptance only of *verae causae*, and of no more of them than required. Taking a true cause to be one 'already supposed to be known by mechanical effects', Whewell maintained that

> the existence of some electric fluid (whether one or two) as a true cause of the phenomena may, on this view, be held to be proved by the facts which suggested such a conception from the first;—by the accompaniments of the discharge,—the spark,—the sound; and especially the mechanical effects,—the shock, the power of striking, breaking and penetrating material objects.[86]

Given this 'physical truth', one should choose the two-fluid theory over the one-fluid because it involved 'the simplest system of assumptions under which the phenomena can be explained'.[87] Until recently, there had been less evidence for the reality of magnetic fluids. There was no magnetic phenomenon like an electric discharge, for example. But that situation had changed. 'More recently, indeed, we have had the magnetic fluid proved to be real, and

[82] Ibid. 111. [83] Ibid. 114. [84] Ibid.
[85] Ibid. 140. [86] Id. (1835*b*), 6. [87] Ibid.

connected with the electric fluid by most curious and unforeseen relations, but of these it is not our business now to speak.'[88]

In the second place, Whewell wrote to Faraday in September of 1835 that he supported Ampère's unified view of electricity and magnetism. He thought that Ampère 'did well to introduce the word *electrodynamic* action, to include electro-magnetic, voltaic, and magnetic, after he had shown that the three are identical'. He urged Faraday to adopt the word *electrodynamic* and the theory it embodied.

According to his theory the *convertibility* of electricity and magnetism, of which you speak (1114) as a problematical point, is a fundamental fact. And it seems to me that taking his theory, and supposing the electric fluid to have inertia, the only point in this part of the subject, which remains obscure, is the law of electrodynamic induction properly so called, that is the force by which a current produces a current in a neighbouring substance, whether a current be a vibration, a strain, or anything else.[89]

Faraday replied that he still had 'a little reserve' about Ampère's theory and so clung to 'the relations of Electric & Magnetic forces' as a simple and convenient means for an experimentalist to refer to phenomena 'readily distinguished in practise'.[90] Looking forward to an eventual agreement between them, Whewell's answer to Faraday introduced the chemical context of Faraday's work. He regarded Faraday's electromagnetic researches 'as the greatest event which ever happened in the history of chemistry'. Indeed, Ampère's 'beautiful' and 'well established' theory was 'only a small part of your magnificent subject'.[91]

In 1835, therefore, Whewell appeared to accept a *real* electric fluid as a *unifying* agent for much of physical theory. Faraday's experiments on electromagnetic induction greatly extended the scope of Ampère's basic theory and were properly described by his term *electrodynamic*. Moreover, Faraday's electrochemical researches were extending the electrical view even further.

By the time Whewell published his *History of the Inductive Sciences* in 1837, recent experiments had roughly jolted his attitude towards the reality of unseen fluids, thus affecting his views in several areas of the physical sciences. In contrast to its importance in the Bridgewater Treatise only four years earlier, for example, the ether declined in prominence, not only as a unifying factor in optics but also as a possible unifying entity in physics generally.

The crucial experiments were J. D. Forbes's on radiant heat. Before Forbes, the generally accepted view of heat, according to Whewell, was the material theory of heat combined with an emission theory of radiant heat. Heat consisted of a special substance, caloric, and radiant heat consisted of particles of caloric emitted from radiating objects. By demonstrating that heat, like light, could be refracted and polarized, Forbes's work undermined the caloric

[88] Whewell (1835*b*), 12.

[89] Whewell to Faraday, 9 Sept. 1835, Faraday (1971), i. 293–4. Whewell's reference (1114) is to Faraday (1849), sect. 1114.

[90] Faraday to Whewell, 19 Sept. 1835, ibid. 295.

[91] Whewell to Faraday, 25 Sept. 1835, ibid. 297.

theory. Like light, heat seemed to consist of vibrations, not matter. A recent theory by Ampère did indicate how the conduction of heat might be explained by a vibration theory of heat, even though conduction phenomena seemed so closely to resemble the flow of a material fluid. On the other hand, a vibratory theory did not so well explain specific heat and latent heat. Thus, from a rather settled view, Forbes's results had thrown the science of heat into a decidedly unsettled state, one in which there existed no good general theory of the causes of heat. At present, there was no consilience of inductions that could establish any unifying conception of thermal phenomena.[92]

These developments influenced electric theory, as Whewell explained in a section entitled '*Question of the Material Reality of the Electric Fluids*'. The history of caloric should teach us not 'too readily' to trust evidence that seemed to disclose 'the real state of things'.

It is a curious instance of the mutual influence of the histories of two provinces of science, but I think it will be allowed to be just, to say that the discovery of the polarization of heat has done much to shake the theory of the electric fluids as a physical reality. For the doctrine of a material caloric appeared to be proved (from the laws of conduction and radiation) by the same kind of mathematical evidence (the agreement of the laws respecting the elementary actions with those of fluids,) which we have for the doctrine of material electricity.[93]

It would not be 'contrary to precedent', therefore, if the theory of electric fluids, though it represent the laws of electric actions, 'should yet be fallacious as a view of the *cause* of the actions'.[94] Even the electric spark was now insufficient to guarantee the reality of an electric fluid. 'A moment's consideration' would show that none of the mechanical effects accompanying sparks 'can really prove material currents, rather than vibrations, or other modes of agency'.[95] The vibrations of sound, for example, could produce the mechanical effect of breaking glass.

Nor did the magnetic fluid survive the jolt to Whewell's realism. 'If we have seen reason to doubt the doctrine of electric fluids as physical realities, we cannot help pronouncing upon the magnetic fluids as having still more insecure claims to a material existence'.[96] Like the hypothesis of an electric fluid, that of a magnetic fluid explained many phenomena, but there was no magnetic phenomenon like the electric spark to suggest mechanical effects caused by a fluid. Its claims to reality had thus traditionally been weaker than those of the electric fluid, and clearly could no longer be upheld.

As in his correspondence with Faraday two years before, Whewell supported Ampère's reduction of magnetic effects to electric currents. Ampère, Whewell

[92] Whewell (1837*a*), ii. 524–34. [93] Ibid. iii. 39–40. [94] Ibid. 40.

[95] Ibid. 41. In a footnote to his discussion of electricity Whewell cited his report to the British Association in 1835, writing: 'I may seem there to have spoken more favourably of the Theory as a Physical Theory than I have done here. This difference is principally due to a consideration of the present aspect of the theory of heat.' Ibid. 42 n.

[96] Ibid. 61.

wrote, had 'clearly established . . . that magnetic action is a peculiar form of electromotive actions . . . Electricity and magnetism are identical.'[97] Again extending Ampère's theory and his term *electrodynamic* to Faraday's researches of the early 1830s, Whewell described electromagnetic induction as 'electrodynamic reaction', Faraday's results being seen by Whewell (in some disagreement with Faraday, he pointed out) as instances in electricity of Newton's mechanical law of action and reaction.[98] It was a reduction to electricity, not magnetism; it was electrodynamics, not magnetodynamics.

Moreover, electrical studies provided grounds for another convergence—a 'convergence of mechanical and chemical theory'.[99] Relying heavily on Faraday's recent researches in electrochemistry, Whewell approvingly quoted Faraday's conclusion that 'electricity is only another mode of the exertion of chemical forces'.[100] To be sure, others—notably Faraday's mentor Sir Humphry Davy—had speculated that such was the case, but it was due to Faraday that 'the great principle of the identity of electrical and chemical action was completely established'.[101] Faraday's confirmation of Davy's guesses was like Newton's of Hooke's regarding gravity or Young's of Huygen's regarding the undulatory theory of light.[102] Hence, the mechanical effects of electricity (including magnetism) converged with chemistry through Faraday's reduction of electric force to chemical affinity. Electricity and magnetism were, in Whewell's phrase, the 'mechanico-chemical' sciences.

Unlike some of Whewell's earlier writings, his *History* did not speculate on the ether as the seat of converging physical theory. So prominent earlier, the ether now received less attention altogether. Whewell did still portray light as transverse vibrations in a fluid ether and took care to state that the concept of transverse vibrations in a fluid violated no mechanical principle. He also explained that hypotheses about the distances between particles of ether offered hope that dispersion—a problem for the undulatory theory—might be explained. But the unifying role in optics went more to the transverse vibrations themselves. Whewell wrote that at that stage in the history of the subject when phenomena relating to polarization and double refraction appeared so intractable, 'the conception of transverse vibrations came in, like a beam of sunlight, to disclose the possibility of a mechanical connexion of all these facts'.[103] With the theory of transverse undulations, when 'the phenomena [become] more numerous, more various, more strange:—no matter: the theory is equal to them all'.[104] Whewell formalized the significance of transverse vibrations three years later in his *Philosophy*. Though light still consisted of undulations in a fluid ether, Whewell's 'Inductive Table of Optics' displayed a consilience of inductions depending on vibrations more than the ether.[105] Or,

[97] Whewell (1837*a*), iii. 91, 95. [98] Ibid. 110–11. [99] Ibid. 6.
[100] Ibid. 169; Faraday (1849), sect. 918. [101] Whewell (1837*a*), iii. 169.
[102] Ibid. 169–70. [103] Ibid. ii. 419. [104] Ibid. 428.
[105] Whewell (1840), i. 302–4; ibid. ii, between 282 and 283.

as one of Whewell's additional notes to the 1847 edition of his *History* stated about consilience in optics:

> The phenomena of fringes of shadows and coloured bands in crystals *jump together* in the Theory of Vibrations. The phenomena of polarization and double refraction *jump together* in the Theory of Crystalline Vibrations. The phenomena of polarization and of the interference of polarized rays *jump together* in the Theory of Transverse Vibrations.[106]

All of this impressive unity effected by vibrations was limited, however, to optics. In neither the *History* nor the *Philosophy* did Whewell suggest the convergence of optics with other physical sciences. And in his additional notes to the 1847 edition of the *History*, he explicitly rejected chances to do so. Light, for example, had connections with both chemistry and heat. Light produced chemical effects, and the colours of bodies were associated with their chemical composition. But these chemical connections lay outside the province of the undulatory theory of light. Demonstrated analogies between light and heat were much more extensive in 1847 than a decade earlier. Such analogies suggested an undulatory theory of heat according to which radiant heat was 'propagated by the same machinery' as light. Unlike Forbes who affirmed such a theory, Whewell was wary:

> Yet such a Theory has not yet by any means received full confirmation. It depends upon the analogy and the connexion of the Theory of Light, and would have little weight if those were removed. For the separation of the rays in double refraction, and the phenomena of periodical intensity, the two classes of facts out of which the Undulatory Theory of Optics principally grew, have neither of them been detected in thermotical experiments.[107]

Indeed, the undulatory theory of light 'professes only to explain the phenomena of *radiant visible* light'. It was neither a theory of heat nor of chemical composition.[108] Faraday's discovery in 1845 of magneto-optical rotation had 'shown that the magnetic polarity is directly connected with that optical polarity by which the plane of polarization is affected'. However, it evidently did not suggest to Whewell, as it did to Thomson, a way of conceiving the convergence of magnetic, electric, and optical processes of nature.[109]

Consequently, over the years ether and electricity had rather opposite fortunes in Whewell's continuing inspection of physical theory for signs of convergence. Gravitational theory set the criteria by which such fortunes would be determined. It was a theory supported by a consilience of inductions in which one true cause explained many diverse phenomena. An even greater convergence, resulting from an even wider consilience, would require a cause no less real than gravity, but one that combined more branches of physical theory than gravity did. In these terms, the luminiferous ether was a true cause, but one whose unifying potential Whewell estimated in different ways at

[106] Whewell (1847), ii. 507.
[107] Ibid. 601–2. [108] Ibid. 510. [109] Ibid. 511.

different times. In 1830, the ether unified optics but was rejected for a possible wider unifying role. In 1833, the ether was prominent for its connections to many other parts of nature, and Whewell allowed it at least the possibility of a wider unifying role. By 1837, however, Forbes's experiments had weakened the ether's position in optical theory and evidently caused Whewell to ignore it as a source of broader convergence. The ether retained its reality but declined as a potential unifier. Conversely, as the electric fluid declined in reality, electricity more and more became the focus of a general unification. If Whewell had concluded that Faraday's magneto-optical rotation heralded further convergence, one senses that he would have predicted the reduction of light vibrations somehow to electricity, not electrical and magnetic phenomena to the ether. At any rate, there was no satisfactory convergence on a grand scale. The ether was a real cause that pervaded the universe, but it was not a source of unity. Significant unity did seem to exist in electricity, where, however, no real cause was identifiable. There may have been promise, but there certainly was nothing yet that warranted formal presentation in an Inductive Table.

3. Conclusion

By the mid-1840s, Whewell had been grappling with modern physical theory for a quarter of a century. His ideas on convergence resided within a particular historical context and a particular metaphysical framework. In noting connections and speculating on possible unities among areas of physics, Whewell continued the legacy of his Cambridge association in the 1820s with Lunn and Cumming, and he responded to the experimental findings in the 1830s of Forbes and Faraday. The primary importance of convergence in physical theory flowed from his metaphysical view of human knowledge as emulation of God's logical thought. In the end, however, physical constraint severely circumscribed metaphysical pleasure. If Whewell did envisage a 'great comprehensive theory', he did not claim to have found it by 1847. The considerable convergence he regarded as established fell far short of an all-embracing, all-simplifying unification. Occasionally, where others supported convergence, he hesitated. Though metaphysics was necessary, it was hardly sufficient for scientific knowledge. As Whewell declared in defining the importance of metaphysics:

> The cultivation of ideas is to be conducted as having for its object the connexion of facts; never to be pursued as a mere exercise of the subtilty of the mind, striving to build up a world of its own, and neglecting that which exists about us. For although man may in this way please himself, and admire the creations of his own brain, he can never, by this course, hit upon the real scheme of nature. With his ideas unfolded by education, sharpened by controversy, rectified by metaphysics, he may *understand* the natural world, but he cannot *invent* it.[110]

[110] Whewell (1840), ii. 526.

Thus, if God did create the world around an idea of a single cause whose effects permeated all of physical nature, man had not yet clearly conceived the idea.

Given Whewell's own careful thoughts, we should ourselves be wary of too readily assuming that early Victorian Cambridge guided her students towards great comprehensive theories. Consider James Challis, senior wrangler in 1824 and, as Plumian professor from 1836, the chief Cambridge lecturer of the time on physical subjects. Though he did eventually develop a unified conception which tried to reduce physical phenomena to motion in an underlying fluid ether, that view was present neither in his lectures nor his publications of the 1830s and 1840s. William Hopkins, dominant coach at that time for the mathematical tripos, did not drill his charges in such subjects as electricity and magnetism and their connections. Challis and Hopkins reflected the main thrust of the tripos itself. Its questions displayed no special concern for unity—not even in those few on electricity and magnetism, subjects that Whewell had urged into the examination. Nor was unity a concern of Whewell's questions for the Smith's prize examinations, where examiners were less bound by Cambridge custom than on the tripos. Instead tripos and Smith's prize questions on electricity and magnetism asked for presentation of mathematical theory, not deliberation over physical unity. In overwhelming proportion, publications of tripos graduates dealt neither with chemistry, electricity, magnetism, and heat nor with possible convergences among them. To be sure, Whewell's prominent writings discussed these subjects and conveyed a desire to unify them. But, it was a desire carefully hedged, in Whewell's own writings, and it was a desire that failed to penetrate tripos studies. To the extent, therefore, that Whewell's works may have helped direct William Thomson's young mind into fields that ought to have been cultivated in that season, one would guess that Thomson's mind, unlike those of most other Cambridge undergraduates, had been prepared to go beyond the standard bounds of tripos studies by his previous exposure as a Glasgow undergraduate to the subjects of chemistry, heat, electricity, and magnetism.[111] Whatever the exact influences on his thinking, however, his bold attempt at unification stood in sharp contrast especially to the official Cambridge scheme, but also to Whewell's ideas about convergence.

Returning to Maxwell's retrospective account of the 1840s, it is instructive to compare Grove's *Correlation of Forces* not, as Maxwell did, to Somerville's work, but to Robert Chambers's *Vestiges of Creation*, published at about the same time as Grove's book. Writing in the 1870s about the intellectual climate within which his *Origin of Species* was published in 1859, Darwin observed:

It has sometimes been said that the success of the *Origin* proved 'that the subject was in the air,' or 'that men's minds were prepared for it.' I do not think that this is strictly

[111] This significance of Glasgow in shaping Thomson's ideas is also argued in D. B. Wilson (1985) and (1987).

true, for I occasionally sounded not a few naturalists, and never happened to come across a single one who seemed to doubt about the permanence of species.[112]

After the success of Darwin's theory, the presence in the 1840s of a book like *Vestiges* may understandably have made it seem that evolutionary ideas had been more in the air or more favourably received than Darwin had actually found to be the case. We should not insist too literally on the analogy with biological thought. But we can remark not only that leading scientists, including Whewell, harshly criticized *Vestiges* but also that Grove's optimistic belief in correlations contrasted significantly with Whewell's viewpoint and, even more so, with the standard Cambridge outlook of the time. Just as Chambers's book enthusiastically presented an extensive evolutionary theory, Grove optimistically spoke of extensive physical correlations: 'Light, Heat, Electricity, Magnetism, Motion, and Chemical-affinity, are all convertible material affections; assuming either as the cause, one of the others will be the effect.'[113] However Grove's work may have influenced Maxwell or Thomson, it did not reflect Whewell's assessment of physical theory in the 1840s. If Whewell's and Grove's works indicated that a comprehensive convergence were desired or in the air, Whewell's, at any rate, certainly did not point to such a theory as a foregone conclusion. In fact, in biology and in physics, respectively, Whewell's views conflicted with Chambers's and, to a lesser extent, Grove's to define what Whewell obviously regarded as responsible scientific positions. And they were. From his perspective of the 1840s, the ultimate success of a conceptual unification of physics must have appeared, if not so outrageously implausible as that of evolutionary theory, nevertheless difficult to predict.[114]

[112] Barlow (1958), 123–4.

[113] Grove (1862), p. xii. This is a quotation from a lecture that Grove gave in 1842.

[114] And, one might add, difficult for Whewell to recognize when it came along. In his 'Additions to the Third Edition' in Whewell (1857), Whewell notably did not latch on to the new energy theory as a principle of unification. In fact, he restricted his comments in this area to the dynamical theory of heat which, though promising, was not quite established in Whewell's judgement: 'It is difficult to make these experiments [by Joule and Thomson] so as to be quite satisfactory; for it is difficult to measure *all* the heat gained or lost in any of the changes here contemplated.' Whewell (1884), ii. 608–9. Instead of unification through energy conversions—which Grove thought his idea of correlations of forces had foreshadowed—Whewell continued to emphasize 'the Connexion of co-existent Polarities', ibid. ii. 622. Presumably, Whewell would have said that, despite his high regard for unifying theories, his proper appreciation of standards of evidence kept him from prematurely jumping to conclusions on the question of energy, as Thomson had.

10. *The History of the Earth, Life, and Man: Whewell and Palaetiological Science*

by M. J. S. Hodge

1. Introduction: The Older and the Newer Historiographies

ANY part of Whewell brings us to confront the whole. But Whewell as a whole is a daunting prospect. He was a very big man was Whewell, perhaps a great one; and they do not make them like that any more. Only by grafting Cassirer, the historical philosopher, on to Conant, the scientific educator, could we have in our day a comparable combination of intellectual stride and institutional clout, of polymathic savant and academic supremo. But no such reflections can help us as historians. Great man theories of history, let alone big man theories, are in no better repute than myths of giants who walked the earth, or at least the quadrangles, in distant days.

What can help us is the Whig tradition in the writing of history, and the reactions that tradition has provoked in recent decades; especially we can be helped if we can learn to be as critical of the fashionable, newer reactions as we are of the unfashionable, older tradition itself.[1]

Whig history has taken many forms, obviously; but most of them require the science of the early nineteenth century to be inspired by the same spirit of rational progress that inspired the partisans of reform in religion, education, and government. Whewell has, then, always seemed easy to place within such a historiography. Even the dutiful volumes of Todhunter and Douglas make implicit concessions to the Whig tradition in the muted, apologetic hagiography they offer for their Tory, Anglican, Idealist hero.[2] Early triumphs are seen to give way to the failings of a later loser: most conspicuously a loser to Mill and to Darwin, and even to young men bent on transforming Trinity College itself. Meanwhile, more willing guardians of the Whig heritage, from Huxley on,

My understanding of Whewell and his views on palaetiological science has benefited greatly from suggestions made by Geoffrey Cantor, John Christie, Menachem Fisch, Larry Laudan, Rachel Laudan, Michael Ruse, and Simon Schaffer.

[1] A good introduction to the Whig tradition is provided by Chris Wilde in the entry on 'Whig history' in Bynum (1981).

[2] Todhunter (1876); Stair-Douglas (1881).

have found in Whewell a natural foil for the vaunted Victorian vindications of reason and reform in empiricism, evolutionism, and liberalism.[3]

It is here that we can so easily go wrong. A strong sense of the indefensible preconceptions in all Whig views of Whewell can make it hard for us to see the limitations in those interpretations of the last three decades, such as Cannon's, that have been adopted in deliberate opposition to the historiographies of Huxley and the rest.[4] Anything, we may presume, anything so resolutely opposed to something so thoroughly unacceptable must itself need no reappraisal. However, the fallacy in this presumption, once the presumption itself is caricatured quite tactlessly, is obviously resistible. There should, then, be agreement that three decades of anti-Whig historiographies for Whewell ought now to feel too long for uncritical comfort.

One reason for taking a wary stance, at this moment, is that the invaluable achievements of the last three decades may otherwise become indelible idols of a nascent profession. Historical, philosophical, and sociological studies of science have exhibited many of the customary concomitants of professionalization only since the 1950s. As a corollary of this timing, any viewpoint dating from this period is associated with the new, enhanced status of these disciplines; while, conversely, any expression of dissatisfaction with such a viewpoint may be taken, once again quite fallaciously, to be an invitation to abandon what has accompanied professionalization, and to return to the unsubtle errors once perpetrated by naïve amateurs.

Such fallacies are especially pertinent to any analysis of Whewell's writings on the history of the earth, life, and man. Because of their relevance to an understanding of what some have called the Darwinian revolution, these writings have been given close attention since 1959, the Darwin centennial year. On this ground, all of us who write today are massively indebted to Cannon and his ally, Hooykaas, and to those others who have learned from them.[5] What we have to learn now is to combine that indebtedness with an appreciation of what they cannot do for us, especially in understanding Whewell in relation to Darwin and, earlier, in relation to Darwin's mentor Lyell.

Cannon was not only repudiating the Whig tradition in general, but also, specifically, the views of Lyell and Huxley about the history of geology and biology. Unsurprisingly, Cannon had no trouble in refuting Lyell's legends about how geology had developed, nor in discrediting Huxley's contrasts between Lyell and Whewell. However, Cannon went on to propose that it was 'catastrophists' such as Whewell, and not the 'uniformitarians' led by Lyell, who were being reasonable and empirical rather than dogmatic and arbitrary about the course and causes of change at the earth's surface. What is more, Cannon argued that the orthodox Anglican, natural theology of a catastrophist such as Whewell made him more a 'precursor' of Darwin than any early

[3] See, especially, Huxley (1887).

[4] W. Cannon (1960); id. (1960*a*); id. (1961*a*); id. (1978).

[5] Hooykaas (1959).

'evolutionist' such as Lamarck. Faced with this historiographical alternative to Huxley, we have to conclude, then, that Cannon did not move completely away from distinguishing good guys, who were undogmatic and empirical and were precursors of Darwin, from bad guys who were not. Hats have been switched around, but the old questions, as to who should wear the black and who the white, are still being asked and decided on the old criteria.[6]

Now, Cannon was perfectly aware that these dialectical switches were often yielding not a radical Tory history, much less a Socialist one, but only mirrored inversions of the old Whig epics. For a principal aim of Cannon in overturning longstanding wisdom about the Whewells and the Lyells of the 1830s was to prepare us for the irony in the tragedy, as Cannon saw it, of what happened in 1859. On Cannon's account, Darwin was eventually to confront a venerable consensus about the coherence of science, morals, and religion, as upheld by the likes of Whewell; and Darwin, according to Cannon, was to confront this consensus with its very own universe, but now completed, in accord with its very own constitutive principles, by natural selection; it was an addition, however, that made this universe abhorrent to its original possessors because manifestly fatal to the coherence cherished in their consensus.[7]

There is, then, a further reason for being wary of this comprehensive Cannonical thesis as to the inception and reception of Darwinian science. For it continues the old Huxleyan habit of concentrating on those issues in the 1830s that are supposed to teach us most about the debates of the 1860s. Regardless, therefore, of whether we think Cannon right—about the relationship between Darwin's *Origin of Species* and Whewell's Bridgewater Treatise on *Astronomy and General Physics, Considered with Reference to Natural Theology* (1833) a quarter of a century before—we should still pause, before following an approach to the 1830s that is so often in the service of a thesis about what was to happen a generation later.

There are equally compelling reasons, although of a very different kind, for being cautious concerning any interpretations of Whewell deriving from his recent rehabilitation by philosophers of science. For this rehabilitation has often been encouraged by developments in philosophy of science itself, developments that can easily prompt inappropriate assumptions about the character of Whewell's work.

According to a familiar stereotypical Anglophonic narrative, three decades ago the philosophy of science was dominated by a combined commitment to logic and to experience marked by the label logical empiricism. Within this commitment there was, it seems, no place for history and much more sympathy for Hume than for Kant. Not surprisingly, therefore, we find Whewell benefiting, in attention and evaluation, from the widespread disenchantment, dating from the 1960s, with logical empiricist views. It is a disenchantment often associated, as for instance in Butts, with Kantian sympathies; or, as in

[6] W. Cannon (1961*a*). [7] S. Cannon (1978).

Curtis more recently, with a neo-Popperian concern for the rationality of progress as exhibited in the historical development of science.[8]

The trouble is that such philosophical rehabilitations of Whewell may bring inappropriately problematic contrasts—history versus formal logic, Kantian rationalism versus Humean empiricism—to the characterization of his work. These, and other contrasts, as currently influential, can divert us from what was concerning Whewell as he took up the issues his philosophy of science was designed to resolve. For the project consummated in the two volumes on the *Philosophy of the Inductive Sciences, Founded upon their History* (1840) was not originally aimed at deploying Kant and history to oppose any Humean legacy, nor was it aimed at dislodging the formal logic of the day from the analysis of science.

Quite generally, then, as we in the 1980s go back a century and a half to the 1830s, we should beware of perpetuating uncritically too much from several rewarding legacies from the sixties—the 1860s, that is, and the 1960s.

2. The French Revolution, the Industrial Revolution and a Cambridge Nationalist

One way to integrate our reading of Whewell with an understanding of his epoch is to start our thinking about the man with our knowledge of his strikingly successful career, and to start our thinking about his epoch by recalling what two revolutions, the French and the Industrial, entailed for England as a state and a nation. Such an integration has to be instructive because Whewell's career was very much an individual success conditioned and facilitated by national institutions and cultural ideals that were distinctive of England (where England excludes Scotland). For recall Whewell at his career apogee in 1841. With the five volumes of the *magnum opus* (the *History* and the *Philosophy*) now published, he marries, and is elected Master of Trinity College, a Crown and therefore a national appointment and second to none in its power and prestige in the English academic realm. It is the year he is President of the British Association for the Advancement of Science, thus becoming ringmaster of the annual state circus of science assembled that year at Plymouth. Consider, too, Whewell in the following year, writing a most revealing letter to a friend, a letter that moves, quite naturally, from opening remarks about examinations, to talk of God, truth, and the Church, the Nation and its Constitution. These six, including that is the examinations, although not necessarily in that order, were for Whewell nothing less than what he believed in as objects of belief essential to civilization in its highest form. Nor does he leave us in any doubt as to the leading connections among them. Most familiarly, he was one of those defending the distinctive traditions of English

[8] Butts (1968); Curtis (1986); id. (1987).

university education as essential to those established relationships between the Church and the Nation that were enshrined in the Constitution, and thereby made essential to the continued securing of liberty under King or Queen and Parliament.[9]

Now it is easy to see that such commitments served, for Whewell and many another like him, as alternatives to the unreformed Church and revolutionary politics of France, and so served to contain the challenge, the threat from that quarter. What is perhaps less easy to see is that they served no less to contain challenges and threats arising from the industrialization of England. Here it helps to view Whewell as the hyperassimilated emigré within his own country. For, in the course of his progressive career successes, he became assimilated entirely to the older national culture of metropolitan England, as entrenched in London, Oxford, and Cambridge, a culture that was—thanks to the burgeoning strengths of its agricultural, financial, and imperial capitalism—still able to marginalize the provincial alternatives associated with the new manufacturing interests represented most vividly by the exemplar of Manchester's transformation. Whewell, in rising from scholarship winner to grandee don, had indeed circumvented Manchester and all it stood for, in his passage from a Lancaster carpenter's household to the Master's lodge. It tells us much about England then—and before and since—that success, in becoming the accomplished academic and ecclesiastical *arriviste* that Whewell was, required such circumvention as a condition for hyperassimilation.

That nationalism, in the sense here indicated, should be a decisive element in the ideology of anyone with Whewell's career trajectory allows us, therefore, to make sense of two recurrent sources of difficulty for a savant in his position: science itself and Scotland. The difficulty with science was that, abstractly considered, it was professedly free from nationality. The difficulty could be resolved however by respecting that ideal while at the same time insisting that different nations could make distinctive contributions to science, so that, for instance, there could be a case made for a particular foreign, not to say alien, import such as French physical astronomy, if it could be represented as fulfilling earlier Newtonian hopes or improving existing tripos practices.

The difficulty posed by Scotland could not be so resolved, obviously. Scotland as viewed from Oxford and Cambridge was habitually disrespectful of the distinctively English commitment to the National Church, the National Constitution and the contribution made to the maintenance of both by a system of university teaching that made classics and mathematics primary, and had the clergy not in special seminaries but as the most prominent presence in a scheme of higher education equally adapted to the landed gentry. It is no mere happenstance at all that Whewell was as disconcerted by his friend the smooth deist Lyell praising American universities in comparison with Whewell's own,

[9] The letter, to J. G. Marshall, is printed in Stair-Douglas (1881), 279–83. Morrell and Thackray (1981), *passim*, is excellent on Whewell's career.

as by his enemy, the abrasive and pious Brewster.[10] Both were Scots knowingly continuing decades of Edinburgh chastisement of Oxford and Cambridge. To reduce such conflicts to personality clashes or regional rivalries is to miss how they involved an issue made unavoidable, for anyone with Whewell's outlook, by the very existence of Scotland within Britain: the issue of reconciling the ultimate sources of authority for the British people with the presence of another national tradition within the nation's bounds.

A main suggestion here will be, therefore, that the character of Whewell's encounter with geology is often distinctive for the same reasons that his entire career was distinctive: namely that the goals he was pursuing as an individual were largely set for him by a context of nationalistic notions as to what counted as success in advancing civilization and so, too, science.

3. Locations for Geology

To understand the various locations geology had for Whewell, including the location, within his career, of his own encounter with this science, one has to recognize the broad scope and deep significance the science held for him. One might think his views on the origin and duration of the earth, including its stocking with plants and animals, man among them, would be spread over several sciences. However, for Whewell, all such matters are directly connected with geology if not subsumed within it. For these are all matters of historical causation, of causes, that is, acting occasionally and successively in past times. All fall, then, within the palaetiological sciences, the sciences of causes active in the past. Now, although several other sciences—glossology, for example, or philology as others were already calling it—count as palaetiological, it was geology that was Whewell's original exemplar of the genre, and the only one that he treats of at any length. Its structure and development are, moreover, exemplary in that the structure and development of any other palaetiological science are understood by comparison and contrast with geology. So geology, broadly construed as Whewell construed it—so as to comprehend climate changes, biogeography, and the origin of species, including man, no less than earthquakes and mountain formation—is what has to be considered.

Whewell's life and his writings both exhibit sufficient restless movement that anything within them has a location *en route*, a location where his various trajectories take him through whatever the subject is and on to something else beyond. There is the movement apparent from any survey of his writings over three decades from the 1830s: a movement from mathematics, pure and mixed, to morality, theoretical and practical. Geology here takes its place in the 1830s, the decade dominated, from his review of Herschel's *Preliminary*

[10] K. Lyell (1881), especially ii. 127. On Scottish critiques of Oxford, see Rupke (1983).

Discourse (1830) to the *Philosophy* in 1840, by writing on induction and the inductive sciences. In this context, the principal challenge was to understand geology as an inductive science, and to adjudicate between the rival accounts given, by Lyell and Herschel on the one hand and by Sedgwick on the other, of what it is for geology to have a secure position among the inductive sciences. There is another movement, in the trajectory that is Whewell's career climb from undergraduate to Vice Chancellor, via a tutorship in Mathematics and Professorship in Mineralogy (1828). Within this movement, there is a progress from one election to another; the first is the election, in 1827, to Fellowship in the Geological Society of London, an election appropriate for one who was already a dedicated, publishing mineralogist; the second is the election to President in 1837, an election appropriate not because of any pretension to original geological research, but because Whewell had come to be seen by many Fellows as an excellent alternative to less progressive elder statesmen who might otherwise gain the office.

Engaging geology in the course of these several trajectories, Whewell delineates various placings for the science. In the *History* and the *Philosophy*, both adopting the same ordering, geology finds its place at the very end of a movement that begins with mathematics and then proceeds through the inductive sciences, much in accord with the distinctions embodied in the BAAS sections. Geology provides, then, the transition to topics that lie beyond the 'physical sciences', topics such as classical studies, civil history, the First Cause as an object of theological science, and 'tradition', including the Scripture narrative. Although this transitional siting, for palaetiological science in general and for geology in particular, may seem natural, there was no inevitability about it. As Whewell remarks in the *Philosophy*, at the end of the previous section on Biology, the palaetiological sciences were not typical when considered as instances of the scheme of ideas, the scheme that the *History* and the *Philosophy* are designed to elaborate. For these sciences do not have their own Fundamental Idea or Ideas, only a conception—historical cause—a version of the Idea of Cause. When he has done with biology, he has then done with the circle of sciences as founded on distinctive Ideas, as he admits. What he does not admit is that geology, if not other palaetiological sciences, might well have been treated much earlier on under the heading of causal science. But such a decision would have forfeited the positioning of geology as a conduit to the sciences, the moral sciences, beyond the physical.[11]

As so positioned, geology is delimited by Whewell, in accord with the position and boundaries of its assigned object: a post-nebular and pre-human positioning between, that is, the incipient earth speculated upon in nebular hypotheses and that same earth inhabited finally by man, and so coming within the history of God's providence as revealed in the special dispensation for our species.

[11] Whewell (1847). See i. 635–6.

These positionings are, therefore, Whewell's own work. They are not given by unavoidable prior intellectual or institutional precedents. There was no consensus concerning geology as a site on the map of knowledge or as a division in the classification of the sciences. Nor was there any customary curricular niche for the science. In Whewell's life at Cambridge geology was familiar, even prominent, from early on, following the appointment in 1818 of his Trinity friend, Sedgwick, to the Woodwardian chair. But its prominence was due to the conjunction of Sedgwick's energy and reputation and the current topicality of the science, not to the statutory claims it could make on undergraduates' time and effort. For curricular purposes it was no closer to being joined with the mathematics, classics, and divinity required for a degree than was botany, say, or mineralogy. Until late in Whewell's life, geology at Cambridge was largely living in the limbo of professorial lectures, outside the rigmarole of examinations and set texts overseen by the Schools and Senate and serviced by the colleges and tutors, private and public. Where Cambridge is reflected in Whewell's positionings of geology is not through any curricular role, but in the prospect of advancing physical geology by bringing mixed mathematics to it, and in the promise that geology, in so far as it studies purposes, will lead to Final Causes, and in so far as it studies origins, will lead to the First Cause.

Within geology itself, when exhibited as an inductive science in the *History* and the *Philosophy*, the main movement is one typical of all induction: from facts to laws without causes, and so onward and upward to laws of causes. For, within this inductive ascent, Whewell distinguishes the phenomenal from the aetiological members in any palaetiological science. Here the exemplar of Newtonian astronomy is both emulated and transcended. It is emulated in that there are, Whewell insists, laws and causes of geological dynamics corresponding to the equivalent branch of the science of mechanics. It is departed from however, in that geology is not a science of mechanical causes, but rather of historical causation. It is not a science of forces that are permanent causes of motion, such as gravitational attraction is, causing motion whenever not opposed by other forces; but of the actions of causes that have worked their effects in a temporal succession, as the causes have, for instance, that have raised mountain ranges of different ages at various times.[12]

Such comparisons and contrasts among distinct sciences are decisive for the higher order inductive science of science—a moral science of physical science—that is pursued throughout the *History* and *Philosophy*. Only phenomenal geology has been completed in passing through an inductive epoch to its sequel. The perfecting of geology as a science that will make it yield true knowledge, aetiological no less than phenomenal, has hardly begun; requiring as it does much more inductive progress than has so far been made in geological dynamics and in the application of that dynamics in arriving at the true theory

[12] Whewell (1837*a*). See especially iii. 481 ff. Also id. (1847), i. 637–8.

of the earth. The marks of truth as manifested in a consilience of inductions and exhibited in inductive tables are not demonstrable here; not because geology comprises counterinstances to such higher order conclusions from the history of other sciences, but because the necessary inductive progress has not yet been made.

This historical and philosophical judgement was reached by Whewell at a particular moment in the history of geology and in his own life, the very early 1830s, and he never felt a need to retract or even revise it. Whewell's treatment of geology is not, then, like his treatment of most sciences; for it is not an exercise in generalizing from the history of attained success; nor, conversely, of applying to the particular case of geology the general doctrines constituting a philosophy of science designed to fit all inductive sciences as such. His philosophy of geology is not, therefore, a straightforward outcome of bringing together a particular science and a general philosophy of science. It was, rather, a philosophical response to geology conditioned by contingencies of timing at its inception, around 1830, before it was further constrained by the structure of exposition given it in 1837 and 1840.

4. Philosophical and Theological Outcomes

Whewell's arguments in his treatment of palaetiological science in general, and geology in particular, are sufficiently complex and convoluted that we do best to invert the order of his own exposition, and to begin with his conclusions. There is a clear advantage in beginning with his conclusions, because we can then avoid one drawback to most of the accounts historians have given of Whewell as a writer on geology. For most historians, understandably enough, have gone to Whewell to see where he stands on the contested issues within the geology of the time, most obviously the debate that Whewell himself taught the world to call the uniformitarian–catastrophist debate. But to read Whewell mainly for instruction on such matter is often to miss the significance for Whewell of the higher lessons of philosophical principle that he is explicitly and insistently concerned to draw from his examination of palaetiological science, lessons that were peculiarly of Whewell's own formulating and that did not become elements in any public discussion.

The first thing to notice about these lessons is that they are nearly all negative. Most evidently is this true of his two leading maxims: the first, 'That no palaetiological science has been able to arrive at a beginning which is homogeneous with the known course of events', which is to say, '*No Natural Origin discoverable*'; and the second—explicitly a corollary of the first—that '*Science tells us nothing concerning Creation*'; for since 'science can teach us nothing positive respecting the beginning of things, she can neither contradict nor confirm what is taught by Scripture on that subject; and thus, as it is unworthy timidity in the love of Scripture to fear contradiction, so it is ungrounded presumption to look for confirmation, in such cases'.[13]

That Whewell is travelling a *via negativa* throughout his examination of palaetiological science is evident to anyone who reads to the end of his presentation. What is less evident is that his negational ambitions are conditioning all of his prior discussion of the development and nature of this genre of science. For, on the face of it, Whewell seems often to be writing in a very hospitable, welcoming, and so positive way. All sorts of proposals are introduced respectfully, even sympathetically, including contested current speculations such as terrestrial cooling from decline of primitive central heat, nebular hypotheses, the transmutation of species, and so on. What is more, the uniformitarian and the catastrophist positions are likewise both given extended, respectful, and sympathetic hearings.

However, if one asks what, for Whewell, do all such scrutinies of more or less general proposals and positions yield, as overall conclusions, the answers are consistently negative. The extent of this negativity in the outcome is easy to miss, because Whewell often seems to be taking sides as he discusses rival views, at least in so far as he indicates that, in his judgement, one rather than another has the better arguments going for it. But this apparent decisiveness over particular matters should not mislead us. For if we ask, quite simply, what does Whewell show himself to believe, to accept, about the earth's past and the changes in life upon it, we notice at once that is very hard to discern the answer to this seemingly straightforward exegetical question.

One reason for this difficulty is that Whewell is adamant that there is as yet no true physical theory of the earth, no true theory, that is, of the course and causes of the changes undergone by the earth and its inhabitants. There is much secure knowledge of geological phenomena, to be sure, and some acceptable, if not certain, knowledge of geological dynamics. But when it comes to the deployment of that dynamics in constructing a true physical theory, there is no candidate worthy of discussion. There are the 'fanciful' theories of Burnet and others to recall, and there are the 'premature' theories of Werner and Hutton also; but there is no successor to these that is neither fanciful nor premature but true. Indeed, a main thesis of Whewell's is that there cannot be, because there is no resolution available, as yet, of the issues dividing the uniformitarian from the catastrophist doctrines, and those doctrines consist principally of incompatible views as to how one should move from geological dynamics to physical theory.[14]

Whewell's attitude here is telling in its implications for our whole interpretation of his place in the history of geological science. For it would be easy, but also mistaken, to couple him with Sedgwick, and to see him, ultimately, as another leading critic of Lyell's uniformitarian synthesis of geology, another critic who responded by joining in developing the case for an alternative to Lyell's

[13] Whewell (1837*a*), i. 679, 687–8.

[14] Ibid. iii. 594–605.

synthesis; the case, that is, for what has been called the directionalist (or progressionist) synthesis such as one finds it in Sedgwick, or more elaborately, in De la Beche, especially his *Researches in Theoretical Geology* of 1834.[15] That this view of Whewell is mistaken is apparent as soon as we notice that there is no moment in Whewell's writings where he defends, even by implication, several central elements in that progressionist synthesis. Most decisively, there is nowhere in Whewell any defence of the thesis that a progressive decline in the activity of the physical agencies affecting the earth's surface, especially igneous agencies, has made the earth progressively more fit for higher and higher types of life, which have accordingly been introduced as a progressive organic succession adapted to the progressive increase in habitability consequent upon progressive physical cooling and calming.

To be sure, Whewell does accept that a palaeontological research has disclosed a succession, if not a progression, of faunas and floras; he does accept that there are in the fossil record various evidences of climate change, notably a general cooling; and he does accept that there has been a decline in igneous causes of physical change. But he is not seeking to give coherence to his views on geology as a whole by seeing how these various beliefs can be brought together in any integrated synthesis; even less is he knowingly allying himself with one particular synthetic stance.

There is, ultimately, for Whewell, as he explains in his closing palaetiological chapter, 'On the Conception of a First Cause', a synthetic integration of palaetiological science, including geology; but it is not an achievement of any of these sciences themselves, taken singly or even in concert, for it is a theological integration.[16] All the origins that these various sciences point to, but cannot explain by reference to any known course of events, all these different origins are creditable to one God as the First Cause; the origination of the earth, the origination of life, the origination of language may all be presumed due to one and the same originating cause, the same First Cause. This positive outcome, from the palaetiological sciences brought together with theology, depends on the negative outcomes in each of those sciences. What the negative outcome comprises, in any one case, depends on the explanatory successes and failures exhibited by that particular science. There is, then, in Whewell's thinking on palaetiological science, a highly distinctive blend of two ingredients. On the one hand are his quite general and abstract views as to how any such science can be scientific at all; on the other hand are his specific judgements as to what any one such science has or has not done so far.

5. Laws, Causes, Induction and History

Whewell's concern with these issues goes back to the period 1830–2, when he prepared his reviews of Lyell's first two volumes; and when he prepared

[15] On the progressionist synthesis, see Rudwick (1971); de la Beche (1834).
[16] Whewell (1847), i. 700–8.

himself to take up where Herschel had left off in the characterization and vindication of the physical sciences as grounded in an inductive philosophy descending, before all other ancestors, from Bacon.

By July, 1831 he was already tabulating '*Types of Progress of Sciences*' in a manuscript notebook.[17] Later, in the same notebook, and in conformity with that tabulation, he took up the question of how to characterize geology as an inductive science that makes inductions of the highest sort, that is, inductions to causes. Within a few sentences the discussion spreads and escalates so as to include a range of very general issues such as would be canvassed in the palaetiological science chapters in the *History* and *Philosophy*.

Whewell begins by insisting that causes are of two kinds: 'the cause of a thing continuing as it is, and the cause of its beginning to be.' There are, that is, 'laws of force' and there are 'accounts of origin'. In other words, there are 'permanent' causes and there are 'progressive' causes; there are 'perpetual' and there are 'primeval' causes. Examples of the first kind are, Whewell says, the force of gravitation, the luminiferous ether, the cooling of bodies as the cause of dew, refraction as the cause of the rainbow. Examples of the second are 'the *true* geological theory of the condition of the earth—Laplace's account *if true* of the formation of the solar system', and 'The true origin of the geographical distribution of plants and animals'.[18]

Now, says Whewell, the second kind appear to be 'as good subjects for science as the other'. And they are 'at least as attractive to the scientific intellect'. However, they do not seem 'to have succeeded at least not on any great scale', there being 'no large instance of an established theory of this kind', the 'parallel roads' of Glen Roy (a geological case) being 'perhaps one of the best'.

In making sense of this very general but not total lack of success, Whewell cites Cuvier's designation of the geologist as an antiquarian, and goes on to relate geology first to our knowledge of man and then to our knowledge of God. When studying the action of physical causes within the human period, as in the case of the ruins of the Temple of Jupiter Serapis, geology may indeed overlap with human history. Here, then, we 'include among our *causes* the acts, habits, institutions, history of man'; these being once again, 'very different . . . conceptions from what we have in permanent causes'. Whereas, at the near end of the past, geology borders on the history of man, at its far reaches it leads to God, to God viewed as an antiquarian views ancient men. In the 'investigation of primeval causes we are led to a nearer view of the operations of the Deity'. It is like 'making out the character of a nation from its architecture', the ancient Egyptian nation for instance.

[17] Whewell Papers, Trinity College Library, Cambridge University, R18 17^{15}. All the passages quoted in this section are from pp. 43^r ff.

[18] Note that Whewell does not identify what the true geological theory is. He is in fact convinced that it does not yet exist. Indeed, he will eventually write (and never retract the judgement) that, in all the palaetiological sciences, there is '*No sound palaetiological theory yet extant*'. See Whewell (1847), i. 664.

Geology is therefore a causal and inductive science; but it is like history in its conceptions of cause. Pursuing this twofold characterization, Whewell reflects that geology, as it has been pursued so far at least, and indeed in regard to the questions the phenomena suggest, 'differs somewhat from the sciences so called'. For the 'object of science is to determine general *laws*', that is, propositions that are 'universally and constantly true, from the nature of their subjects, and independently of any one particular exhibition of them in fact'. The object is not, as it is in history, to ascertain particular events that have happened.

To that extent, Whewell finds he must distinguish 'reasoning' that is 'inductive' from reasoning that is 'historical'. For consider the example he develops, of an eclipse as an object of inductive reasoning by an astronomer and that eclipse as investigated by the historian. An astronomer, says Whewell, can determine that, according to the laws now regulating the motions of the sun and moon, an eclipse must have happened at a certain remote period, and have been visible in certain places. But whether this is the eclipse mentioned by some classical authors and associated with certain ancient traditions, the astronomer does not decide. On such questions, rather, he is examined as a witness by the historian or antiquarian. For he 'gives to the historian or antiquarian the laws which he has determined'.

Now Whewell goes on to emphasize that although the eclipse example is instructive, it is unrepresentative in that astronomers are exceptional in being able to reason back into past time and forward into future time, so as to reach, inductively, events very remote from present ones. To be sure, in a complete science such reasoning can be conducted as exact calculation; and perhaps astronomy can so proceed; but 'in no other case does it belong to the inductive philosopher alone to recall into existence the series of events which form the past'. So, Whewell concludes, when the 'inductive philosopher' goes 'beyond those times in which he can unite his efforts with those of the historical reasoner it is his duty to arrest his narration'. In other words, outside astronomy, the occurrence of past events cannot be inferred in the absence of any records of them, records studied by historians. So where records do not exist, there is usually no ascent to past events and no narrative possible.

Whewell is now in a position to discern the limits of geological knowledge. From this last reflection, it appears, he says, 'that the geologist cannot in accordance with inductive principles' tell us 'what happened when the present or any former condition of the earth began to obtain'. The geologist 'cannot descend from his proper region of necessary and constant connexion', to engage the 'accidents of individual and transient change, or of origin which is the greatest of such changes'.

The ultimate drift of this argument is plain enough, but the rationale of its component steps may be far from obvious. What makes for the difficulty is that Whewell has here a quite different argument from the one that will eventually appear in 1837 and 1840; but it is an argument with much the same conclusion:

the geologist cannot tell us about origins. To see how the present argument works, one does best to go to its closing passage, where Whewell allows that geology can achieve some explanatory causal narrative conclusions, even while being precluded from the beginnings of things.

Whewell closes by insisting that, although 'confining the geologist to his proper field', he is still leaving him able to decide 'many or most of the questions' prompted by 'speculations on the origin of the earth'. For there are cases where the geologist, as inductive reasoner, can move from inductive generalizations that include no causal narrative to confident assertions of that kind. For instance, the geologist may begin with the generalization that 'from trap rocks proceed veins into the adjacent rocks *as if* they had been injected in a state of igneous fusion', and that there is 'a *gradation* without interruption from newest lavas to trap rocks of all kinds'. Then, by coupling this universal appearance of igneous origin, with a refutation of all proposals implying 'aqueous production', one will have established 'the *fact* that trap rocks *were produced* in a state of igneous fusion'. Likewise, to take another example, the 'superior rocks being shown to contain fragments of the inferior, it will be believed that they *were formed* after them'. In this way, says Whewell, 'we may state any of the theories ill or well supported'. For, in such cases—where a theory about the facts as to what has happened in the past is inferred from some generalization about present facts—all that is 'requisite is to arrange the inductions (legitimate or otherwise) which are included in the theory, so that the *narrative* which it includes may be as orderly and clear as it is possible to make it'. For conversely, if the theory is one that is very 'loosely' constructed, there will be in it a noticeable 'hiatus' that will show the 'insufficiency of the structures so elevated'. Exclusion of alternatives and continuity as secured by coherence are, therefore, the grounds of truth in a causal, narrative theory that is constructed when the geologist moves beyond inductive reasoning that is not historical to inductive reasoning that is. Such are the ways whereby geology can be both historical and inductive and so scientific.

Whewell has already set out the limitations that these conditions impose on the geologist's ambitions. If his series of facts takes him back only a finite time into the past, he is 'not at liberty to place a hypothesis in the previous period thus left blank'. Only by including this series in a wider generalization can he pass 'beyond the barriers'. His 'boldness of assertion' is not to increase with the difficulty of investigation, in the 'dim and shadow districts', but is to 'diminish with the scantiness of means of proof and the insecurity of his accumulated pile of inferences'. It being no longer 'in his province to tell us of the beginning of things', he would leave that to those philosophers who have 'imagined that they were authorized' from present facts 'to speak of chaos, and creation, of deluges and conflagrations', to construct cosmogonies and to fix the 'times and circumstances of cataclysms and ecpyroses'. With this modesty about the very remote would go diffidence about the most proximate events. The geologist would leave to others any investigations of the individual events 'preserved by record or tradition, such as have excited the fears and wonders of men or influenced their moral destiny'; for that is a 'province' where the judicious among such men will undertake the interpretation of the evidences by appropriate 'peculiar rules'. Moreover, the 'indistinctness and number of the historical mass of such geological changes in early times is so great', and 'our ignorance' of the details of the 'circumstances' of those that 'geology indicates to be in the order of time' also so 'obvious', that, Whewell concludes, 'an irreconcilable contradiction between history and science is scarcely possible'.

6. The Delimiting of Palaetiological Science

It has been well worth looking at Whewell's arguments and conclusions in this manuscript text, for they can alert us to his most persistent preoccupations as a writer on geology and so on palaetiological science generally. Especially they exhibit his consistent strategy of deploying epistemological explications in the service of delimitational adjudications. The proper duties of those enquirers who work in this or that province are to be explicated, so that the nature of their actual and possible achievements can be identified correctly and so related appropriately to the actual and possible achievements of others. The principle case of such adjudication is that between geology and Scripture.

That the adjudication is an epistemological one may lead us to overlook that it is an adjudication at all. We are used to finding geology and Scripture related to one another on corresponsive grounds, whether particular or general. Buckland upholds the correspondence of a particular geological event to a particular Biblical moment, the Flood; Sedgwick upholds organic progression, over vast eons before man's arrival, as analogous to the progressive revelation in Scripture of the progression from Creation to Judgement. We are then inclined to presume that the rationale for Whewell's philosophy of palaetiological science is not adjudicational, because it is not grounded in a corresponsive thesis, not even a quite general and abstract one. But Whewell's conviction is that there is adjudicational work to be done through epistemological delimitations, even when no corresponsive thesis is involved—much less adopted.

It was reflection on Lyell's geology, especially on his rejection of Lamarck's views on the origin of species in the transmutation of species, that gave Whewell his eventual delimitational strategy as adapted in 1837 and 1840. For that strategy, although quite general in its form—as expressed in the abstract negational maxims, about the undiscoverability of origins and the uninformativeness of science concerning creation—is a generalization of the lessons Whewell drew from Lyell's encounter with the transmutation of species issue. Lyell, as Whewell saw it, had judged advocates of transmutation, such as Lamarck, unsuccessful in their attempt to find, in the changes of character presently wrought in animal and plant species by changed conditions, causes adequate to have brought new species into existence in the past. These changes that species are subject to at present are changes within the course of events as now known to us. So, as Whewell represents it, this failure is a failure to find a beginning for species that is homogeneous with the known course of nature.

Now, as Whewell saw it, in insisting on this failure, Lyell was admitting one major exception to his own claim that all past events recorded in the rocks could be successfully explained as the effect of causes still at work at present, and acting with no lesser or greater intensity than in the past. So Whewell's understanding of the failure of transmutation rests on his understanding of Lyell's most general commitment to the adequacy of existing causes. Whewell understood Lyell's claim as one of causal adequacy on behalf of causes existing today, and presumed to be acting with uniform intensity throughout past, present and future. For anyone committed to this claim, Whewell urged, there were at least three challenges to be met. First, causes must be found for the mechanical effects indicated by such phenomena as tilted and elevated rock strata. Second, past climate changes must be traced to causes now in action. Third, it will have to be shown that changes from one set of animal and plant species to another are explicable in this way.[19]

Now, Whewell welcomes Lyell's comprehensive marshalling of the evidence of causes now in action, and he proclaims that in doing so Lyell has contributed decisively to geological dynamics. However, when it comes to the first of the three challenges in moving beyond dynamics to geological history, Whewell declares himself entirely unconvinced. The extent and form of the elevation of ancient strata, the relations of successive strata to each other and the shapes and connections of river valleys are all beyond the competence of existing causation. In such cases, one has to conclude that the changes belonging to the past stages of the earth's existence, as disclosed by geological inquiry, 'cannot be considered as forming a continuous and homogeneous series' with those 'which are now taking place upon its surface'.[20] As to the second challenge, Whewell hails as ingenious Lyell's novel speculation referring climate changes to the effects of changes in the position of land and sea relative to the poles and the equator; but he declines to accept or reject it. On the third challenge, postponed by Lyell until his second volume, Whewell endorses Lyell's conclusion that transmutation cannot provide a resolution, because the evidence is that species are not now sufficiently mutable for new species to arise from old even in changing conditions over long periods of time. The appearance of new species, at successive epochs, indicated by irresistible 'geological evidence to have repeatedly occurred', is a fact '*not* belonging to the operation of that tendency to change in organized beings which we see still brought into play'. So, Whewell concludes, with 'this striking exception' of the origin of species, he is willing to 'assert', with Lyell and others, 'that all the facts of geological observation are *of the same kind* as those which occur in the common history of

[19] Whewell (1831). See especially 193–4.
[20] Ibid. 199–200.

the world'; it being another question whether they are, as Lyell asserts, of the same degree or intensity.[21]

We are now in a position to see how decisive for Whewell's eventual view of the palaetiological sciences was this striking exception. For, in the *History* of 1837, this exception within geology is exhibited as exemplifying the norm for palaetiological science generally. It is in addressing the question of the transmutation of species, as a question within 'organic geological dynamics', that Whewell raises the more general '*Question of Creation as related to Science*', and so the question of the limitations on all palaetiological speculations that try to move from the present to the remote past.[22]

Whewell here formulates two versions of a dilemma, the first within organic geological dynamics itself, the second within physical geology. The first concerns 'only such causes as we know to be in constant and orderly action'; the second concerns 'the facts which have happened in the history of the world'. The first dilemma is that, if species are not transmutable then 'we must suppose' that their variations, which are 'apparently indefinite' are in truth 'bounded by rigorous limits'; while, if we admit a transmutation of species, 'we must abandon that belief in the adaptation of the structure of every creature to its destined mode of being', a belief many would give up 'with repugnance', and one that the best naturalists have been convinced is true.[23] A decisive element in this version of the dilemma is, then, the matter of the destined mode of being, the way of life of each species. For the transmutation hypothesis, as Whewell views it, supposes that a species may change in taking up a new way of life, so that its eventual structure would not be due to its being fitted to one way of life that it was specially destined for, but due rather to any ways of living that it was forced into or fell into through contingent circumstances.

Consider now the second version of the dilemma, as raised by geology. For this science discloses that 'many groups of species' have 'succeeded each other at vast intervals of time'. The dilemma then arises 'anew': either accept the transmutation of species, and 'suppose that the organized species of one geological epoch were transmuted into those of another by some long-continued agency of natural causes'; or else believe 'in many successive acts of creation and extinction of species, out of the common course of nature; acts which, therefore, we may properly call miraculous'.[24]

Now Whewell postpones discussion of this second version of the dilemma until he has examined the first. His examination not only concludes that species are limitedly mutable, but also that the additional hypotheses are false that transmutationists have thought it necessary to join with unlimited mutability, in order to explain geological and other phenomena; particularly the hypothesis that traces new structures to changes in 'wants', and the hypothesis of a continual spontaneous generation of the simplest organisms.[25]

[21] Id. (1832). See 103–9, 126.
[22] Id. (1837*a*), iii. 580.
[23] Ibid. 574–5. [24] Ibid. 574. [25] Ibid. 575–80.

It is after this examination that Whewell raises the question of creation in relation to science, and asks whether 'since we reject the production of new species by means of external influence', do we 'accept the other side of the dilemma', and 'admit a series of creation of species, by some power beyond that which we trace in the ordinary course of nature'?[26]

It is at this point that Whewell finds himself with a response, if not an answer, that is supplied by his most general conclusions regarding palaetiological science. For he says that 'the history and analogy of science' teach us to 'reply as follows' to this question.

> All palaetiological sciences, all speculations which attempt to ascend from the present to the remote past, by the chain of causation, do also, by an inevitable consequence, urge us to look for the beginning of the state of thing which we thus contemplate; but in none of these cases have we been able, by the aid of science, to arrive at a beginning which is homogeneous with the known course of events. The first origin of language, of civilization, of law and government, cannot be clearly made out by reasoning and research; and just as little, we may expect, will a knowledge of the origin of the existing and extinct species of plants and animals, be the result of physiological and geological investigation.[27]

This reply to the question leads, then, to the very issue of science and creation. Although 'philosophers have never yet demonstrated, and perhaps never will be able to demonstrate, what was that primitive state of things in the social and material worlds, from which the progressive state took its first departure', nevertheless, they can 'go very far back' in all these 'lines of research'. They can 'determine many of the remote circumstances of the past sequence of events', they can 'ascend to a point which, from our position at least, seems to be near the origin', and they can 'exclude many suppositions respecting the origin itself'. But it is 'difficult to say' whether 'by the light of reason alone, men will ever be able to do more than this'. It is then 'no irrational opinion, even on grounds of philosophical analogy alone', that in all sciences that seek origins, 'we may be unable to arrive at a consistent and definite belief, without having recourse to other grounds of truth, as well as to historical research and scientific reasoning'. If we are to 'apprehend steadily the creation of things' we will have to 'summon up other ideas than those which regulate the pursuit of scientific truth', and to 'call in other powers than those to which we refer natural events'. Not surprisingly, then, 'in this part of our inquiry, we are compelled to look for other than the ordinary evidence of science'.[28]

Having invoked these general conclusions, Whewell elaborates their implications for geology. This science, together with other palaetiological sciences, leads us to a single 'focus of being', a single source not only of animal and plant life, but also of social and rational life, arts and language, law and order. Proudly, then, Whewell insists, this 'reflection concerning the natural scientific

[26] Whewell (1837*a*), iii. 580–1. [27] Ibid. 581–2. [28] Ibid.

view of creation' has not come from 'any wish to arrive at such conclusions'. Rather, 'it has flowed spontaneously from the manner in which we have had to introduce geology into our classification of the sciences'; this classification being 'framed from an unbiassed consideration of the general analogies and guiding ideas of the various portions of our knowledge'. He goes on to warn, however, that although geology, like any other palaetiological science, leads to theology, geology and theology must not be confused or mixed. Theology cannot supply a supernatural geological dynamics to supplement natural dynamics; nor can geology tell us how to interpret any narrative of the providential history of the world and man. Moreover, Whewell insists, we should not expect a 'full insight' into the consistency that we presume exists between 'the results of true geology or astronomy' and the 'statements of true theology'; after all finite minds cannot expect to 'comprehend the infinite'.[29]

7. The Natural and the Supernatural

Whewell saw, therefore, no contradiction or even tension where we may sense one in his thinking about palaetiological science. On the one hand, he seems to consider seriously the possibility of a complete palaetiological scheme, with the origins of the solar system, the earth, life, man, civilization, and the rest all credited to natural causes in a comprehensive, progressive series of changes wrought by a series of successive causal actions over vast stretches of past time. Here the living would be made continuous with the lifeless, the human with the animal, and the civilized with the barbarous. And yet, on the other hand, Whewell is insistent that such natural continuities are not established by any science, and so are not conclusions imposed on any general reflections concerning palaetiological science.

The reason Whewell sees no contradiction here is that he is entirely confident that the more comprehensive in its ambitions is any palaetiological scheme, the more will it be evident that the natural continuities it implies will be mere suppositions, and so entailing no exclusion of God, acting as First Cause, in the successive originations of those quite distinct series, of quite different changes, that are the proper object of the various particular palaetiological sciences.

His confidence in this outcome is most plainly expressed in his treatment of the nebular hypothesis. A chapter is devoted to this hypothesis in the Bridgewater Treatise, where indeed it seems to have been first given that name. Whewell's position there is that the natural theologian need not worry over the prospect of the hypothesis turning out to be true. For God will still be evidenced in two ways. First of all, however far back the hypothesis goes, one can still ask how there originated the state and powers of diffuse matter that it takes as its initial condition; and since such an initial condition is not traced by the hypothesis to

[29] Ibid. 583, 586.

any natural causation, it may be evidence for God's wisdom and beneficence. Second, the progression of cause and effect, leading away from that initial condition, and including eventually the formation of a terraqueous globe, its stocking with life and man, will include productions that evidence God rather than the state and powers of the nebular material itself.[30]

For suppose a planet to be produced in accord with the hypothesis; it may be presumed to resemble only 'a large meteoric stone'. How, then, does it come to be 'covered with motion and organization, with life and happiness'? The rhetorical questions and their answer follow easily:

> What primitive cause stocked it with plants and animals, and produced all the wonderful and subtle contrivances which we find in their structure, all the wide and profound mutual dependencies which we trace in their economy? Was man, with his thought and feeling, his power and hopes, his will and conscience, also produced as an ultimate result of the condensation of the solar atmosphere? Except we allow a prior purpose and intelligence presiding over this material 'primitive cause', how irreconcilable is it with the evidence which crowds in upon us from every side.[31]

The same line of argument, now quite generalized, can be used by Whewell, in 1840, to support the same conclusion, that in order to form the whole hypothetical series of events—from nebular condensation to human civilization—into a connected chain of causation, supernatural influences may have to be considered as part of the past series of events. In contemplating such a series, we are led, says Whewell, 'by a close and natural connexion, through a series of causes', all the way from those that regulate the remotest heavenly nebulae, to those determining 'the diversities of language, the mutations of art, and even the progress of civilisation, polity and literature'.[32]

However, regarding the earth's formation and stocking with its first and subsequent life, Whewell says he has spoken 'hypothetically' of these events as 'occurring by force of natural causes', only so that 'the true efficacy of such causes might be brought under our consideration and made the subject of scientific examination'; to raise, that is, but not to settle, the question of whether there are natural causes adequate to such events.[33]

> It may be found, that such occurrences as these are quite inexplicable by the aid of any natural causes with which we are acquainted; and thus, the result of our investigations, conducted with strict regard to scientific principles, may be, that we must either contemplate supernatural influences as part of the past series of events, or declare ourselves altogether unable to form this series into a connected chain.[34]

We should now be able to appreciate why Whewell felt no tension, much less contradiction, between two themes in his writings on natural science in

[30] Whewell (1833*c*). See Book 2, Chapter 7, 'The Nebular Hypothesis'. On Whewell and the nebular hypothesis, in general, see Schaffer (1989) and the references given there to early historians of the topic, such as Jaki and Merleau-Ponty.

[31] Id. (1833*c*), 184–5.

[32] Id. (1847), i. 657–8.

[33] Ibid. 658.

[34] Ibid.

relation to natural theology. On the one hand is the theme, dominating much of the Bridgewater Treatise, that God acts in the physical world through law. One enunciation of this theme is especially explicit. In Book 3 ('Religious Views'), chapter 8 ('On the Physical Agency of the Deity') begins with yet another Whewellian negation. 'We are not', he says, 'to expect that physical investigation can enable us to conceive the manner in which God acts upon members of the universe.' Science, even more clearly than 'everyday reason', shows us 'at what an immeasurable distance' we are from being able to conceive '*how* the universe, material and moral, is the work of the Deity'. However, regarding the 'material world'—as distinct, that is, from the moral world—at least we can 'go so far as this;—we can perceive that events are brought about, not by insulated interpositions of divine power exerted in each particular case, but by the establishment of general laws'. This is the 'view of the universe proper to science, whose office it is to search out these laws', Whewell declares.[35]

On the other hand, however, there are Whewell's explicit declarations that among the events inquired into by the palaetiological sciences are some that manifest God, acting as a power beyond the known, lawful course of nature. For instance, he expects Lyell to agree with him, he says, in thinking it undeniable that 'we see' in the transition from one set of species to another, as shown by palaeontological studies, 'a distinct manifestation of creative power, transcending the operation of known laws of nature', geology thereby 'lighting a new lamp along the path to natural theology'. The more closely this transition is examined, indeed, the more it seems to Whewell out of reach 'of any other power than that of which the Creator has confined the regulation and manifestation to the depths of his own bosom'; it is a power that 'appears to belong, not to what we are accustomed to speak of on the laws of nature, but to that Supreme Will, which is their source and foundation'.[36]

Now, there is no inconsistency, for Whewell, in these two lines of thought. For, in judging the origin of some series of events beyond the known laws of nature, he does not need to assert that it has been originated by an act of God that involved breaking or overriding any laws, in the way that the virgin birth of Christ was sometimes represented as brought about by a special, exceptional Divine action in violation or circumvention of a known law. Indeed, for Whewell, the origination of any series of events, such as a new set of species, is a matter of God's introducing new laws rather than either breaking or conforming to laws already in force. He seems nowhere to have said explicitly that these originations are to be construed in that way, as additions to the constitution, additions to the laws of nature; but such would seem very much in accord with his overall stand; for he would then be arguing that from the way

[35] Id. (1833*c*), 356–7.

[36] Id. (1831), 194; id. (1832), 125. In following Whewell's views on these matters, I have benefited from an unpublished paper written some years ago by Michael Ruse: 'William Whewell on the Origin of Species'.

a law is known to be conformed to now in the course of nature, there can be no inferring how it was that the law first came to be instituted. But regardless of whether this was precisely Whewell's position, it is plain that he saw the originations discerned, but left unexplained, in palaetiological science, as moments when additions were made to nature. So his argument is that from the way things go, lawfully, now, after these additions, one cannot arrive at an explanation of how the addition was made in the first place. One may know that it has been made but not how. When Whewell says that God does not interpose his power, on isolated occasions, to produce a particular event, he sees that as consistent with his account of God's originating action in making these additions to the world; for such additions are not particular events, they are initiations of series, lawful series, of events, a set of organic species, for instance, or a new language, or whatever.

8. The Doctrines of Uniformity

It will already be plain that, although an admirer of Lyell's geological dynamics, Whewell's motivational preoccupations as a writer on geology contrast strikingly with Lyell's own. Where Whewell was resigned, even ultimately resolved, to conclude that geology, like any palaetiological science, was inevitably striving after what it could not be expected to achieve, Lyell was wanting to frame the business of geology in such a way that success could be ensured, and the ill reputation of the science—for vague speculation and divisive disagreement—scotched. Whewell is rightly known as the most acute and enlightening critic of Lyell's proposals for the foundations of geology. In reading Whewell on Lyell, we do well to keep in mind this contrast in their goals.

Whewell eventually made his critique and rejection of Lyell's uniformitarianism largely turn on two issues, that may ultimately be seen as one. For the one issue is that of where in geology we can know, and where by contrast we must retreat to suppositions that are more or less arbitrary, and so more or less unsatisfactory supplements to what we do or can know.

Lyell's own position almost comes down to a single twofold thesis: namely, that the only knowable causation is that active in the literate human period, the present in the sense of the last three thousand years through today and tomorrow; and that the present is not special in any other way than in the knowability of its causation, where knowability is pretty much a matter of observability. The agencies active in the prehuman past are the same, and they have worked in similar circumstances, so that the sorts and sizes of their effects have been similar too. Conversely then, a cause that is confined to the prehuman period would be not knowable because not observable, even potentially, by man. The assumption about the uniformity of prehuman and present causation, about the stability and permanence of the system of causation, is then demanded by the requirement that geology attempt always to confine itself to explaining past changes as the effects of causes that are or could be known causes. That requirement derived in turn from the *vera causa* ideal first enunciated in Reid's explication of Newton's first rule of philosophizing. For that ideal specified that one's explanatory causes should be causes that are known to exist—to be real and so not hypothetical—from the evidence of facts other than the facts they were used to explain.

According to Lyell's argument, therefore, that ideal for geology entailed a strong and strict acosmogonical reproductive causal adequacy presumption for geology. For past events are to be presumed to be the effects of causes that are still undiminishedly active today and tomorrow, in that they could, and indeed eventually will, produce, (and so reproduce) in the future effects of the same sort and size as they have in the past. Accordingly, no past effects recorded in the rocks are to be explained as residues from an original state of a nascent earth—an early molten fluid earth or an early landless panoceanic earth for instance—a state that no causes active today could ever produce even in an unlimited stretch of future time. Cosmogony, in the sense of geogony, is therefore excluded from geology. The science has no concern with whether, much less how, the earth did once originate from some preplanetary, prototerrestrial beginnings. The reproductive adequacy thesis, as required by the *vera causa* ideal, requires a presumption of uniformity that precludes cosmogony.[37]

To this Lyellian proposal, Whewell replied with three countering arguments. First, he argued that present causation, in Lyell's sense, is no more knowable than past causation; second that the *vera causa* ideal of explanation by known causes is often too restrictive for any physical science; and third that there is more than one way to subsume the earth's past within a presumption of uniformity in natural causation, and that there are reasons not to prefer Lyell's way, but to prefer another that leads to entirely different expectations as to what geology may or may not establish.

In developing the first argument, Whewell insists that it is arbitrary either to assume a uniformity in the course of nature that precludes catastrophe, or to assume that catastrophic rather than uniform change has occurred. We should look to the rocks themselves to discern whether or not past change has been uniform or catastrophic. This declaration may seem to beg the whole question as Lyell had framed it. For Lyell had been adamant that what is seen in the rocks is always indeterminate, in regard to the issue of whether it has been caused slowly by causes no more violent than today's, or suddenly in catastrophic events. But Whewell thinks that there are plenty of cases where the first option can be confidently eliminated. The physicist in him, especially, would insist

[37] For a more extensive interpretation of Lyell along these lines, see: R. Laudan (1982), and id. (1987), ch. 9. For Whewell's critique of Lyell's appeal to the *vera causa* ideal, see Ruse (1976*a*). A manuscript entry from the early 1830s reveals, tellingly, that Léonce Elie de Beaumont, whose theories were embraced by Sedgwick and rejected by Lyell, was Whewell's exemplary catastrophist: 'Catastrophists (E. de Beaumont) Uniformitarians (Lyell)', Whewell Papers, R18 17^{12}, p. 22^{r}. The significance of Elie de Beaumont's theorizing, especially for English discussions, is explained in Rudwick (1971), Greene (1983), and R. Laudan (1987).

that certain configurations of effects can only be wrought by certain constellations and intensities of forces. Not that Whewell is less sceptical about our ability to know causes than Lyell is. Indeed, he maintains that causes are never known except from their effects, and that this goes for present causes no less than causes acting in the past. So we know the causes of the elevation of the Alps no less, and no more, than we know the causes of an elevation of land now going on; for we know each only from their effects. Lyell's equation of knowability in causes with action in the human present is, therefore, unacceptable.[38]

Moreover, even if we did agree to try to move from the present, as more knowable, to the past, as less so, we could not know what we would need to know concerning the present. Most decisively, we could not know the extent of any fluctuations in the intensity of causes acting in the present. What we have observed so far, in the way of violence, may be exceeded tomorrow for all we know. So if we cannot set limits to violence in the short run, even less can we for the longer run, including the prehuman past. Nor, if we take in a longer run of the future, can we sustain any presumption of stability in the causal system. Lyell's attempts to justify a perpetual subterranean, combustive motion notwithstanding, the heat of the earth must be presumed to dwindle eventually. There will be an end to the earth, then; just as Whewell had argued, in his Bridgewater Treatise, that there would be an end for the solar system, thanks, eventually, to the resistance of the ether through which the planets make their orbits.[39]

Lyell has, therefore, misunderstood the precedents set by astronomy for geology. A combination of the *vera causa* ideal, with a presumption of stability in the system of causation, is by no means the only proper way to model this young science of geology upon astronomy. Rather, we should 'consider our knowledge of the heavens as a palaetiological science', and, specifically, consider the nebular hypothesis. For there we have 'the beginning of a world'. Whewell stresses that he does not, for his argument, maintain that this hypothesis is 'true'; but if geologists would borrow 'maxims of philosophizing from astronomy', then 'speculations' such as have led to that hypothesis 'must be their model'.[40]

Again, consider other provinces of palaetiological speculation, says Whewell, in 'the history of states, of civilization, of languages'. We may presume 'some *resemblance* or connexion between the principles' determining 'the progress of government, or of society, or of literature, in the earliest ages, and those which now operate', but no one has 'speculated successfully' by 'assuming an *identity*' in these causes.[41]

Where do we now find a language in the process of formation, unfolding itself in inflexions, terminations, changes of vowels by grammatical relations, such as characterize the oldest known languages? Where do we see a nation, by its natural faculties,

[38] Whewell (1837*a*), iii. 616–17; id. (1847), i. 668–9.
[39] Id. (1847), i. 674; id. (1833*c*), 207–9.
[40] Id. (1837*a*), iii. 619. [41] Ibid. 619–20.

inventing writing, or the arts of life, as we find them in the most ancient civilized nations? We may assume hypothetically that man's faculties develop themselves in these ways; but we see no such effects produced by these faculties, in our own time, and now in progress, without the influence of foreigners.[42]

In the *Philosophy* in 1840, Whewell will give these examples, especially glossology (or philology), more extended treatment. But they never take on a life of their own.[43] Geology continues to dominate the understanding of palaetiological science, and the other instances of the genre never require any revisions in the way that genre is conceived. It is only as instances of a genre that also includes palaetiological astronomy, which is to say hypothetical nebular astronomy, that they can help discredit Lyell's mistaken appeals to unpalaetiological astronomy as a precedent for geology. He continued, in the *History*:

Is it not clear, in all these cases, that history does not exhibit a series of cycles, the aggregate of which may be represented as a uniform state, without indication of origin or termination? Does it not rather seem evident that, in reality, the whole course of the world, from the earliest to the present times, is but *one* cycle, yet unfinished;—offering, indeed, no clear evidence of the mode of its beginning; but still less entitling us to consider it as a repetition or series of repetitions of what had gone before?

Thus we find in the analogy of the sciences, no confirmation of the doctrine of uniformity, as it has been maintained in geology.[44]

This argument against Lyell is so typical of Whewell that we may overlook how characteristic it was of his time. There were in his day not only more distinct sciences than ever before, but more self-conscious efforts than ever before to distinguish and advance novel sciences. Whewell is the child of his time in taking this kind of effort as exemplary for the history of science as a whole, making his history a history of sciences, plural but comparable. Arguments from the analogy of science in this sense, arguments from comparing comparable, but only comparable, sciences, are what his writing on the palaetiological sciences is all about.

9. Whewell, Chambers, and Darwin

Whewell may well have sensed, fearfully enough, that sooner or later someone would have a go at putting the nebular hypothesis together with spontaneous generation and the transmutation of species, to construct just the kind of complete palaetiological scheme that he had insisted was only to be considered in order to see that the gaps in it could not be closed by natural causes. Whether or not he did sense this, when the time came he was indeed exasperated, the more so as the job was done by an anonymous author (it was Robert Chambers)

[42] Ibid. 620.
[43] Id. (1847), i. 649–51; 661–3.
[44] Id. (1837*a*), 620.

in a book obviously designed for the general reader (*Vestiges of the Natural History of Creation*, 1844) so taking the issues it raised to a wide public, in circumvention of the natural guardians of epistemological rectitude in such matters, Whewell himself and his Christian, savant friends. Whewell responded publicly by having printed a collection of extracts of his writings on physiology, comparative anatomy, and geology, under the title *Indications of the Creator* (1845). Within a year, he was writing a preface for a second edition of this anthology. In that preface he engaged Chambers's own criticisms of his writing, criticisms made, anonymously again, in Chambers's *Explanations: A Sequel to Vestiges of the Natural History of Creation* (1845).

As with his treatment of the transmutation issue in the *History*, Whewell's response was twofold. On the specific question of whether *Vestiges* made its cases empirically, for spontaneous generation and transmutation, Whewell argued that the facts, as assessed by the best authorities, were not in favour. Here, Whewell had to take up, especially, the state of discussion surrounding various generalizations in comparative embryology that were decisive for the *Vestiges*' attempt to establish that the development of life on earth proceeded, naturally, in accord with powers and tendencies exhibited in higher animal embryos.

However, Whewell was no less concerned to counter *Vestiges* by arguing that it presumed to have established what we had no reason to expect could be had, and that is a palaetiological scheme that finds natural origins for such series of causes and effects as are presented by the history of life and man. In arguing this theme, Whewell had to insist, against Chambers's accusations, that he, Whewell, was not saying that we should not try to seek scientifically after such a scheme; rather, he was saying that we should try, but without expecting to succeed.[45] This general rebuttal of Chambers is combined by Whewell with the specific criticisms, to provide his answer to a direct challenge: 'It may be said', Whewell acknowledges, 'if the hypothesis of creation' in the *Vestiges* is 'not the true one, what then is?' To this question, says Whewell, 'men of real science do not venture to return an answer'. Moreover, as he eventually conceded, 'I must necessarily confess that we cannot obtain from science a complete view of the history of the universe.' There is 'no surprise or humiliation' in our inability to 'fathom and comprehend the acts of the Creator and Governor of the World'. He closes his second edition preface by recognizing that he represents 'the results of the History of Science as being in a great measure negative' regarding all such questions. He quotes Chambers's plea that science should explain to us the great ends of the Author of Nature and our relations to him, to good and evil, to life and eternity; and Chambers's disappointment at the emptiness of any answer from science when it tells only

[45] Id. (1846), 9. For the immediate context of Whewell's response to the *Vestiges*, see Brooke (1977*a*), and Yeo (1984). See further Brooke (1977); Raub (1988). I cannot join Raub in identifying Whewell as the reviewer of Chambers in *Fraser's Magazine*: see n. 9 of her article.

of classes of animals, of electrical and chemical agencies, and so on.[46] To this Whewell retorts as his parting shot:

> Strange caprice of the human mind! that one who feels such answers to be worthless, should think that he has remedied all their deficiencies, and supplied all that the soul can desire, when he has told us that *life grows out of dead matter, the higher animals out of the lower, and man out of brutes*![47]

To Whewell it was both reprehensible and incomprehensible that anyone could want to conduct his theodicy publicly through a palaetiological scheme that included such theses.

As viewed by Whewell, Darwin's *Origin* was the transmutation of species coming round for the third time. But Darwin himself, an eminent scientist, a friend, and the author of a book that made no appeal to popular favour, could not be countered with such hauteur as one would indulge in countering the anonymous composer of a book like *Vestiges*. For us, as Whewell's and Darwin's historians, the matter is further complicated because there were interactions between the two men that can instruct us about both.

The subject of Darwin's relations with Whewell, relations of many kinds, personal, institutional, ideological, and so on, is such a complicated one that it cannot be resolved by any sweeping generalization on the one hand, nor, on the other, by fastening on to any particular text, moment, or transaction as a single, privileged key to the whole story.

A comprehensive and critical account of the two men's relations with one another must wait for another occasion. However, it may be appropriate to indicate here how the subject could be advanced beyond where the literature now leaves it. Several points, of rather disparate sorts, may be worth making in particular. First, there are motivations from social historiography and from intellectual historiography that have conspired, mostly unwittingly, to make many people wish that Whewell will turn out to be the brains behind Darwin. On the social side there is the neo-Namierite presumption that individuals are best understood as products and protégés of some informal, invisible school, network, or college that fostered and raised them. Cannon's Darwin, as child of a Cambridge network having Whewell as a dominant presence, is very much in this tradition.[48] On the intellectualist side, there is a neo-Kantian presumption that behind many a scientist is a philosopher who has taught him how to think about the universe as a whole and about science as the way to explain its contents. Ruse's Darwin, with his indispensable debts to Herschel and Whewell, as teachers of philosophy, is very much in this tradition.[49] Even more generally there are, perhaps, hopes in many a scholarly heart, that two giants, Whewell and Darwin, will turn out to be, ultimately—and if it is an irony, then all the better—collaborators in a single revolution of thought. So,

[46] Whewell (1846), 21–3, 29–30.
[47] Ibid. 30.
[48] W. Cannon (1961); S. Cannon (1978).
[49] Ruse (1975*a*). For a more recent paper in the same vein, see Recker (1987).

in a word, the challenge to critical scholarship is not to get carried away by a tide of assumption that would tend a priori to greatly overrate the importance of Whewell for Darwin.

Secondly, the truths about Whewell and Darwin, it is now becoming plain, are to be found in the details. Thanks to recent editions of Darwin's early notebooks and correspondence, especially, we can reconstruct with some confidence what was going on and when, between the two men, at least in the decisive years from 1836 to 1839, in letters, in conversation, and in reading each other's published work.[50] When this evidence is put together with other relevant documentation, notably Darwin's marks and annotations in his copy of Whewell's *History*, then it is apparent that Darwin learned much from Whewell, had many debts to him, was much influenced by him, if one prefers; but that this indebtedness was far more plural and fragmentary and much less coherent, integral, and synoptic than either the Cannonical or the Rusean presumptions would suggest. For an itemizing of Whewell's contribution to Darwin's life and work would form a long list of things that Darwin either accepted, was challenged to reject, or was concerned to modify and reinterpret and subsume within his own thinking. Thus, to cite only some instances, there is Whewell's version of Kant's teleology that Darwin rejected;[51] there is Whewell as aid in understanding the physical geology of wave motions in rock material;[52] there is Whewell on necessary principles as elements in all scientific knowledge, to be reinterpreted as instinctive, ancestral, animal legacies for humans as capable of success in pursuit of scientific knowledge;[53] there is Whewell, very acceptably to Darwin, distinguishing formal from physical laws;[54] there is Whewell on the nature of a natural system in systematic botany and zoology, to be partly accepted, partly rejected;[55] there is Whewell on the teleology of sleeping and daylight, to be scorned;[56] and so on and so forth. Later, in writing the *Origin*, there is Whewell on God working through law to be used as an epigram at the front of the book; and later still, there is Whewell on hypotheses, to be cited in legitimation of a 'provisional hypothesis' of pangenesis, a hypothesis about generation or reproduction and so inheritance and variation; and so on and so forth.[57]

[50] Barrett (1987). Darwin's annotations on the *History* are discussed, for instance, in Manier (1978) and in the essay review of this book by Schweber (1979). The most recent, enthusiastic advocate of Darwin's debts to Whewell is Curtis. See his two articles cited above in n. 8. It will be apparent that I can accept some but not all of Curtis's proposals.

[51] Cornell (1986).

[52] Darwin to Whewell, 18 June 1837, in Stair-Douglas (1881), ii. 24–5. Whewell's two presidential addresses to the Geological Society show him to be knowledgeable concerning Darwin's work at the time. They were printed, respectively, in Whewell (1838*a*), 624–47, and (1839).

[53] Curtis (1987).

[54] Ruse (1975*a*).

[55] See Darwin's 'Notebook D', p. 26 in Barrett (1987).

[56] See 'Notebook D', p. 49, ibid.

[57] Thagard (1977), 353–6. See the introductory section of Chapter 27 of Darwin's treatise on *The Variation of Plants and Animals under Domestication*.

What all the evidence, taken together, makes quite implausible is that Darwin owed to Whewell either his most general convictions concerning the universe or his most general convictions concerning science. Central to Darwin's understanding of the universe is his view of the relationship between the physical world—of the stars, planets, earth, ocean, climate, and so on—and the organic world of plants and animals, including man. On the physical side, Darwin never dissented in any fundamental respect from Lyell's assumptions concerning the stability of the physical system of terrestrial causation, with its corollary the exclusion of cosmogonical theorizing from geology. His view of the organic world, directly and knowingly in disagreement with Lyell's view of the organic world, was nonetheless accommodated to Lyell's view of the physical. A principal challenge for Darwin was to explain how progress, although not invariable, would be inevitable in some lines of descent at least, in the adaptation of life to a terrestrial surface that was changing only unprogressively à la Lyell. In his understanding of this challenge and how it could be met, Darwin shows no signs of indebtedness to Whewell's views on the physical world and the relation to it of the organic. In working out his earliest conclusions concerning organic progress Darwin drew on Lamarck (as expounded and criticized by Lyell) on Owen and on Carus, among others, but neither then nor later did Whewell contribute. Nor is that surprising; Whewell's writings do not convey a consistent, positive, detailed teaching on such matters.

As for Whewell's teachings on the evidential and explanatory ideals appropriate to inductive science, there is no reason to think that Darwin's early work was at all directed by them. Most especially is this true of Whewell's leading proposal, the consilience of inductions. No one had access to this proposal in printed form until Whewell published it in 1840 in the *Philosophy*. Contrary to legend, there is no exposition of it in the *History*; indeed, it may well be that Whewell had not worked it out when he was writing that book. Nor is there any sign that Whewell was likely to have communicated the consilience doctrine to Darwin in conversation at the Geological Society, or indeed in any other setting. What is more, when Whewell did defend consilience in 1840, he offered it as a replacement for the *vera causa* ideal, which, he insisted (as he had in 1837), was too restrictive.

Now Darwin's argument for his theory of natural selection was conformed to that ideal and not to Whewell's consilience as a replacement for it. Certainly there is a part of Darwin's argument that would have been appropriate had he been presenting his theory as consilient. But that part is as much required by the *vera causa* ideal.[58] So its presence provides no ground whatever for reading that part as a trace of Whewell's influence; the place of that part in the whole of that argument is the place it should have, given that Darwin is conforming to the ideal Whewell had sought to discredit and replace. What is more, in July 1837, nearly a year and a half before Darwin had first come to natural selection,

[58] Hodge (1977), 237–45; id. (1989).

he was arguing—for an earlier theory of adaptive species formation in branching descent—along very much the same lines that he would give in his argument for natural selection in 1839, 1842, 1844, and 1859.[59] So there is simply no ground at all for finding any moment in Darwin's life as a moment when, from reading in Whewell's writings, receiving letters from him, or talking with him, he learns to embrace the consilience of inductions as a decisive new ideal for any theory of the origin of species.

Perhaps, then, if Darwin's idea of the universe and his ideal of science are not derived from Whewell, may his concern with man, mind, and morals not be traced to that source? Again, the evidence all suggests otherwise. That any theory of transmutation should include man, his mind and morals, was a conviction Darwin clearly had in July, 1837, when he opened his 'Notebook B'. There is every sign that Darwin's reading in Lyell's account of Lamarck, and in his grandfather's books, contributed to this conviction. There is no sign that Whewell had anything to do with it. A year later, Darwin opened a 'Notebook M', devoted to 'metaphysics', that is to say, mind and morals in men and animals. It has been suggested that in opening this notebook, Darwin was acting on a resolution to counter Whewell's views on such subjects.[60] Again, the trouble with this suggestion is that there is no sign that Whewell played this role in Darwin's life. His earlier 'Notebook C' and his contemporary discussions with his family show that he had many good reasons, at that time, for paying even more attention than he had before to metaphysical matters. Nothing suggests that Whewell provided Darwin with those reasons.

But if Whewell is not a decisive resource on broad issues for Darwin in the younger man's London years, was he not such a resource in the Cambridge undergraduate years before the *Beagle* voyage? One should doubt it, and for two reasons. First, a great deal that Whewell could have taught later and Darwin could have learned later was not available to them at that time. The Whewell of the *History*, let alone the *Philosophy*, was largely a child of the years when Darwin was away (1831–6); and Darwin himself at Cambridge had no specific ambitions, let alone achievements, as a scientific theorist, that Whewell could have helped him pursue. As to general educational culture, Whewell may well have been an inspiring and instructive presence at soirées. However, to go beyond this conclusion, to make Darwin primarily a product of informal Cambridge culture as exemplified by Whewell, is to risk, with Cannon, continuing the tendency that traces to Darwin's own century (and especially to the biographical efforts of his son Francis, a don at Cambridge), to claim Darwin for Cambridge and England rather than Edinburgh and Scotland. That this tendency now needs critical reappraisal has become apparent with enhanced understanding of the influence on the early Darwin of two Scottish and Edinburgh legacies: his Lyellian geology and his Grantian zoology.[61] To

[59] Hodge (1985); Hodge and Kohn (1985).

[60] W. Cannon (1976), 377–84; Curtis (1987); Barrett (1987).

[61] Hodge (1985); Sloan (1985); Hodge (1990).

persist in making Darwin descend from Whewell may be to persist, in effect, in English and Cambridge efforts to make Scotland and Edinburgh peripheral to the nation's understanding of its own heroes and history.

10. Whewell and Darwin's Theory

Moving on from the question of Darwin's debts to Whewell, to the older man's response to the younger, one notices right away that although the Whewell who writes about Darwin's *Origin* in the 1860s is that much older than the author of the *History* and the *Philosophy*, the view taken of palaetiological science remains the same. One might expect it to be so from the discussion of geology and of the nebular hypothesis in Whewell's writings on the plurality of worlds in the 1850s. For, despite geology being deployed in a new argument there, against the plurality of worlds inhabited by man, the account of geology itself contains nothing not in the *History* and *Philosophy*.[62]

Eventually, in a new preface to a seventh edition of his Bridgewater Treatise, Whewell explained, in 1864, why he found Darwin's *Origin*—he left book and author unnamed even when quoting them—unpersuasive.[63]

Whewell takes Darwin to be engaging the old choice between chance and design, as that choice was first formulated by the Greek atomists and those, like Cicero, who opposed their claim that the world has been produced by a 'fortuitous concourse of atoms'. There is, says Whewell, a 'new approach' to the fortuitous concourse doctrine 'in those who maintain, in its extreme form, the Nebular Hypothesis'. And he directs his readers to his old arguments against anyone who puts that hypothesis in opposition to the belief in an Intelligent Creator. The same arguments, he proposes, can be brought to bear on a current version of a proposal concerning animal structure. 'In opposition to the view of the Design shown in the structure of animals, as manifested by the adaptation of their parts each to its purpose', that structure has been credited to the 'operation of external circumstances and internal appetencies'. This is Whewell harking back to Lamarck as encountered in Lyell; but he goes on to say that 'very recently' the argument has been stressed that 'all those structures of animals which are less adapted to their mode of life, cease to be continued in the species, precisely because of this want of adaptation', so that 'no structures remain for us to consider' except those where 'the organisation answers the purpose of life perfectly'.[64]

'This doctrine', says Whewell, 'again taking the extreme form of it', makes the universe the 'result of an infinite number of trials, the failures being constantly rejected by the fact of their being failures'. So, in its extreme form, it says 'the world' is the 'work of chance'. But the same arguments that refute the Nebular hypothesis refute also 'this hypothesis in this form'. The arguments

[62] Whewell (1854*a*). [63] Id. (1864). [64] Ibid. pp. xv–xvi.

are that, if any animal structure has been produced by 'necessary causation' from some earlier state of the animal world, then 'there must have been in that previous state the germs of the adaptations' later produced; and 'these germs and possibilities of adaptation must, in reality, contain evidence of design'. The argument holds, moreover, no matter how far we go back.[65]

Whewell does not think, however, that this argument is needed. For he thinks one faces 'a mere work of fancy' in attempts recently made 'to show that the organs of animals, as they now exist, with at least a seeming adaptation to their purpose, may be derived by necessary causation from a state of things in which there was no such adaptation'. Darwin, he argues, in confronting cases such as the structure of the mammalian eye, has to resort to two assumptions that deprive his reasoning of all force. The first is that the mere possibility of imagining the required transitional steps is a reason for thinking them to have taken place; and the second is the assumption that to confirm that they have done so, one needs only to suppose that there has been an unlimited number of generations over an indefinite time. Beyond his resorting to these two assumptions, Whewell urges, Darwin has not given us any new insight to replace the teleology offered us by Cuvier and other authorities earlier in the century. There is then, in Darwin's speculations, nothing that calls for any revision of the natural theology of the first edition of his own Bridgewater Treatise.[66]

Whewell had already explained, in writing to an Aberdonian Professor of Theology in 1863, that recent discussions in 'geology and zoology' did not seem to him to have 'materially affected the force of the arguments' given in his *Indications of the Creator*. In 'tracing the history of the world backwards, so far as the palaetiological sciences enable us to do', it seemed that 'all the lines of connexion stop short of a beginning explicable by natural causes; and the absence of any conceivable natural beginning leaves room for, and requires, a supernatural origin'. Darwin's 'speculations' do not 'alter this result'. He still has 'an inexplicable gap at the beginning of his series'. Whewell's God of the gaps is, therefore, still defensible. Furthermore, most of Darwin's hypotheses 'are quite unproved by fact', says Whewell. 'We can no more adduce an example of a new species, generated in the way his hypotheses suppose, than Cuvier could.' Darwin still has to admit that the 'existing species of domestic animals are the same as they were at the time of man's earliest history'. Nor can unlimited geological time rescue Darwin. The 'advocates of uniformitarian doctrines in geology' persist in helping themselves to 'additional myriads of ages', but the 'best and most temperate geologists' still accept 'great catastrophes'; so the 'state of the controversy on that subject' is not 'really affected permanently'. What Whewell has written in the 1830s is, still, then 'a just representation of the question between the two doctrines'.[67]

A few months later, he wrote to his friend James Forbes about the antiquity

65 Whewell (1864) pp. xvi–xvii.
66 Ibid. pp. xvii–xviii.
67 Todhunter (1876), ii. 433–4.

of man. The 'course of speculation' on this subject has 'somewhat troubled me', he admits.[68]

> I cannot see without some regrets the clear definite line, which used to mark the commencement of the human period of the earth's history, made obscure and doubtful. There was something in the aspect of the subject, as Cuvier left it, which was very satisfactory to those who wished to reconcile the providential with the scientific history of the world, and this aspect is now no longer so universally acknowledged. It is true that a reconciliation of the scientific with the religious views is still possible, but it is not so clear and striking as it was. But it is still a weakness to regret this; and no doubt another generation will find some way of looking at the matter which will satisfy religious men. I should be glad to see my way to this view, and am hoping to do so soon.[69]

It was not to be. Whewell never, it seems, fulfilled this hope, at least not in print. Elderly men's retrospects can be misleading, but Whewell's recurrent invocation of Cuvier in these years is not a false note. Cuvier always represented many things for Whewell. Most obviously, he was a leading authority on natural history and on palaetiological science in the years when Whewell was forming his own views on these subjects. But Cuvier was also the enduring, magisterial, autocratic, Protestant careerist who had survived to maintain, throughout successive French political upheavals, a stance that the Tory, Anglican Whewell could find reassuring even if he did not identify himself quite unequivocally with the Frenchman. Whewell had been largely content in the 1820s with Cuvierian science, including its successful resistance to the challenges from Lamarck and Geoffroy St Hilaire. It would have been easier for Whewell if the later challenges, from Lyell, Darwin, and others, had been taken care of in advance. Whewell often lived his life, as an analyst of palaetiological science, as if they could have been; as it was, it often fell to him to find a level of generalization about this genre of science from whence those challenges could be represented as leaving the original stance, when abstracted from dispensable details, essentially inviolate.

11. Conclusion

In his reflections and writings on palaetiological science, Whewell lived and died by history. His views constitute a highly instructive response to the state of the art at one moment, the early 1830s. After that, Whewell was defending conclusions drawn from that momentary encounter. There may be a certain hubris in his apparent presumption that he could make a foray into this area, and leave it again having derived lessons that he would be unlikely to have to

[68] Ibid. 435.

[69] Ibid. 435–6. See also three letters to Lyell (pp. 429–32), concerning languages and the antiquity of man, that also show Whewell resting his case on the conclusions he had drawn nearly three decades before.

revise. His influence in geology, among the palaetiological sciences, seems to be very limited. To be sure, he contributed through the BAAS and the Geological Society to the high standing the science then enjoyed. But that standing was an achievement already secured by others before he came on the scene in the 1830s. On the inorganic side, he did join Sedgwick and others in promoting the pursuit of a distinctive Cambridge commitment to physical geology, as exemplified by the work that William Hopkins began to publish in the middle of that decade.[70] On the organic side, especially in the 1840s, his support probably advanced the cause of morphology as developed by his friend from Lancaster, Richard Owen, but only after he had convinced himself that this morphology did not displace teleology. Beyond such particular influences, there remains the possibility that he may have provoked Chambers into pursuing the hope of a comprehensive palaetiological scheme that relied only on natural causes.

In his encounter with palaetiological science, Whewell shows himself to be the judicious critic and cautious reformer that he is usually made out to be. Never subversive, never reactionary, he is not a radical or a visionary either, not a Bentham or a Coleridge. The reasons for his particular blend of judiciousness and caution do not have their sources in anything peculiar to geology or palaetiological science generally. That blend is characteristic of the kind of national institutional success story that is Whewell's career, in his triumphant early maturity in the 1830s, a career so exemplary of his generation and so exemplary of such motivations as the English nationalist Tory containment of the disconcerting consequences of the French and Industrial Revolutions. Whewell on palaetiological science is, therefore, a case where context has not been conditioning text topic by topic or argument by argument, but rather in the overall outlook and attitude. Nothing that is taken seriously by prominent respectable savants is to be lightly dismissed, and nothing that is cherished by moderate, intellectual divines is to be directly challenged. The authority of the formal and informal institutions committed to the congruence of these is to be cultivated, as a condition for the well-being of the nation and its contribution to civilization.

[70] For Sedgwick, Whewell, and Hopkins, see the full account in Smith (1985).

11. *Antithetical Knowledge*

by MENACHEM FISCH

1. Introduction

AS we have seen, Whewell's *Philosophy* of 1840 was the culmination of a process of intense philosophical deliberation that spanned the best part of the two decades prior to its publication. In its finished form the *Philosophy* did not recapitulate the process of its formation, but rather (and typical of the outcome of a Whewellian investigation) argued back down, in semi-deductive fashion, from what one might term axioms of the Fundamental Idea of antithetical knowledge, to a detailed account of the emergence, the structure, and the nature of truth of a well-formed inductive science.[1] In the *Philosophy* Whewell had accomplished what he had initially set out to achieve, namely, an empirically grounded and unitary theory of the emergence of antithetically structured science. He had attempted to put the theory to test, in part at least, in the *History* (neglecting, however, to update the work in the light of subsequent philosophical findings) and seems to have been largely satisfied with the results. No doubt the *Philosophy* was put together all too hastily, and, as Todhunter complained,[2] its overall set-up left much to be desired. Nevertheless, the two subsequent editions of the work and the long and varied list of Whewell's successive philosophical writings left the basic system of the *Philosophy* virtually intact.[3] Whewell evidently felt no need to further schematize or modify the *Philosophy* after 1840. In the third edition of 1858–60 the work was taken apart at its natural seams and was republished in the form of three separate works; a conceptual history of the sciences entitled *The History of Scientific Ideas* (1858); a critical history of the philosophy of science entitled *On the Philosophy of Discovery* (1860); and Whewell's own philosophical system tellingly dubbed *Novum Organum Renovatum* (1858). However, nothing of importance was changed or added to the original philosophical system.

After 1840 Whewell seems no longer to have reassessed his philosophy of science in the light of new problems. His philosophical efforts, so it seems,

I wish to thank Joseph Agassi, Zev Bechler, and Gerd Buchdahl for their criticisms and comments on an earlier version of this paper.

[1] See my 'A Philosopher's Coming of Age: A Study in Erotetic Intellectual History', this vol. Ch. 2, sect. 2, and my forthcoming *William Whewell Philosopher of Science*, sect. 2.1.

[2] Todhunter (1876), vol. i, ch. 10.

[3] Cf. Whewell (1847), (1845), (1848*b*), (1849*c*), 1860), (1852*b*), and (1854). Although the *Philosophy* was not superseded by any of these works, his (1851*a*) analysis of the transformation of hypotheses in the history of science certainly goes a step beyond the *History*.

were directed thereafter towards demarcating his views from those of others (the German idealists in particular), criticizing the shortcomings of other philosophical systems past and present, and following up the consequences of his views into areas other than the philosophy of science (notably ethics, education, and natural theology).

In a sense Whewell's attitude is understandable. In so far as it sought to account for the formation of the sciences of its day; new scientific advances posed the *Philosophy* no new problems. Whewell even seems to have had no difficulty combating the threat posed by Darwin's theory of evolution.[4] Subsequently, with the possible exception of one passage in Mill to which we shall return shortly, neither was he faced with any serious philosophical challenge by rival philosophers. Herschel, who reviewed the *Philosophy* at length, noted their differences, but seems to have been reluctant to engage in serious philosophical debate.[5] Others parted early with Whewell on epistemological fundamentals, going their separate ways though without seriously calling him to task. The *Philosophy* thus went virtually unchallenged. And Whewell, rather than reflect further on his position, as he had done continuously prior to 1840, chose now to direct his efforts towards the criticism of rival philosophies and the exploration of the ramifications of his views beyond the confinements of the philosophy of science. This was unfortunate since his system was by no means free of difficulties.

The aim of this chapter is to continue, as it were, where Whewell chose to break off, and to assess critically the system of the *Philosophy* in the light of its philosophical presuppositions. The focus will be on Whewell's theory of scientific truth and its warrants, upon which, as we have seen, the entire superstructure of the *Philosophy* was made to bear.

2. Whewell's Theory of Scientific Truth: A Schematic Overview

In the thirteenth of his 'Aphorisms Concerning Science' Whewell cryptically formulated his notion of scientific truth. A true induction, he says, 'is . . . a true Colligation of Facts by means of an exact and appropriate Conception'.[6] Since, according to Whewell, all knowledge comprises two antithetical elements, one factual and one conceptual, his theory of scientific truth is correspondingly twofold.

> Ideas are the *Form*, facts are the *Material*, of our structure. Knowledge does not consist in the empty mould, or in the brute mass of matter, but in the right moulded substance . . .

[4] See especially his introduction to the 1864 edition of the Bridgewater Treatise.

[5] Whewell responded to Herschel (1841) in a letter dated 25 July 1841, in which he argued that Herschel's position was far nearer his own than Herschel was willing to admit. To the best of my knowledge the debate was taken no further.

[6] Whewell (1858), 70.

Therefore, continues Whewell:

> our knowledge is then most complete, then most truly deserves the name of Science, when both its elements are most perfect;—when the Ideas which have been concerned in its formation have, at every step, been clear and consistent; and when they have, at every step also, been employed in binding together real and certain facts.[7]

Science at its best, Whewell maintained, comprises, in its finished form, an entire body of fact perfectly colligated by means of a well-formed Euclidian conceptual scheme. A prudent Whewellian researcher will hence strive to perfect his hypotheses in both antithetical respects. Aiming to produce hypotheses that are at once both true colligators of the facts under consideration, and, at the same time, well-formed theorems of a well-formed deductive system of ideas.

On the face of it the two ideals of perfection are unrelated. A proposition could be a well-formed theorem of a 'pure' deductive system without in principle colligating anything; much of pure mathematics, for example, deals with nothing that can be said to inhabit the outside world. Similarly, statements can be found empirically true (on a Whewellian showing) regardless of their participation in a wider deductive scheme. As for example what, in 'the Nature of the Truth of the Laws of Motion', he took to be the purely empirical finding that velocity does not change merely as a function of time.[8]

One could hence, at the outset, talk of two independent criteria of perfection for Whewellian science, which, in keeping with his own terminology, might be termed: the *colligatory adequacy* and the *conceptual appropriateness* of scientific hypotheses. Together, he maintained, they exhaust the notion of scientific truth. In striving to perfect the colligatory adequacy of a hypothesis, science, he held, could in principle arrive at empirical truth, and indeed had on occasion already done so. Whereas in the course of perfecting the conceptual component of a science, scientists could in principle discover what he called necessary truth, and indeed had on occasion similarly done so.

In what follows, I shall attempt to reconstruct and evaluate the two prongs of what I shall refer to as Whewell's antithetical theory of scientific truth. Former commentators have by and large tended to regard his notion of experiential truth as basically unproblematic, but have sharply criticized his contention that science has the power to recognize, let alone actually warrant, necessity. As we shall see, a closer examination of the texts makes plain that quite the opposite is the case; namely, that given the epistemological presuppositions of his system, his notion of necessity, subject to certain qualifications, is sound, whereas his claim that science is capable of reaching and recognizing empirically true knowledge turns out to be faulty.

[7] Ibid. 72.
[8] Id. (1847), ii. 216–17.

3. Colligation

First we turn to the less controversial yet far more problematic issue of factual truth.

Whewell's notion of colligation is one of empirical fit, akin to what one would today call *truth by correspondence*. Both the warrants of a perfect colligation, and the status of its truth, were for him wholly empirical: 'all received theories in science, up to the present time', he wrote,

> have been established by making up some supposition, and comparing it, directly or by means of its remoter consequences, with the facts it was intended to embrace. Its agreement, under certain cautions and conditions, . . . is held to be *the evidence of its truth. It answers its genuine purpose, the Colligation of Facts.*[9]

However, before turning to examine what Whewell held were the 'cautions and conditions' that ensure the colligatory adequacy of a hypothesis, we must first enquire what it could mean for him to *say* of a hypothesis that it is empirically true. For such a claim seems highly problematic in the context of the Fundamental Antithesis.

Central to Whewell's theory of knowledge is the claim that all order and design in the scientific world-picture is contributed by the mind. The purely empirical component of knowledge (in so far as it can be analysed philosophically) is in the last analysis *totally* devoid of form.

> Our sensations, of themselves, without some act of mind, such as involves what we have termed an Idea, have no form. We cannot see one object without the idea of space; we cannot see two without the idea of resemblance or difference; and space and difference are not sensations . . . Our sensations, from their first reception, have their form not *changed*, but *given* by our ideas. Without the relations of thought . . . the sensations are matter without form. . . . By the very act of being perceived as perceptions, they have a formative power exercised upon them, the operation of which might be expressed, by speaking of them not as *transformed*, but simply as *formed*;—as invested with form, instead of being the mere formless material of perception.[10]

If that be the case, what can it mean for a conceptual structuring of such 'formless material' to be empirically true? If colligatory truth is taken as pertaining to something like a correspondence relation, as implied in the earlier quotation, what exactly are Kepler's ellipses, for example, supposed to correspond *to* if all there is outside them is but 'formless material'? In what sense can it be said, to take Whewell's Aristotelian imagery a step further, that this or that form be *true* of a given lump of formless matter? The answer is clearly that it cannot. But then, so it could be argued, no Whewellian investigator ever confronts base formless matter directly—just as in Aristotle's world primordial base matter is never encountered as such. On a Whewellian

[9] Whewell ii. 45–6.
[10] Ibid. i. 36.

showing scientists, at all stages of their investigations, confront *facts*, that is, previously informed, rather than formless sensation. And hence, so the argument would go, whatever form already resides in the given factual materials under consideration determines the possibilities of their further informations. Thus, to use the Aristotelian metaphor once again, a block of marble, by virtue of its 'marbleness', shape, weight, etc. possesses only a limited range of potentialities. it cannot be appropriately informed so as to become a loaf of bread for instance. But even so, even if former informations limit, even drastically, the scope of further informations the facts will admit to, in what sense can *one* such legitimate potential form be considered *more true* of the facts than another? Even the most elaborate forming will in most cases not limit future information to one. Hence the truth of an interpretation or forming of facts cannot be elucidated in terms of its correspondence to the one single interpretation dictated by their previous formings, at least not in principle. What then, according to Whewell, can an interpretation of facts be said to be true *of*?

The answer, I suggest, lies in the fact that Whewell's ontological commitments transcended considerably the presuppositions of his theory of knowledge. Indeed, although the peripatetic imagery of matter and form worked well for his epistemological claims, it can be very misleading with respect to his ontology. Here theology too had a role to play.

In a word, what Whewell maintained was given to human knowledge by no means exhausted what he held reality to consist of. The world without, he believed, was created purposefully and meaningfully by an omnipotent and benevolent Creator. Yet when presented to man via his sensations, nothing of the Divine plan, he held, was disclosed. The book of nature, though initially composed with a definite meaning and by means of a definite 'grammatical scheme', presents man with no more that the 'formless material' of its bare and uninterpreted characters. Reality, Whewell held, resembles a meaningful yet foreign text. And it is up to man to read it lacking any preknowledge of its grammar: 'When we read a page, we have before our eyes only black and white, form and colour . . . the letters are nothing in themselves; they contain no truth if the mind does not contribute its share . . . If we do not know the language in which the words are written.'[11] 'Induction', he concluded, 'may be conveniently described as the interpretation of phenomena.'[12] The book of nature is hence a meaningful text, but its meaning is not, as it were, *in* the text.

Viewed thus, the truth of an interpretation can be meaningfully spoken of. A true interpretation is one that fully coincides with what the text was initially intended to mean. In science the world picture we frame is obtained by the reading of our conjectured self-fabricated 'grammars' *into* fragments of the 'text'. Still, the (epistemically inaccessible) 'grammar' by which the 'text' was originally written is something of which our conjectured readings may be said

[11] Ibid. ii. 617–18. [12] Ibid.

to be true or false. Epistemic fabrication can, *in principle*, be said to correspond to ontological reality, and as such be true or false of it.

In this respect, the Platonism of Whewell's ontology can coexist in principle with the modified Kantism (if one may call it so) of his epistemology.[13] Whewell, however, went one unfounded step further. Not only did he claim that there was something 'out there' to which a scientific hypothesis could be said to correspond, but, he urged, it is within our epistemic capabilities to *verify* such correspondence when it happens to occur. This though was to go a step too far. To claim that truth as such is *recognizable* is a blunt violation of his own epistemology. For if, as he claimed, the only outworldly input we enjoy is formless sensation, if the Lord's master-plan is forever concealed, and if, as he insisted, all form superimposed in knowledge upon unstructured sensation is truly mind-made, then even if certain of our hypotheses happen to dovetail with the Creator's original plan, there can be no way for us to *know* that they do. If all we have to go by is the bare text and our wits, and if the meaning is not in the text, there can be no way for us to confirm the truth of an interpretation.

John Stuart Mill put the point to Whewell forcefully:

> If we give ourselves the licence of inventing the causes themselves as well as their laws, a person of fertile imagination might devise a hundred modes of accounting for any given fact, while there are probably a thousand more which are equally possible, but which for want of anything analogous in our experience, our minds are unfitted to conceive.[14]

From Whewell's response (to this very passage) it is evident that he either misunderstood or deliberately chose to ignore Mill's criticsm, for he did not take it as a criticism of his epistemology, but as a contradiction of what he was sure science had in fact achieved.

> I reply, that the question is about accounting for a large and complex series of facts, of which the laws have ascertained: and as a test of Mr Mill's assertion, I would propose a challenge to any person of fertile imagination to devise any one other hypothesis to account for the perturbations of the moon . . . besides the hypothesis by which they have actually been explained with such curious completeness. This challenge has been repeatedly offered, but never in any degree accepted; and I entertain no apprehension that Mr Mill's supposition will ever be verified by such a performance.[15]

Whewell hence, deliberately or not, chose to avoid the main issue. By allowing for the recognition in principle of colligatory truth, he had rendered his position inconsistent. Whether or not Mill or anyone else could in fact match Newton's achievement, was of course quite beside the point.

Bearing all this in mind, let us now turn to examine what Whewell believed *vouched* for such recognition.

[13] On the problem of reconciling Whewell's Kantism and Platonism, see Butts (1967) and his references to Marcucci (1963).

[14] Mill (1846), ii. 22–3.

[15] Whewell (1860), 271.

4. Good Versus True Whewellian Science

Whewell's methodology (or 'heuristic' as he dubbed it) falls under three major headings corresponding to the stages preceding, during, and following scientific discovery. On the stage preparatory to discovery I have enlarged elsewhere.[16] It consists of the twin processes he termed 'The Decomposition of Facts' and 'The Explication of Conceptions' designed to disclose, prior to discovery itself, the type of concept sought for.

Discovery itself—that is, coming up with the right concept for the job—Whewell maintained, boiled down to a 'happy guess', inarticulatable and unanticipatable in principle. His methodological writings concerning this stage of an investigation foreshadow those of Popper writing almost a century later. Like Popper, Whewell called for hypothesizing freely, fancifully, and frequently (however, unlike Popper, all within conceptual boundaries previously disclosed by the decomposition of the facts). 'A facility in devising hypothesis', he urged, 'is so far from being a fault in the intellectual character of a discoverer, that it is, in truth, a faculty indispensable to his task'. Moreover,

> The tendencies of our speculative nature . . . perpetually show their vigour by overshooting the mark. They obtain something by aiming at much more . . . profound sagacity is combined with fanciful conjecture not rarely . . . but commonly, and in most cases; probably in all, if we could read the thoughts of discoverers as we read the books of Kepler.[17]

Bold speculation, however, is not enough, for it will lead to real knowledge only if firmly coupled to an 'ingenuity and skill which devises means for rapidly testing false suppositions as they offer themselves' ('a talent', he added, 'Kepler did not possess').[18]

The attitude to be taken towards one's speculations, he urged, should be at all stages one of critical impartiality; never to allow 'the labour . . . spent upon any conjecture to produce any reluctance in abandoning the hypothesis as soon as [one has] evidence of its inaccuracy'.[19]

But what should count as success? The positive content of Whewell's theory of theory-appraisal comprises two distinct components: the first, prospective, turning essentially on the merit of the researcher, the other essentially retrospective and wholly governed by hindsight.

[16] Fisch (1985), 284 f.; id. (1986), sects. 3.31 and 3.32.

[17] Whewell (1847), i. 54–5. Whewell stressed the virtues of daring to be wrong. In a Royal Institution lecture delivered January 1848 he declared (echoing *Philosophy*, Vol. ii, Bk. XI, ch. 5, sect. 2, and paraphrasing Bacon's *Novum Organum*, Bk. 2, aphorism 20): 'To try wrong guesses is with most persons the only way to hit right ones'. To which an enthusiastic Faraday responded later that year (interestingly quoting the Baconian version): 'I can testify for myself that the truth you laid down that truth can more easily emerge from error than from confusion, has been to me practically useful, and a source of continual pleasure in observing that in other things besides those I meddle with, it is so . . . ' (cited by Todhunter (1876), i. 170–1).

[18] Whewell (1847), i. 435.

[19] Id. (1858), 81.

First and foremost the hypothesis to be selected must 'explain the phenomena which we have observed'.[20] It must have survived attempted refutations in the light of the facts in hand. Moreover, since scientific hypotheses aim at comprehensive colligation, they should prove true of all phenomena of the kind investigated. Consequently a good hypothesis

> ought to *foretell* phenomena which have not yet been observed;—at least of the same kind as those which the hypothesis was invented to explain . . . Because the rule prevails, it includes in all cases; and will determine them all, if we can only calculate its real consequences. Hence it will predict the results of new combinations, as well as explain the appearances which have occurred in old ones. And that it does this with certainty and correctness, is one mode in which the hypothesis is to be verified as right and useful.[21]

Not 'true', Whewell was careful to observe, but 'right and useful'. Many *good* yet nonetheless false hypotheses, he pointed out, had in the past proved effective in their ability to predict within their intended domains of application.[22] Again, similarly to Popper, Whewell the (theologically motivated) realist yet epistemological conventionalist, goes on to outline in broad strokes what might be termed a theory of verisimilitude:

> Men cannot help believing that the laws laid down by discoverers must be in a great measure identical with the real laws of nature, when the discoverers thus determine effects beforehand in the same manner in which nature herself determines them when the occasion occurs . . . such a coincidence of untried facts with speculative assertions cannot be the work of chance, but implies some large portion of truth in the principles on which the reasoning is founded.

And in the metaphorical language of hermeneutics:

> To trace order and law in that which has been observed, may be considered as interpreting what nature has written down for us, and will commonly prove that we understand her alphabet. But to predict what has not been observed, is to attempt ourselves to use the legislative phrases of nature; and when she responds plainly and precisely to that which we thus utter, we cannot but suppose that we have in a great measure made ourselves masters of the meaning and structure of her language.

'The prediction of results even of the same kind as those which have been observed, in new cases', he concludes, 'is a proof of real success in our inductive process.'[23]

By prudently eliminating refuted hypotheses, and by constantly attempting to refute those devised to replace them, as for Peirce and Popper so with Whewell, we gradually approach the truth. The *best* hypothesis we can construct is one that survives the test of unknown cases; the one that proves an

[20] Whewell (1847), ii. 62.

[21] Ibid. 62–3.

[22] 'The *Epicyclic Theory* of the heavens', Newton's hypothesis of *Fits of Easy Transmission and Easy Reflection* in the particles of light, and even the doctrine of *Phlogiston* supplied the means of foreseeing the consequence of many combinations of elements even before they were tried' (ibid. 63–4).

[23] Ibid. 64–5.

equally adequate colligator of the kind of facts it was designed to accommodate in yet previously untried circumstances. That is the best a discoverer can do. It is not necessarily truth itself, though, he maintained, it is evidently close. Epicycles and Phlogiston, he observed, yielded true predictions and adequately colligated the facts in hand. And yet came a time when even they were refuted.

But rather than stop here, and view science, as would Popper, as an ongoing and essentially open-ended asymptotic approach to truth, Whewell chose to go one interesting yet unfounded step further, claiming that when actual truth happens to be hit on, science is capable of detecting it as such. We have already pointed out how such a position inevitably jarred with the presuppositions of his epistemology, but there is something further to be learnt from his error.

True hypotheses, Whewell claimed, distinguish themselves from merely good ones on two accounts. First, in exhibiting repeatedly what he termed 'consilience of inductions', and second, in generating 'progressive simplification' as they were applied to different kinds of phenomena.

By 'consilience of inductions' Whewell meant *explanatory surprise*. 'When the hypothesis, of itself and without adjustment for the purpose, gives us the rule and reason of a class of facts not contemplated in its construction,'[24] Such instances, he maintained, as the surprising explanation of the precession of the equinoxes by the theory of gravitation, initially constructed to explain 'the perturbations of the moon and planets by the sun and by each other', or the surprising explanation of the positions of the planes of polarization of the two refracted rays by the undulatory theory of light initially constructed to explain the laws of double refraction in calcspar.

> impress us with a conviction that the truth of our hypothesis is certain. No accident could give rise to such an extraordinary coincidence. No false supposition could, after being adjusted to one class of phenomena, exactly represent a different class, when the agreement was unforeseen and uncontemplated. That rules springing from remote and unconnected quarters should thus leap to the same point, can only arise from *that* being the point where truth resides.

The point was not entirely original. John Herschel, for example, pointed in his *Preliminary Discourse* (1830), to the special significance of what he elsewhere dubbed, 'successions of felicities'.[25] Whewell, however, unlike anyone before him, elevated the lesson to be learnt from repeated explanatory surprise to the status of an actual truth-criterion. More important perhaps, it was he (as in so many cases) who coined its suggestive and catchy name.

[24] Ibid. 67–8. There is some disagreement regarding Whewell's precise meaning of the term. For an assessment of the contemporary literature on consilience see Fisch (1985*a*).

[25] Herschel (1827), 359 (quoted by Whewell in his discussion of the reception in England of the undulatory theory of light (1857) ii. 349 f., and mentioned in connection with consilience (1847) ii. 67). In the *Discourse* Herschel makes much of the special virtues of 'verifications [springing] up, as it were, spontaneously, in to notice, from quarters where they might be least expected, or even among instances of that very kind which were at first considered hostile to them', Herschel (1830), 170 ff.

Original or not, the point is intriguing. Are the unexpected successes of a hypothesis more significant to its appraisal than its expected ones? Does the fact that a theory is found to provide an adequate explanation of phenomena radically different from those it was initially intended to explain *in itself* lend it extra support? (Let alone vouch for its truth!) I have attempted elsewhere to analyse Whewell's intuition with reference to latter-day theories of inductive support.[26] Whewell himself, however, did little if anything to argue for his two truth criteria.[27] He offered no argument whatsoever to explain *why*, if a hypothesis exhibits consilience or brings about a dramatic simplification of a scientific discipline, it should be regarded as duplicating the Divine plan of creation. But in view of the sheer impossibility of producing such an argument, I am led to understand him as really meaning something somewhat different.

The only argument Whewell did offer by way of a justification of his claim that consilience and a tendency to simplicity lend a hypothesis 'a stamp of truth' is that: 'No example can be pointed out, in the whole history of science, so far as I am aware, in which this Consilience of Inductions has given testimony in favour of an hypothesis afterwards discovered to be false.'[28] Undoubtedly, as David Wilson (in this volume) and others have pointed out, Whewell expected the truth, that is God's actual plan, to be both simple and all-embracing. And consequently, simple and seemingly all-embracing hypotheses should, upon this view, be regarded as approximating truth as such, at least to an extent. They could arguably be regarded, if you wish, as *necessary conditions* of true theories. But as *sufficient* conditions, as actual criteria of truth, all that can be said of Whewell's consilience and progressive simplification is that they are *descriptive characteristics* of theories he was *otherwise sure* were in fact true, namely, universal gravitation and undulatory optics (and at a later stage Faraday's theory of polarization.[29]) His criteria of truth, I suggest, were not intended and indeed could not explain *why* he believed these theories were true. Rather, *given* that indeed such was the case, they registered what he maintained demarcated the two with respect to all other hypotheses to date. (Which perhaps explains his weak response to Mill's criticism cited above.) In a word, the normative force of Whewell's twin criteria of appraisal rests primarily, I suggest, not on a philosophically grounded theory of truth, so much as on their being *descriptions of the best*. His mistake, both with respect to his own epistemological claims and certainly with respect to the historical record since his time, lay in his insisting that the best—namely Newton's

[26] Fisch (1985*a*).

[27] Lakatos also never saw the need to explain *why* research programmes exhibiting progressive problem shifts were preferable to those that equally successfully adjusted themselves *ad hoc* to new situations.

[28] Whewell (1847), ii. 67.

[29] In an essay Whewell drafted during the mid-1840s he sketched an inductive table of voltaic and chemical phenomena, similar to those of wave optics and physical astronomy incorporated in the *Philosophy*, leading up (down) to Faraday's discovery that 'Electric polarity and chemical affinity are identical'. See Whewell Papers, Trinity College Library, Cambridge University, Add. MS a. 217[4].

gravity and Young and Fresnel's wave optics—was demonstratively empirically true. However, and this should be emphasized, his thinking so, though no doubt an inconsistency on his behalf, should not be taken as indicating anything seriously wrong with his overall theory of inductive science (that is, his epistemological presuppositions). The four theses comprising his philosophy (the Fundamental Antithesis, his theory of induction, his account of excellent science, and twofold theory of truth) taken in conjunction with the *prospective* elements of his methodology, do not imply that science can knowingly reach empirical completion. All one has to do is to replace 'empirically true' with something like: 'as fine as gravitation and wave-optics' as what consilience and progressive simplification are supposed to warrant. I do not believe Whewell really meant anything else, despite the fact that he unfortunately and wrongly equated the two. Hence, as far as the empirical side of Whewell's theory of truth is concerned, Popper's theory of corroboration as appraisal fits Whewell's meaning better than do his own texts.

Indeed, at the time Whewell was writing, the theories of gravity and wave-optics formed a class of their own, which, notwithstanding his confusion regarding truth, he understood clearly. Unlike other theories, he pointed out, both perfectly embraced an entire class of phenomena, and reached their lofty status in an impressive series of what Lakatos would call (and would similarly never attempt to justify), progressive problem shifts, and neither theory required *ad-hoc* adjustments *en route*. Unlike Lakatos, Whewell, at least in the *Philosophy*, did not offer his criteria of excellence (as I prefer to call them) merely as criteria for the *choice* of the better of two rival 'research programmes', but rather as absolute measures of the truth or falsity of single 'programmes'. Evidently, although he regarded the two theories as equally verified, it was Newton's gravity, which once formed went virtually unrivalled, that served as his paradigm of excellence when writing the *Philosophy*, rather than the far more Lakatosian story of the ongoing rivalry between particle and wave optics.

Finally, it is interesting to note that Whewell's criteria of excellence are wholly retrospective. A hypothesis will be good by virtue of what goes into it; that is, if the existing portion of the science's conceptual scheme be properly exposed, if the hypothesis be formulated in keeping with it, and if the trial and error process of discovery be conducted prudently, there is a good chance that a good hypothesis will eventually be constructed, that is to say, one capable of adequate prediction within its intended domain.[30] But whether it will rise to an

[30] Whewell, however, never maintained that his heuristic for *good* science was foolproof (as, for example, Agassi has suggested (1981), 389), as his delightful example of the weather indicates: 'It by no means follows that when we have thus decomposed Facts into Elementary Truths of Observation, we shall soon be able to combine these, so as to obtain Truths of a higher and more speculative kind. We have examples which show us how far this is from being a necessary consequence of the former step. Observations of the weather, made and recorded for many years, have not led to any general truths forming a science of Meteorology: and although great numerical precision has been given to such observations by means of barometers, thermometers and other instruments, still, no general laws regulating the cycles of change of such phenomena have yet been discovered.' (1847), ii. 34.

excellence similar to that of the theories of Newton, Young, and Fresnel, only time will tell. Nothing a discoverer can do in the process of setting up his hypothesis can, according to Whewell's philosophy, ensure its eventual yielding of explanatory surprises or dramatic simplification when applied to new domains. Whewell's 'heuristic' is hence only capable of ensuring *good* science, excellent science (true science in his mistaken view), is, however, quite another story. Its achievement is not a question of sound method, but in a deep sense one of good luck. Such science cannot be contrived; at most, on a Whewellian showing, it may be recognized in retrospect.

From a somewhat wider perspective, Whewell's treatment of colligatory adequacy can be seen to exhibit three features characteristic of the nature of his entire philosophical enterprise. First, the manner in which his theory of theory-appraisal is found to hinge on an argument *from* an understanding of existing excellent science *to* a normative system for the future formation of similar science, echoes the manner in which, as we have seen, his entire philosophy took form. Second, in the clear divide it exhibits between anticipatable good science and totally unanticipatable excellent (or as he thought, true) science, the essential catastrophism of his account of the emergence of science again comes to light. And finally, his criteria of appraisal of both good and excellent science are in a profound sense historical. What determines how good a hypothesis will be, is to a large extent a function of how well the discoverer is able to acquaint himself with the conceptual history of his science to date. Subsequently, the future history of a science will alone determine its excellence. Indeed, his 'inductive tables' which he insisted captured the 'logic of induction',[31] were essentially schematic genealogical charts of the history of the sciences in which consilience, simplification, and the lack of them were clearly exhibited.

So much for colligation. Let us now turn to the more controversial area of conceptual appropriation, for it is there that necessity takes form.

5. Truth *of* Versus Truth *within* Systems of Ideas

There are two different senses in which one can talk of truth in connection with the conceptual scheme of a well-formed Whewellian science. One can talk, on the one hand, of the truth *of* the system *in toto*, in the sense of it being or not being, as a whole, best suited as the ideal component of the science in question.

[31] Whewell (1847), ii. 74–94. 'Such Tabular arrangements of propositions as we have constructed may be considered as the *Criterion of Truth* for the doctrines which they include. They are the Criterion of Inductive Truth, in the same sense in which Syllogistic Demonstration is the Criterion of Necessary Truth—of the certainty of conclusions, depending upon evident First Principles' (p. 93). See also his exchange with De Morgan following the latter's criticism of the *Novum Organum Renovatum*. Whewell wrote back saying of his methodology that: 'I am disposed to call it *Discoverer's Induction*; . . . I think my formulae are pretty near the mark, and my *Inductive Tables* a good invention. But I do not wonder at your denying these devices a place in Logic; and you will think me heretical and profane, if I say, so much the worse for Logic.' Whewell to De Morgan, 18 Jan. 1859, cited in Todhunter (1876), ii. 416–17.

Alternatively, one can talk of the quite different notion of the truth of a proposition *within* the system, in the sense of which, say, 'the sum of the angles of any triangle is the sum of two right angles' is true, and 'the sum of any two sides of a triangle is always smaller than the third' is false within Euclidian geometry. Although Whewell obviously differentiated between the two notions of ideal truth, he nonetheless conflated them by referring to them collectively as necessary. In order to assess precisely his usage(s) of the term we shall have to go beyond him and draw the line more carefully.

Of the difficulties involved in talking of the truth of a system *in toto* we have spoken above. Whewell maintained that for each class of phenomena there exists one unique wholly adequate interpretative scheme, and mistakenly held that once construed its truth could be vouched for by repeated consilience and its overall colligatory convergence. We shall return to the problem of uniqueness from a different perspective below. But beforehand let us turn to the second of Whewell's notions of ideal truth, namely, truth within systems of ideas.

The conceptual scheme of a Whewellian science comprises concepts and propositions concerning concepts which, though borne out by experience, are not learnt *from* experience. In its finished form such a system should resemble, both in its structure and inner logic, what Whewell called a 'pure' mathematical science. Hence, truth *within* the conceptual scheme of a natural science was, for Whewell, analogous to mathematical truth.

Whewell's views on mathematics and the nature of mathematical truth were initially forged, I have argued,[32] in an attempt to counter Lagrange's version of the calculus as advocated by the Analytical Society. In their finished form, however (first published in *Thoughts on the Study of Mathematics as Part of a Liberal Education*, 1835), they went far beyond the calculus to embrace all 'pure sciences'. Vehemently opposing Dugald Stewart and later John Stuart Mill, Whewell held that the propositions of a well-formed mathematical system were necessarily true, not simply because they followed necessarily from a set of axioms, but because the axioms they followed from were themselves necessarily true. Truth within systems of ideas was for Whewell a matter of more than mere coherence.

The axioms of geometry, he wrote,[33] are the 'necessary conditions which the Idea of Space imposes'. But what is the nature of such necessity, and in what sense can it be said to be 'imposed' by the idea of space? The fundamental idea of space, he maintained, was not, as William Hamilton and others had argued, simply defined by Euclid's five axioms. Rather, he conceived of the idea of space as a conceptual construct of which the five axioms were somehow *found* to be true. In the course of contemplating and attempting to articulate the idea of space, he held, it becomes gradually apparent that one cannot conceive clearly, for example, of two different points connected by more than one

[32] See my 'A Philosopher's Coming of Age', this vol., Ch. 2, sect. 3.3

[33] Whewell (1858), p. 9, aphorism 26.

straight line *within Euclidian space*. The axioms, he urged, are found true *of* the idea of space, their negations found inconceivable *within* it.

5.1 *Applicative necessity*

Whewell did little to clarify his epistemic claims regarding the realm of concept. He never really explained how he conceived of the process by which the mind supposedly examined a fundamental idea and acquainted itself so confidently with various of its features. He dubbed the process 'progressive intuition', but did little more than give it a name. The details of his epistemology, however, are not our concern at present so much as the precise nature of the truth claims he ascribed to knowledge of the conceptual. Time and again he hailed the truth of the axioms of a well-formed system of ideas (and consequently of all that follows from them) as necessary, but he never really took time out to explain the inevitability they were supposed to convey. He certainly did not hold that the axioms were *empirically* necessary, in the sense Leibniz viewed them as obligating each and every possible world. For Whewell, they did not express truths to which all possible worlds were subject. One could easily conceive of worlds governed by a different set of fundamental ideas of which quite different axioms were true. In fact, as I have shown elsewhere,[34] this is a point he makes explicitly in the Bridgewater Treatise. The laws of motion, he argued there, we know for certain are true features of the world God created, but (contrary to Michael Ruse's reading of Whewell)[35] not constraints on creation itself. In short, they are not true a priori of all possible worlds.

Yet they are true *irrespective* of creation, simply because they 'correspond' in Whewell's scheme not to features of reality, but to features of ideas. What is not at all clear is in what sense they are *necessary*. Whewell put it thus: 'The character of axiomatic truths seen by intuition is, that they are not only seen to be true, but to be necessary;—that the contrary of them is not only false, but inconceivable.'[36]

Whewell seems to be arguing that axiomatic truths purport to state relations among concepts which are not merely contingent on the fundamental idea in question, but for some reason cannot be otherwise. Not merely that 'Every

[34] Fisch (1985), 290 f.; id. (1986), 286–8.

[35] Like the majority of Whewell's critics, Ruse ascribes to him a true-in-all-possible-worlds notion of necessity, wrongly equating the inconceivable and impossible. Thus he writes 'assuming that God cannot choose the inconceivable (and Whewell as yet gave no reason to think otherwise[!]) . . . God, if such there be, no longer had an entirely free hand with many laws of science. The kind of Designer that Whewell was arguing for seems hardly the overwhelming powerful Being that Christians normally locate at the centre of their religion . . . God had no more to say on the matter than we have.' (1977), 251–2. Ruse notes no discrepancies between this reading of Whewell's notion of necessity and that expounded upon in the Bridgewater Treatise, although he does stress the fact that such a view of scientific necessity seriously vitiates the argument from design advocated in the Treatise.

[36] Whewell (1860), 338.

effect *is* produced by a cause',[37] but that 'every Change *must* have a Cause'.[38] Now what is it to say that *A*s are *necessarily B*s, as opposed to saying that *A*s merely *are B*s? The former is to say that being *A determines* being *B*, which is to say one of two: either that *A*-ness *causes B*-ness—which clearly cannot be the case in the realm of concept—or that *B*-ness is implied *analytically* by *A*-ness. But that too cannot apply to Whewell. Whewell went to great pains to argue that mathematical truth was more than analytic. The truth of the axioms, he urged, turned on more than *B* merely being included in the definition of *A* or on the logical form of the statement. He viewed fundamental ideas as complex conceptual constructs *of which*, in time, well-trained minds recognize certain statements to be true. Their truth is hence unaffected by, and can in no way obligate reality, since it belongs wholly within the realm of concept. And yet, to use Kant's idiom (so prudently avoided by Whewell), they are *synthetic*, in that, to use Ayer's definition, their 'validity is determined by the facts of experience'[39]—'experience' in the realm of concept that is. But if that be the case, in what sense are they necessary?

As I see it, within the realm of concept they are not. The idea of space, upon a Whewellian showing, cannot be said to necessitate Euclid's five axioms, they are simply found to be true of it. (The fact we cannot clearly conceive their negations does not impinge on their contingency any more than the failure to perceive the negation of a factual statement implies anything more than that it so happens to be the case. Conception with respect to the ideal is as perception with respect to the factual.)

However, and this to my mind is the crucial point, whenever such a system of ideas is *applied* there is an important sense in which necessity does come in to play.

The idea is the following: When a tool is applied to do a job, its contingent features *necessarily* impose constraints. For example, although a thin lens's focal length is certainly a contingent feature of the lens (determined by the lens's radii of curvature and index of refraction), it nevertheless dictates *necessarily* the location and size of the images produced each time the lens is used. It is necessarily true that one cannot climb higher than two metres on a ladder that just happens to be two metres high. The same, I suggest, applies to Whewell's notion of fundamental ideas. A fundamental idea functions in science, according to Whewell, as an interpretative apparatus by which an entire corpus of fact is structured. Independently of its special role in scientific knowledge, certain of its features come to light, he believed, in a process of 'progressive intuition' and are expressed in a set of axioms. The axioms thus construed, we have seen, are only contingently true *of* the fundamental idea, but apply *necessarily* each time the fundamental idea itself is employed. It just so happens that *within* Euclidian space only one straight line can be drawn between any two points, therefore in situations in which we confront phenomena

[37] Id. (1847), ii. 573. [38] Id. (1860), 362. [39] Ayer (1946), 78.

by means of the idea of Euclidian space, the five axioms apply *necessarily*. For want of a better term let us call such necessity 'applicative'.

Take for example the way in which Whewell analyses the nature of the truth of the laws of motion. He first points to what he calls the three axioms of the fundamental idea of cause; namely, that every effect is produced by a cause; that causes are measured by their effects; and that action is always accompanied by an equal and opposite reaction.[40] It is when the fundamental idea of cause is applied to the special case where change in the motion of a body is taken as an effect, that we obtain the necessary component of Newton's three laws, namely, that when no force acts, the properties of motion (velocity and direction) will be constant; that when force acts, its quantity is measured by the effects produced; and that when one body acts upon another, there will be reaction, equal and opposite to the action.[41] One does not have to assume, as Whewell did, that the three axioms are themselves necessarily true *of* the idea of cause, in order to establish, as he does, that when applied to changes of motion they are 'necessarily and infallibly true'.[42] They are indeed true a priori, but only contingently so, yet once applied to experience, they are rendered *applicably necessary* in the sense outlined above. Whewell failed to draw this line with sufficient precision leaving his notion of truth within systems of ideas unclear. What he did do masterfully was to differentiate sharply, following W. R. Hamilton,[43] between the necessarily and contingently true components of the laws of motion themselves. The former being the three axioms, *qua* axioms of a pure conceptual scheme, restated for the special case of changes of motion, the latter being the very same three propositions *qua* colligators of actual matters of fact.[44]

> The form, and even the language of these laws, is of necessity what it is; but the interpretation and application of them is not possible without reference to fact. We may imagine many rules according to which bodies might move (for many sets of rules, different from the existing ones, are, so far as we can see, possible) and we would still have to assert—that velocity could not change without a cause—that change of action is proportional to the force which produces it,—and that action and reaction are equal and opposite. *The truth of these assertions is involved in those notions of causation and matter, which the very attempt to know anything concerning the relations of matter and motion presupposes.* But according to the facts which we might find . . . we should settle in a different way—what is a cause of change of velocity—what is the measure of the force

[40] Whewell (1847), ii. 573 f. For a more detailed exposition of his analysis of the laws of motion see Fisch (1985*b*), 294–8, and (1986), 156–63. The analysis of Whewellian truth in terms of 'applicative necessity' proposed here, goes beyond the discussion of what I termed 'semantic necessity' in my (1985) and thereby supersedes it.

[41] Whewell (1847), ii. 567 ff.

[42] Ibid. 577.

[43] For a partial discussion of the exchange between Hamilton and Whewell on the nature of the truth of the laws of motion, see my 'A Philosopher's Coming of Age', this vol., Ch. 2, sect. 4., and forthcoming *William Whewell Philosopher of Science*, sect. 2.331. For an assessment of Hamilton's position in the debate see Hankins (1980), 172–80.

[44] See Fisch (1985), 296–8.

which changes motion,—and what is the measure of action between bodies. The law is necessary, if there is to be a law; the meaning of its terms is decided by what we find, and is therefore regulated by our experience.[45]

For a retrospective analysis of the nature of scientific truth, such as was attempted in his 1834 classic, no more was needed. The applicative necessity of an applied well-formed theorem of a system coupled to the empirical truth of the colligatory function it performs with respect to the facts, exhaust the complementary nature of truth to which a finished piece of antithetical knowledge pertains. In the *Philosophy*, however, Whewell's concern was with the *prospective* aspects of the formation of such knowledge, and here a third notion of truth presented itself.

5.2 The Nature of the Truth of Whewellian Transcendental Reasoning

I am referring to Whewell's peculiar use of transcendental reasoning. To recall,[46] to be able to preserve both a unitary theory of the growth of science and ensure that both components of antithetical knowledge—the factual and the ideal—form in a meaningful manner, Whewell had to show that certain axioms of the system somehow become known fairly early in the game. For if that were the case, inductive moves could be viewed at once as both colligations of fact up the 'ladder of axioms' *and* meaningful derivations of theorems down from axioms. His solution to the problem, we have seen, took the form of a methodology of pre-discovery geared to unearth, prior to discovery itself, as much as possible of the science's conceptual scheme deployed already in the knowledge we already possess. The idea being to be able to appropriate the type of concept sought for with reference to that part of the conceptual scheme already at work. The process, to recall, involved 'decomposing' the known facts, and rendering explicit, by a process of reasoning akin to a transcendental deduction, the general presuppositions of workers in the field.

In the course of empirical investigation, he stated, certain central axioms of a science's conceptual scheme come to light, the truth of which is gradually realized despite the fact that 'they are not results of experience, nor can be'. Rather, he argued: 'they are [found to be] necessarily assumed as the interpretation of experience. They are not proved in the course of scientific investigations, but brought to light as such investigations showed their necessity. They are not the results, but the conditions of experimental sciences.'[47] But what is the nature of such 'necessity'? It is not an applicative necessity in the sense outlined above. Applicative necessities, we found, hinge on a prior realization of the truth of the axioms with respect to a fundamental idea, which in turn belongs

[45] Whewell (1847), ii. 588 (my italics). Note that Whewell's modal argument with respect to other possible worlds applies only to the *application* of the fundamental idea of cause. The axioms of cause hold necessarily in each and every world *to which* the idea of cause is made to apply.

[46] My 'A Philosopher's Coming of Age' this vol., Ch. 2, sect. 6.2

[47] Whewell (1860), 349.

wholly *within* the realm of concept. Whewell's transcendental reasoning supposedly discloses various universal laws tacitly presupposed by scientists, but, so it seems, does not bear witness to a realization that they represent true features of a fundamental idea. Such a realization, he stated elsewhere, 'is not the first, but the most complete and developed condition of our conceptions which enables us to see what are axiomatic truths in each province of human speculation'.[48]

What type of necessity, then, *is* conveyed by transcendental deduction in Whewell's system? It is certainly not necessity in a Leibnizian true-in-all-possible-worlds sense. Even if our minds are doomed to contemplate the world only in certain ways, that in itself could not in any way limit the variety of possible worlds (although it could limit the variety of *knowable* ones). And in any case, Whewell never went as far as Kant in claiming that his fundamental ideas (a gloss over Kant's forms of intuition and categories) described the ways in which we were *obliged* to confront reality (let alone that the list he presented exhausted our epistemic capabilities).[49] Hence the necessity ascribed by him to the fruits of transcendental deduction cannot even be regarded, as in Kant, as pertaining to epistemic inevitability.

It would seem then that Whewell's particular use of transcendental deduction does not really convey a necessity at all, at least not in the applicative or epistemic senses of the word. In a sense it conveys a necessity on the higher philosophical level—namely that of the Fundamental Antithesis itself—by demonstrating that all knowledge, however low-level, comprises elements not learnt from experience. But in terms of its positive content, all a Whewellian transcendental deduction can do is to report that such and such a universal assumption is *in fact* presupposed by the experts in all experiential investigations of a certain type. In no way does it *warrant* the universal assumptions but only exposes them; it gestures, as it were, towards a science's as-of-yet unarticulated system of ideas, by pointing out possible candidate axioms. But it cannot vouch for their truth—neither applicative nor colligatory.

And yet, despite all this, in Whewell's mature system, applicative and transcendental necessities (whatever the latter might mean) are conflated. Time and again he sounds as if he is claiming that transcendental reasoning not only points to candidate axioms, but can actually vouch for their necessity: 'that this doctrine of the indestructibility of Substance is a primary axiomatic truth, is certain; both because it has been universally taken for granted by men seeking for general truths; and because it is not and cannot be provided by experience'.[50]

A rather odd statement, coming three pages after the following definition of axiomatic truth: 'The character of axiomatic truths seen by intuition is, that they are not only seen to be true, but to be necessary;—that the contrary of

[48] Whewell (1847), ii. 633.
[49] For an assessment of the extent of Whewell's Kantism see Fisch (1985), nn. 19, 46*a*.
[50] Whewell (1852*b*), 14.

them is not false, but inconceivable . . . the test of axioms is that the contrary of them is inconceivable.'[51] Similarly, with respect to the axioms of cause, in the last analysis it is not the test of the inconceivability of their contrary that seems to count, so much as the assent of experts: 'we accept and adopt . . . the existence of necessary truths concerning causes as there exist necessary truths concerning figure and number. *We find such truths universally established and assented to among the cultivators of science, and among speculative men in general.*'[52] Similar remarks can be found in his discussions of the truth status of other fundamental ideas.

Now it is certainly the case that Whewell never took time out to fully analyse his two notions of necessity. He never differentiated properly between them, and his treatment of both—applicative necessity in particular—left much to be desired. Nevertheless, his conflation of the two, such as in the passages quoted, serves to highlight an important aspect of his philosophy; that to do with what he claimed *warranted* necessity. An aspect in relation to which the applicative and the transcendental are found not to be as alien to each other as one might think.

Knowledge, claimed Whewell, comprises both an empirical and an ideal element; the former supplied by the world without, the latter actively constructed by the mind. Scientific knowledge at its very best is knowledge found true on *both* antithetical accounts. It must correspond perfectly to the facts on the one hand, and do so by employing, as far as possible, only fully 'explicated' conceptions on the other. In so far as knowledge achieves the former it can be said to be (as he maintained mistakenly) empirically true, and in so far as it achieves the latter it can be said to boast what I have termed applicative necessity. Truth of the facts is warranted, according to Whewell, by empirical test procedures—prediction within the intended domain, simplicity and consilience outside it. But what warrants truth *within* systems of ideas?

On a Whewellian showing the answer is simple: since the entire ideal component of knowledge is given by the mind, and since truth *of* a fundamental idea belongs wholly within the idea, it follows that the mind itself is the sole legislator with respect to such truth. There is no higher authority, nor can there be. The highest authority is the best minds available, that is, the opinion of the experts. The Fundamental Antithesis itself warrants the *possibility* of there being applicably necessary truths in a given body of knowledge. But the *justification* of the necessity of any one specific statement rests ultimately upon the judgement of the scientific élite.

The Idea of Space must be quite clear in the mind, or else the Axioms of Geometry will not be seen to be true: there will be no *intuition* of their truth; and for a mind in such a state, there can be no Science of Geometry . . . The Idea of Space is much more commonly clear in the minds of men than the other Ideas on which science depends, as Force or Substance. It is much more common to find minds in which these latter Ideas

[51] Ibid. 10–11. [52] Id. (1847), i. 175 (my italics).

are not so clear and distinct as to make the Axioms of Mechanics or of Chemistry self-evident . . . Hence, though the Axioms of Mechanics or of Chemistry may be, in their own nature, as evident as those of Geometry, they are not evident to so many persons, nor at so early a period of intellectual or scientific culture. And this being the case, it is not surprising that some persons should doubt whether these Axioms are evident at all;—should think that it is an error to assert that there exist in such sciences as Mechanics or Chemistry, Fundamental Ideas, fit to be classed with Space, as being like it, the origin of Axioms.[53]

This kind of élitism with regard to the truth of axioms casts new light on Whewell's use of transcendental reasoning. For a transcendental deduction not only demonstrates *what* universal laws are assumed by the experts, but bears witness to the fact that indeed they *are* assumed by the experts. Which means in effect that in so far as expert assent is the sole warrant of truth within conceptual schemes, transcendental deductions, in disclosing such assent, bear directly on the *truth claims* of the laws they yield as well as exposing them. Full-blooded truth within systems, it is true, requires conscious, deliberate, and well-debated cultivation of the system,[54] whereas, as a rule, transcendental deduction only bears witness to a tacit agreement among experts.[55] Nonetheless, by the end of the day, the warrant of Whewellian necessity boils down to a widespread and wholehearted conviction on behalf of the élite. And in so far as transcendental reasoning can point to such sentiment, it can be thought of, within Whewell's system, at least as *foreshadowing* his notion of necessity. Thus, the difference between the two types of necessity referred to by Whewell, emerges as one of degree rather than kind.

To sum up, the role of transcendental deduction in Whewell is hence twofold (*a*) it serves to articulate the tacit non-experiential presuppositions operating at low-level empirical research; and (*b*) by pointing out the universality of their employment, to *warrant* to an extent the applicative necessity of their truth. Whewell's mistake, however, was in claiming further that in doing so the transcendental deduction of a number of axioms suffices to determine uniquely the true conceptual scheme of a science in advance.

Whewell's misconception with respect to the empirical truth of a conceptual scheme was thus twofold. On the one hand, as we have seen, he held that empirical test procedures could vouch for the colligatory truth of single well-formed theorems of the system, whilst transcendental deduction, by exposing and vouching for the ideal truth of various of the axioms of a system, served to locate in advance the appropriate fundamental idea uniquely suited for the job.

[53] Whewell (1860), 337–8.

[54] Cf. id. (1858), 45 f.

[55] Whewell was fond of remarking that the Fundamental Antithesis is not apparent in true knowledge since 'Although in every inductive inference, an act of invention is requisite, the act soon slips out of notice . . . The pearls once strung, they seem to form a chain by their nature. Induction has given them as unity which it is so far from costing us an effort to preserve, that it costs us an effort to imagine it dissolved.' (1858), 76. Hence in that sense, the fact that an ideal presupposition is indeed tacit, might itself vouch for its truth.

Both claims were of course unfounded. Not only are they refuted by the historical record but were, as we have seen, inconsistent with the epistemological foundations of his philosophy which in principle excluded the idea of an empirically complete science.

Historically speaking, his position is, in a sense, understandable. The histories of science and of mathematics had, after all, yet to produce equally successful alternatives to Newtonian dynamics, gravitation theory, and wave optics. And, as a matter of historical fact, Whewell knew nothing of Lobachevsky, Bolyai, and Riemann's work on 'Euclidianized' though non-Euclidian geometries. The specimens of truly excellent physical and mathematical theories that he was familiar with certainly did look unique at the time. But of course, in the light of his epistemology, if not that of the historical record, he should have been far more cautious. Moreover, even if we grant him some justification for viewing science as capable in the long run of articulating a science's uniquely appropriate conceptual scheme, he still had no grounds at all for maintaining that transcendental reasoning was capable of picking it out at such an early stage of an investigation. It is one thing to claim that representations of entire bodies of fact admit in the long run to one and only one system of ideas, but quite another to hold that the early location of one or two of its axioms suffices to pin the whole system down. It is too much of a pre-established harmony.

This considerable flaw in Whewell's mature philosophical position, was, I propose, a direct result of his premature *History*.[56] For lack of a suitable historiography of 'inductive preludes', the examples the *History* offered were largely regimented by hindsight. The early stages of a science, in which some of its axioms gradually come into view, were retold in the *History* not as a critical dialogue between rival anticipations of the final system, but to a large extent as the story of the few who happened to anticipate the system time would eventually establish as the correct one. But rather than amend the premature *History* in the light of a better thought-out philosophy of discovery, the dialectic was reversed. And his historical examples were allowed to dictate to his philosophy. His historiography of hindsight thus bestowed upon the early transcendental glimpses of a system the unfounded prophetic power of determining it in advance, thus curbing to a great extent the truly open hermeneutic quality of Whewellian science as determined by the Fundamental Antithesis.

[56] For a full discussion of this point see Fisch (1986), 218–22.

12. *Deductivist Versus Inductivist Approaches in the Philosophy of Science as Illustrated by some Controversies Between Whewell and Mill*

by GERD BUCHDAHL

I

THE contrast between the two approaches listed in the title has over the years gained a certain prominence in discussions on the philosophy of science. With the knowledge of hindsight it will be of interest therefore to study its incidence during an earlier period, and in particular, in the writings of Whewell and Mill which may yield added significance for a later generation. For the moment I shall not attempt to define the two schools of thought delineated by these labels, partly because at this stage—being controversial—any account of necessary and sufficient criteria would be difficult to come by, and partly because it may be more fruitful in the present context to evaluate the differences by way of a study of the philosophical positions held by Whewell and Mill themselves. So for the moment we may use them simply as convenient reference labels. The situation is in any case complicated by the fact that, on the one side, despite Whewell's apparently explicit attachment to deductivism, he also elaborated a nominally inductive type of methodology; whilst on the other side, Mill's explicitly inductive approach did not prevent him from attributing considerable importance to deductive methods in the development of scientific theorizing, especially in the realm of physics.[1] The great difficulty lies in the fact that not all inductivists agree on their principles or on their

[1] Portions of this chapter were originally published in *Monist*, 55 (1971), 353–67. The present version provides a fuller discussion of Mill's responses to Whewell, such as his emphasis on induction, by contrast with Whewell's notion of 'Fundamental Ideas'; also of the significance of the different uses made by Mill of his notion of 'description', and of his espousal of a realist position, by contrast with Whewell's more conceptualist approach to science. Cf. Mill (1879), i. ch. 21: 'To the Deductive Method . . . the human mind is indebted for its most conspicuous triumphs in the investigation of nature. To it we owe all the theories by which vast and complicated phenomena are embraced under a few simple laws . . . ' ibid. i. 538.

interpretation of the logic of scientific reasoning, and the same is true of deductivism; and the difference can therefore be discussed only in connection with individual writers. When this is done, we do find moreover some considerable variations in the treatment of their respective doctrines by the members of each of our two schools. To define these differences it will be useful to have recourse to a finer structure of methodological elements, indicative of criteria for the acceptance of scientific hypotheses. Moreover, in order to place these components within a proper perspective, it will be useful to refer to the views of some subsequent philosophers of science with which to contrast those of Whewell and Mill.

To mark the relevant distinctions, we will define the following components of the appraisal of hypotheses, with respect to which various philosophers hold a variety of different views. These have to do with what I shall label (*a*) the formulation; (*b*) the explication; (*c*) the probation; (*d*) the consolidation; and finally, (*e*) the validation of hypotheses. The details (and sometimes even the very existence) of these components are of course controversial, so we can at this stage give only very crude and provisional explanations. 'Formulation' appertains to the logic of discovery; to questions about the development of individual theories, the place of imagination, of background knowledge, existing science, and so on. 'Explication' considers problems of meaning, of the clarification of scientific conceptions, of the interpretation of scientific terms, and similar matters. 'Probation' defines principles of empirical support; for instance it marks the obvious feature that a hypothesis has to account for the data which it is introduced to explain. 'Consolidation' refers to the expansion of probative procedures. For instance, it introduces considerations of simplicity, and of what Whewell called 'consilience of inductions', that is the systematic stratification of individual hypotheses, and their systemic integration within a wider context.[2] Finally, 'validation' comprises the consideration of principles that belong to the field of 'justification of induction', including the peculiarly 'metaphysical' theories which can be found to underpin the various positions held by different schools.

Inspection of the doctrines of the two schools represented by Whewell and Mill respectively, will show how different are their appraisals and their 'weighting' of the importance of these various factors. Nor is there usually a single unitary doctrine concerning each of the individual components. Thus in the course of controversy, Mill may sometimes regard 'formulation' as mere description of the initial phenomena, though on other occasions he is willing to concede the importance of invention and imagination.[3] Against this, whilst Whewell at certain times suggests that formulation of hypotheses amounts to 'happy guesses',[4] at other times a more serious doctrine emerges which involves a deep penetration of the historical element in discovery, and the

[2] Cf. Whewell (1847), ii. 62, 65, 67, 77–8, 869; id. (1860), 274.
[3] Cf. Mill (1879), i. 342 with 343, 352; id. (1868), ii. 8–9, 16.
[4] Whewell (1847), ii. 41.

importance therein of the clarification ('explication') of the basic concepts of the science involved.[5] Moreover, where Mill will regard explication, if at all, as only 'subsidiary' to the processes of induction,[6] for Whewell this constitutes an intimate part of formulation. In a similar way, whereas Whewell on the whole tends to treat probation and consolidation together as part of a single process, Mill separates them rather more sharply.

Consider, as a preliminary example, the differences in their approach towards probation and consolidation. Whewell's basic procedure is that of fitting hypotheses to the data. As further data come to light, the original hypotheses may need modification, but as his survey of the history of the sciences was intended to indicate, in the long run special support is usually forthcoming, in the shape of new phenomena of a different kind which then turns out to be capable of being covered by the original hypothesis. Eventually it may be possible to show that the original hypothesis is part of a group that can be made to fit a still higher-level thoery, provided the new theory also leads to an increase in 'simplicity'; a theory which will (as Whewell often says) view the old data from a new point of view. It may then be possible to exhibit the whole hierarchy of hypotheses in what Whewell labels a 'Table of Inductions',[7] exemplifying what I have called 'consolidation'. Nevertheless, the fundamental logic behind this scheme is that of displaying a set of hypotheses to which the vast variety of data has been fitted. The tables of induction do no more than indicate graphically the *deductive* movement from higher-level to lower-level hypotheses, and from the latter towards the data. So far, therefore, we never pass beyond the process of endless fitting and continual testing, however much the architectonic of the tables may represent a stable and completed monument of theoretical achievement.

Clearly, this is very much in the spirit of the kind of methodology espoused in our time by the Popperian dynamic approach, with its emphasis on 'criticism' and attempted falsification of hypotheses. Any 'logical' movement that exists here is entirely from hypotheses to the data which are exhibited formally as their deductive consequences.

Whewell puts this quite succinctly: The formulae of this method

> declare only that the results *can* be clearly explained and rigorously deduced by the employment of a certain Definition and a certain Proposition. . . . The doctrine which is the *hypotheses* of the deductive reasoning, is the *inference* of the inductive process. The special facts which are the basis of the inductive inference, are the conclusion of the train of deduction.[8]

This is as pure a statement of the *deductivist* doctrine as could be desired. Also, though deductivism certainly employs the *expression* 'inductive inference', it seems so far to be employed solely to summarize the fact that verifiable deductions have been successfully fitted to the corresponding hypotheses.

[5] Cf. ibid. 5–16.
[6] Cf. Mill (1868), ii. Bk. IV, ch. 2 pp. 189 ff.
[7] Whewell (1847), ii. 78–9, 82–3, 93–4.
[8] Ibid. 90, 92, italics in text.

Now this emphasis on hypothesis has a methodological and a logical aspect. And in both respects, the deductivist approach is sharply at odds with that of the inductivist, represented in its purity by the philosophy of Mill. As regards the methodological aspects, what we may call 'the inductive *process*' the method of hypothesis is given only a subsidiary role by Mill, and is joined by additional inductive schemas, as for instance the derivation of causal laws from instantial premisses by means of the employment of a number of 'causal methods', the details of which need not concern us here.[9] What is important for the moment—and this concerns the logical aspect of the question, that is, inductive *inference*—is Mill's view that 'probation' is more than mere 'fitting of data to hypotheses'.

Consider the most primitive of cases, direct generalization. It is not sufficient to show that the death of individual men 'fits' the hypothesis that all men are mortal. On the contrary, what remains to be proved, even after all methodological precautions have been taken, for example, elimination of irrelevant circumstances, counteracting causes, and the like—a task of the inductive methods—is that these individual deaths do in fact *argue to the conclusion* that all men are mortal. Or, what for Mill is the same thing, we require to show that from these statistics of individual mortality we can *inductively infer* and thus substantially predict, the future demise of a specified individual.[10] It is not enough to show that the data would fit some given hypothesis, but we require to 'prove' that the hypothesis actually holds, 'is true'. An alternative way of putting this is to say that we require to establish, as a minimum demand, that the hypothesis should possess at least some antecedent probability of being true.

What holds of the 'primitive cases', that is, the formulation of laws, applies even more to the case of the explanatory hypotheses which are meant to account for these laws. Here again (as we shall see in more detail in what follows) Mill demands more of the inductive process than the fitting of laws to theories; we need to show independently of the success of such a method that our hypotheses are true. This is particularly (and perhaps only) clear in cases of hypotheses with putative existential import. It is not enough to show that the phenomena hypothetically imply a causal agent responsible for them; we need to exhibit independently the existence of such a cause.[11] But even where such existential claims cannot be substantiated—and this point concerns the *logical* aspect of inductive inference—we need to establish at least a finite antecedent probability for the hypothesis, for instance by falling back on considerations of analogy and uniformity.[12]

In order to avoid confusion in what follows, we note that we shall in fact have to distinguish three different species that fall under the genus 'induction' as employed by Mill, two of which concern the inductive *process*, (*a*) inferring laws of phenomena from particular instances, employing Mill's methods;

[9] Cf. Mill (1868), ii. 11–13; on the Methods in general, id. (1879), i. Bk. III, ch. 8, pp. 448 ff.

[10] Cf. ibid. i. 352–3.

[11] Cf. ibid. i. 348–9 n.; id. (1868), ii. 20–3.

[12] Mill explains this case by reference to the example of geology; cf. ibid. 25–8.

(*b*) inferring an explanatory hypothesis, given these laws; whilst the third (*c*) concerns inductive *inference* alone, and is identified by Mill with enumerative induction, amounting to the projection of individual cases into the universal.

In general, then, Mill insists on some independent *logical* requirement that has to be satisfied before a hypothesis can be accepted; and no amount of additional 'consolidative' supplements, such as the existence of strong consilience, of analogy, of simplicity,[13] and so on, can take the place of induction in this sense of the term, which is evidently very different from that used by Whewell. As Mill sees it, the deductivist for ever remains at the level of hypothesis.[14] As in the case of our present-day Popperians, he cannot speak of establishing the truth (or even probability) of a hypothesis in any inductivist sense. For him, science simply consists of a programme of continual research, where (as Kant had already formulated this position) we can only 'project' our systematic theories on to nature, but cannot say that anything 'given' *corresponds* to the systematic unities thus projected.[15]

II

This however is not the whole story, for not all deductivists share such a thoroughgoing belief in the unavailability of any further conditions capable of generating some kind of validity for their conclusions, additional to the operational strength bestowed by the methodological process of probation and consolidation.[16] The considerations here operative belong to the component of validation, and since this is brought into play only after the operations of the inductive or deductive process have been exhausted, we clearly dwell here at the level of presuppositions and metaphysical foundations. To discuss the aspect of validation, it will however be convenient first to consider some contrasting inductivist positions.

We have seen already that Mill insists on the necessity for an additional inductive inferential step. As will become apparent later, he, together with most inductivists, holds that there must attach to such inferential reasoning a peculiar feature labelled 'inductive validity'.[17]

This is an old doctrine. It affirms that the condition of inductive validity is in some way present 'in the nature of things', that it has an 'ontological'

[13] The assumption that 'nature always acts by the simplest means', Mill includes amongst his 'fallacies of the simple inspection'; cf. ibid. 320.

[14] Cf. below, near end of Sect. VI.

[15] Kant (1933), 535; cf. Buchdahl (1969), 506–9.

[16] Goodman (1954), 73–7.

[17] When speaking of 'inductivism' in this way, we will for the moment restrict ourselves to the older types of inductivism, and ignore neo-inductive developments in the present century, since the assumptions required for a satisfactory formulation of confirmation are still somewhat debatable. Moreover, we shall soon see that some of the formulations of the 'new problem of induction', such as a satisfactory definition of 'lawlikeness', for instance in the work of N. Goodman, ibid., share some of the problems of the older inductivism, when that is properly understood.

foundation; that there must correspond to inductive inference something that *makes* such an inference valid. Already in Aristotle this appears as the conception of the 'commensurate universal', but the idea goes under many names. In Mill, as will be explained later, we meet this as the doctrine that there is something about the world in virtue of which inductive inference can be said to be 'valid', even though such an inference may be *fallible* to varying degrees.[18] Sometimes Mill seems to say that the underlying assumption involved is that of the 'uniformity of nature' expressed as the fact 'that the universe is governed by general laws';[19] sometimes also as 'the law of causation': 'it is a law that there is a law for everything'.[20]

Now although this 'law of causation', according to Mill, itself depends on the 'validity' of inductive inference, the *significance* of this notion is clearly caught by the term 'uniformity' in the sense of universal lawlikeness. Mill restricted his view simply to that of *affirming* the presupposition of lawlikeness. The 'new problem of induction' as formulated in our own day, will then be seen to take off where Mill ended, that is, in attempting to define this notion of lawlikeness in some more satisfactory way, for instance by means of the concept of 'entrenchment' of projectible predicates, as in Goodman's philosophy of induction.[21] But already earlier classical inductivist writers had characterized foundations of validity somewhat more explicitly than Mill. Thus the early Russell spoke of 'the uniformity of nature',[22] and J. M. Keynes elucidated a number of characteristics, such as the principles of limited variety of groups of qualities, and of atomic uniformity.[23]

There is however an alternative prominent characterization of certain forms of inductivism which again goes back to Aristotle who construes the 'commensurate universal' as expressing something 'necessitarian'.[24] The position which then emerges is that inductive inference is valid only on condition that there should exist a peculiar relation, for instance some form of 'entailment' or necessary connection, between the terms entering into the formulation of scientific laws. This is the kind of non-logical necessity usually referred to as 'nomic necessity'.[25] William Kneale was a prominent exponent of this form of necessitarian inductivism; for him, scientific laws express the ranges of what is possible and what is impossible. True, we can never *know* that a hypothesis which we have formulated (under the guidance of a principle of simplicity) in the light of the best possible evidence, and which has run the gamut of attempted falsification, *is* a principle of impossibility (and hence, by conversion, of necessity);

[18] Mill (1868), ii. 101; cf. below, Sect. III.

[19] Mill (1879), i. 354.

[20] Ibid. 376.

[21] Goodman (1954), 80–3, 89–90. This is a useful way of regarding the matter, since it establishes an intelligible meeting-point between inductivists and deductivists. Kant, for instance, who belongs to the latter school, held that the problem for the 'metaphysics of nature' was that of establishing the 'lawlikeness' of nature. Cf. Kant, (1910), p. 176, ll. 27–8. For this, cf. also Buchdahl (1974). Whewell's doctrine of 'fundamental ideas', to be explored presently, aims in a similar direction.

[22] Russell (1912), 98–103.

[23] Keynes (1921), 249, 260.

[24] Buchdahl (1963), 18–34.

[25] Johnson (1924), 9; Braithwaite (1953), 293.

nevertheless, the policy of choosing a hypothesis on the evidence would be pointless, Kneale argues, unless there *were* principles of necessitation.[26] In the same way, Keynes had already described the lawlikeness of general uniformity (elucidated in his aforementioned various principles as involving a 'necessary connection'.[27]

To be sure, not all inductivists explicitly construe the notion of 'validity' in terms of nomic necessity; Mill, for one, strenuously opposes any interpretation of laws as explicitly necessitarian,[28] although it is a much-debated question whether his analysis of lawlikeness in terms of 'unconditionality' as well as 'invariableness' does not in the end boil down to much the same thing. To this I shall return in the next section. Others, for example Hume, Braithwaite, and Popper, whilst allowing some form of syntactical or semantical sense to the notion of natural necessity, do not employ this as a characterization of inductive validity: Hume, because he has a non-rational theory of inference; Braithwaite, because he opts for a self-correcting notion of inductive inference on the lines of Charles Peirce; and Popper because, being a deductivist, he regards the very concept of inductive inference as otiose.[29] This brings us back to validation and deductivism.

Whilst not all inductivists are explicitly necessitarians, certainly not all necessitarians are inductivists. On the contrary, necessitarianism is also an important aspect of some central examples of deductivism. The most prominent thinkers here are Kant and, influenced by him, Whewell. They differ however from the inductivist necessitarians in disclaiming any ontological significance for nomic necessity. Paradoxically, not unlike the Hume of the *Treatise*, necessity for them is grounded in something 'subjective' or 'mental', though Hume's psychologizing solution is converted into something with transcendental or a priori status. For Kant, the lawlikeness of nature is caught at three different levels, (*a*) the level of 'nature in general', a notion which contains as a necessary ingredient certain categorical concepts (which have for Kant a necessitarian grammar); (*b*) the level of Newtonian particulate nature (what he calls 'the metaphysics of science') which contains these same categories but applied to the empirical concept of matter; finally, (*c*) the level of 'the order of nature' which is, as already remarked, the purely 'subjective' resultant of the scientist's 'projection' of scientific theories on to the concatenation of empirical laws.[30] Theories, on this view, are generated under the guidance of certain

[26] Kneale (1949), 258.

[27] Keynes (1921), 251, 263.

[28] Mill (1868), ii. 419–21.

[29] For Braithwaite, nomic necessity has no ontological significance, but is the resultant of purely logical considerations. Cf. Braithwaite (1953), 11, and ch. 9. For his theory of induction, cf. ch. 8. For K. R. Popper's anti-inductivist views, cf. Popper (1959), 29, 41, 371. His views on 'natural or physical [i.e. nomic] necessity' will be found in appendix *x, especially pp. 430–2. His conclusion is that although Kneale's notion of nomic necessity is perfectly acceptable, 'this doctrine is, in my opinion, incompatible with any theory of induction' (p. 432). On this, cf. also Buchdahl (1969), ch. 2, sect. 2b, pp. 26 ff.

[30] For details, cf. ibid. (1969), ch. 8, sect. 4c; also id. (1982); id. (1983); id. (1986); id. (1987).

'ideas', which may possess specific content, such as the ideas of gravitational force, of the pure elements of chemistry, and similar notions. And these ideas are themselves generated by more abstract methodological or regulative maxims such as those of economy, continuity, simplicity; or of homogeneity, diversity, affinity.[31]

Common to this approach is Kant's doctrine that the categories and ideas are conditions of the possibility of their respective realms, be they nature in general, particulate nature, or the order of nature. It is however senseless to ask whether this 'lawlikeness' *exists* or is 'given'. Its origin is conceptual, and its validation lies in the satisfaction of its transcendental tasks, which are those of defining the possibility of the various levels of nature as just indicated.

This general Kantian approach recurs in Whewell, but the stratification by levels is dropped, and the different tasks assigned to them by Kant are somewhat conflated, often as we shall see, with rather confusing results.[32] So far I have emphasized that for Whewell the scientific *process* exhausts itself in the formulation of hypotheses and hypothetical conceptions; now we must add, it does so under the guidance of certain very general (and necessitarian) notions, called 'fundamental ideas', which in the process of the testing and enlargement or modification of hypotheses become progressively 'clarified'.

Let us first note some general aspects of Whewell's theory of ideas. Whewell holds that 'conceptualization', involved in hypothesis formation, is an extremely general and pervasive aspect of all thinking and knowing. In this connection, he emphasizes the Kantian doctrine that it is not possible to make a sharp distinction between the 'sensational' and the 'intellectual' aspects of knowledge; that the division between 'data' and 'inference' is artificial.[33]

In other words, Whewell emphasizes the 'theory-ladenness' of our data; a view that has become prominent in our day. In general then, his is a subjective and 'dynamic' approach: we begin with relatively narrow conceptions, we try progressively to clarify these, partly through critical argument,[34] partly through empirical testing of corresponding hypotheses; and we enlarge the result via the exhibition of consilience and simplification. But there is not really some independent realm of 'fact', and certainly no underlying ontological basis, against which to check this edifice; and thus the requirement of a separate inferential step does not arise.

For this reason alone we should expect Whewell to lack the notion of inductive inference of the kind espoused by Mill. There is in Whewell no provision for forms of argument which involve the drawing of conclusions, or the *detaching* of such conclusions from the evidence. More extreme even, there is no attempt to assess in any way, quantitatively or quasi-quantitiatively, the

[31] Kant (1933) A654/B673 to A651/B679; A662/B690 to A663/B691; id. (1910), 534–8, 544–5.

[32] Below, latter part of Sect. IV.

[33] Cf. Whewell (1847), i. Bk. I, ch. 2; also Whewell (1860), 250.

[34] Cf. here especially Whewell's discussion of his notion of 'explication of conceptions' and 'the use of definitions', id. (1847), ii. 5–8, 11–16.

inductive probability of any hypothesis. The emphasis is all on 'process', on testing, enlarging, deducing, etc. And the choice of hypotheses is not limited by any abstract estimation of their initial probability.

Yet this is not to say that there are no checks of any kind on the choice of the hypotheses. But they are checks of a conceptual nature; and are presented in Whewell's version by his notion of 'fundamental ideas'. These ideas are very general conceptual forms; from what has been said we shall expect some of them to bear resemblance to Kant's 'categories' and 'forms of intuition', namely, space, time, causation, substance; whilst others will resemble Kant's 'regulative ideas' or 'maxims', namely, affinity, final cause, symmetry, likeness, and so on; still others, such as polarity, will seem more to have been taken from the science of Whewell's day.[35]

Now in Whewell's presentation these ideas emerge in the context of a historical survey of the development of some given branch of science. And such a survey is supposed to show how its corresponding leading fundamental idea came to be 'clarified' through progressive criticism. However, Whewell certainly believes that the appropriate ideas do not emerge 'accidentally' from such a survey, since it is the idea (always potentially present?) that, so to speak, actually 'creates' that particular branch of science whose organizing centre it is. Whewell's approach to these ideas is 'idealist', or 'subjectivist', in the Kantian sense: the idea is a mental function, a spontaneous creature of the scientist's mind—Whewell draws here a comparison with Kant's forms of intuition and categories of thought. Just as Kant had argued that there could be no perception ('intuition') without spatio-temporal forms being 'contributed' by the mind, and that there could be no 'objective judgements of experience', no 'nature', without presupposing the 'subjective' addition of categorical concepts like substance, causation, and mutual causal interaction, so Whewell wants to say that there could not arise, for instance, a science of statics without the logically prior infusion of the concept of cause in the sense of force.[36]

Thus 'ideas' impose a 'form' upon the sensory data, and this form has a necessitarian aspect, in two ways; (*a*) By expressing the universality and necessity of scientific laws, and (*b*) by 'regulating' what are acceptable formulations of the phenomena. So ideas have formal and regulative functions; they aid what we may call 'consolidation'. On the other hand, as already mentioned, Whewell sometimes also regarded ideas and the consequent formulations as the potential end-product of the scientific process of reasoning. For instance, he held that the science of statics had by this time arrived more or less at a final and perfect state, because it expressed the 'idea' of force in 'clear and appropriate' form. (He wrote textbooks on statics and dynamics.) Now at first sight this feature of Whewell's doctrine seems to be in apparent contradiction to his falsificationism with respect to hypothesis. However, as will be shown in more detail later, this is a contradiction without practical substance. For it is

[35] For Whewell's list of 'fundamental ideas', cf. Whewell (1847), i. 80.

[36] Cf. ibid. 164 ff.

always possible to construe the 'final formulation' of the fully-articulated 'ideas' ('clarified' and 'appropriate' in Whewell's sense)[37] as an *ideal*, whilst at the same time holding that we can never *know* whether we have arrived at the final form; a view which resembles Kneale's necessitarianism, where similarly we never know whether we have succeeded in formulating a principle of necessitation.

But though we may never know for certain whether we have arrived at a final formulation, Whewell's necessitarianism, just like Kneale's, implies that finality is at least possible in principle, a point bearing on the question of 'validation'. Let us briefly follow the manner in which Whewell construes the relevance of this necessitarian aspect for his theory of scientific reasoning.

The 'ideas', as we said, when suitably clarified, exhibit a necessitarian aspect through their corresponding axioms. They articulate the form in which scientific hypotheses are eventually chosen. Whewell frequently claims that the necessity of the ideal axioms can be intuited, but this (as will be shown) is a somewhat regressive aspect of his doctrine. Basically, the foundation of the ideas is transcendental: they are the condition of the formulation of lawlike hypotheses which they thus generate in a definite manner (which Whewell describes as 'necessary and universal'),[38] though the appropriate form is only elucidated gradually in the historical development of science; and the question of what empirical content is suitably subsumed under its relevant form (Whewell occasionally admits) is never perhaps completely settled.[39] Nevertheless, and perhaps inconsistently, the existence of the fundamental ideas does seem to be used by Whewell as a means of explaining the possibility of arriving, *at least in principle*, at definitive scientific conclusions; and when thus used, the ideas really provide the mirror-image of Mill's doctrine of inductive validity.

There are thus some considerable inconsistencies in Whewell's dynamic approach towards science, and the five components of scientific methodology outlined at the start of this essay melt into one another with extreme fluidity: formulation of hypotheses, and the progressive clarification of the ideas which they contain, are part and parcel of the processes of probation and consolidation; and the problem of validation is subject to a correspondingly confusing treatment.

Whewell in fact offers us two contrasting solutions, one of which is pragmatic, not to say psychological, the other transcendental or logical. The passage from which I quoted above (n. 8) made it appear as though Whewell only demanded *possible* explanations: it is a prime task of the scientific process to show that the results '*can* be clearly explained' by some proposed hypothesis. In reality, however, he does not mean to stop here, but goes on to ask whether we ought not to demonstrate that these results 'can be clearly explained and rigorously

[37] Whewell (1847), ii. 18–23.

[38] Ibid. i. 245.

[39] Cf. ibid. 250: 'The laws of motion . . . are certainly and universally valid; and the only question for observation to decide is how they are to be understood.'

declared *only* by the adopted hypothesis; with the suggestion—exactly as in Mill—that only in this way do we have a guarantee that we are in possession of the 'true' explanation.[40]

Whewell's reply to his own question appears at first sight to be altogether psychological. His position is that through continual familiarity with the scientific problems at issue we are led to the

> conviction, that the inductive inference [that is, the 'superinduced' hypothesis] is not only consistent with the facts, but *necessary*, [a conviction which] finds its place in the mind gradually, as the contemplation of the consequences of the proposition, and the various relations of the facts [a reference to 'consilience' and 'simplification'] becomes steady and familiar. . . .
>
> We may leave it to be thought, without insisting upon saying it, that in such cases what *can* be true, *is* true.[41]

But I think that this psychologizing defence is a somewhat misleading expression of Whewell's true intentions. For never far from his mind are considerations taken from his theory of 'ideas', as is intimated also by the use of the word 'necessary' in the above passage. And there is indeed another passage where the requirement of 'attention to and familiarity with facts and experiments' is coupled with his necessitarian position concerning ideas.[42] According to Whewell, the axioms flowing from the ideas are 'substantial general truths'.[43] For instance, the various axioms of causation, derived from the idea of cause: 'nothing can happen without a cause',[44] 'effects are proportional to their causes', 'causes are measured by their effects', and 'reaction is equal and opposite to action', have 'universal validity',[45] and express 'universal or necessary truths'.[46] The only problem is that of their correct 'mode of interpretation',[47] the solution of which requires 'reflection' through which we may hope to obtain 'clarity' as to what is truly involved, without perhaps being certain that we can ever arrive at finality; our only hope here is 'attention to and familiarity with facts and experiments'—clearly a relative matter. Thus, in the case of the first law of motion, successive generations had gradually come to grasp that 'cause' should be interpreted as 'external force',[48] that 'happening' means 'change of velocity', and that 'time alone is not a cause of change of velocity'.[49] Some of these notions are purely conceptual matters, others require attention to substantial empirical facts.

Nevertheless, Whewell often muddles the issue, by pretending that the resulting formulations, that is the *interpreted* axioms, are instances of the 'existence of necessary truths'.[50] He will then argue in reverse, and in Kantian fashion, that the existence of such truths can only be explained on condition of our postulating the corresponding abstract axioms, which thereby receive a

[40] Ibid. ii. 90; cf. also ibid. 622. For Mill's similar doctrine, cf. Mill (1868), ii. 11.
[41] Whewell (1847), ii. 90, first italics mine.
[42] Ibid. i. 251. [43] Ibid. 249. [44] Ibid. 217.
[45] Ibid. 250. [46] Ibid. 245. [47] Ibid. 249.
[48] Ibid. 247. [49] Ibid. 221. [50] Ibid. 174.

quasi-transcendental status. The axiom of causation is then thought of as one of the 'conditions of experience',[51] with the convenient slide from causation being a transcendental presupposition of 'our conceiving any real occurrences . . . of a real sequence of things',[52] to the quite different claim of its being a conceptual ingredient in a putative *causal* sequence, such as is represented by the first law of motion. (This is one of the results of conflating Kant's transcendental and 'metaphysical' levels, mentioned before; it corresponds to a mistaken intepretation of Kant, of which Whewell's is an explicit instance.)

I shall return to the question of 'necessary truths' in a later section.[53] For the moment, our result is as follows: Whewell's quasi-transcendental foundation corresponds to the validational component of induction, which in Mill appears as an explicit inferential step. Yet no more than Mill does Whewell claim that validation achieves final empirical certainty, and to the extent that Mill often charges Whewell with such a view, he is mistaken; his misapprehension being due to Whewell's confusing presentation. For Whewell, on one side, validation is mostly a transcendental matter, hovering, so to speak, perpetually behind the scenes; it forms no part of the process of probation as it does in Mill, who assigns to it the explicit mode of an 'inductive inference', nevertheless with a proviso akin to Whewell's that we can never know for certain that we have arrived at an inductive conclusion incapable of being overturned. On the other side, with the loss of the weight of 'validation' as part of probation, Whewell is moved to lay more emphasis on explication (for example, clarification of ideas) and on consolidational procedures. The development of these contrasting positions, opening up the possibility of copious disputes, some apparent, some real, between the contending parties, we must now follow up further. The study of such disputes is not pointless, since for both sides (and for us) they are a means of achieving greater clarity about the general issues involved.[54]

III

The contrast between our two nineteenth-century giants could not be sharper, at least on the surface. As will by now be expected, Mill construes Whewell as claiming that we *can* arrive at statements of necessity; even, that we can know when we have arrived at them. And he holds this claim to entail its own refutation of the whole conception of necessity itself.[55] Moreover, it prevents him from even beginning to grasp that Whewell's necessitarianism was meant to provide an alternative to Mill's own theory of validation.

[51] Whewell (1847), i. 175.

[52] Ibid. 251.

[53] Below, sect. VI.

[54] There are two important discussions of the Whewell–Mill controversy in the literature, although both seem to me to fail to do sufficient justice to the philosophical parallelism as regards 'validation' that runs through this dispute. Cf. Stoll (1929), 91–103; G. Wright (1957), 46, 52, 56–60, 196, 108–200.

[55] Cf. Mill (1879), i. 274–6, 278–81.

Now one might expect that this would involve Mill in falsificationism pure and simple. But here he is on the other side of the fence. Belonging to the 'inductivist' school, he claims that Whewell has altogether omitted to pay attention to the most important aspect of scientific reasoning, which is to provide *inductive proof*. According to Mill, it is not enough to formulate hypotheses, to derive from them verifiable consequences, and so on *ad infinitum*. This only enables us to show that the phenomena have so far not 'disproved' the hypothesis.[56] Nor is it sufficient to say (with Whewell) that science, when faced by exceptions, will modify its hypotheses, and simply transform them into something different (albeit simpler). For by this method we for ever remain in the realm of hypothesis.

Sometimes Mill formulates his opposition by insisting that in addition to showing a hypothesis to account for the data, we have to supply an 'independent proof'[57] of its truth—independent, that is, of its 'power to account' for the 'data' that flow from it deductively. Unfortunately, he is none too clear on the way in which this may be achieved. There are in fact two types of hypothesis in terms of which Mill argues his case, causal or explanatory hypotheses, and colligating laws of phenomena. Although he often conflates these, they present somewhat different problems, and in the course of his arguing for the necessity of independent proof, Mill sometimes switches to the more plausible case of 'transcendent hypotheses', for example those involving what are often called 'unobservables', such as gravitational force, the luminiferous ether, and similar entities.[58] In this case, the 'independent proof' is supposed to be furnished by one or other of his celebrated 'Canons of Induction'. For instance, we must be able to produce experimental situations which indicate that in the absence of the putative 'cause' the 'effect' does not follow, and that it 'only' follows in the presence of the cause.[59]

This is rather confused. Mill is none too clear on what would count as 'independent evidence'. At one extreme, it seems to involve direct observation of 'the cause', at the other, the employment of analogy, as in the case of geological theorizing. Both are an obscure (and never explicitly formulated) way of offering considerations in favour of some kind of hypothetico-deductive or (using Peirce's term) 'abductive' argument. At any rate, the move involved, from hypothetical explanation to hypothesizing inference, a form of 'transcendence' to the hypothesis, was for Mill only a special instance of our second case of transcendence, for him much more basic, since it involves an explicit transcendence in time. Kepler's law of the ellipse is a colligating formula which hypothetically accounts for actually or potentially observable positions of a planet. But as an account of 'induction'—and according to Mill, this is all that Whewell has provided—it, too, is incomplete. For the hypothesis here is a statement in logically universal form, whereas the observational evidence is particular. The proper induction of the law hence requires an inference,

[56] Ibid. 503.

[57] Id. (1868), ii. 15–16.

[58] Ibid. 11–13, 21–3; cf. id. (1879), i. 348–50.

[59] Cf. id. (1868), ii. 11–12.

equivalent to the conclusion, not just that the positions of the planet lie on an ellipse, but that they invariably do so; and it is really this, a 'transition from known cases to unknown, which constitutes induction in the original acknowledged meaning of the term'.[60] 'Induction is proof; it is inferring something unobserved from something observed: it requires therefore an appropriate test of proof.'[61]

What Mill is objecting to in Whewell is hence twofold: (*a*) formulating hypothetical (or, to use Whewell's term, 'colligating') conceptions is *not* induction but only 'description'. (What Mill here means by 'description' is a very complicated question, and we shall try to unravel it as we proceed. This much is clear: colligation being merely description, for example expressing a set of experimental or observational values by means of a formula, is to say, in Mill's terminology, that it does not as yet involve induction.) (*b*) Genuine scientific reasoning however does, or ought to, involve induction, that is, transcendence from known to unknown.

Consider the example of Kepler's hypothesis of the elliptical path of the planetary orbit of Mars. The 'colligating conception' of the ellipse, we might say, 'accounts for' and is thus 'supported by' the data, that is the observed positions of the planet. But this colligation, although it involved Kepler in a considerable amount of inferential reasoning, constructive diagramming, transpositions of co-ordinate origins, etc., according to Mill is still only 'description', in so far as it states either the observed, or the (at some finite future time) potentially observable, positions of Mars. And this is not induction. Induction enters because it is part of the expression of Kepler's law to argue—as Kepler indeed does—that his colligation involves the claim that the planet will *continue* universally (unlimited past and future) to revolve in an elliptical orbit. But how is this 'proved' by showing that the positions of Mars can be fitted to an ellipse?[62] In order to show this, Mill claims, we have to employ something like the Method of Agreement; that is, all the positions of Mars agree in lying on an ellipse; therefore all positions of Mars (timelessly) agree in lying on an ellipse.[63] Here, 'colligation' means to 'describe' the different positions as agreeing in lying on the orbit of an ellipse; the different positions may then be regarded as 'the observable effects' of the 'cause', which is the 'timeless' elliptical motion of the planet. (It was rather eccentric of Mill to call the planet's motion the 'cause' of its individual positions; but let that pass.)

Now according to Mill the Method of Agreement itself assumes, however, that 'every event' or the beginning of every phenomenon must have some cause. Proving this 'law' forms the second component of Mill's reference to 'proof'.)[64] But what about the status of this 'Law of Universal Causation'? Here we come, of course, to the core of Mill's logical doctrine, and its

[61] Ibid. 352. [61] Ibid. 352.

[62] Cf. ibid. 338–40, 350–3; ibid. ii. 190–2.

[63] Id. (1868), ii. 190. Whewell, of course, never faced this problem of 'generalization over particulars', making it simply part of his generalized method of hypothesis, aided and abetted by the support from his theory of 'fundamental ideas'.

[64] In the passage from id. (1879), i. 352.

discussion will have to be relatively short and dogmatic. According to Mill, our knowledge of this law is itself based on an 'enumerative induction', and such inductions, or direct generalizations, are usually very precarious. Mill now offers various considerations,[65] intended to show that in the case of the field of evidence supporting universal causation the precariousness of the conclusions is reduced to a minimum—though *only* a minimum! He certainly realizes and explicitly states that the conclusion cannot transcend the evidence and that our 'reliance' on the law '[can]not hold in circumstances unknown to us, and beyond the possible range of our experience'.[66] (In other words, an inductivist does not have to say that his conclusions hold, come what may, as is so often urged in criticism of such a position. He is only concerned with the question of what conclusions may be reasonably inferred on the evidence.)

However, and here we come to a philosophically crucial point, Mill does make a distinction between *precariousness* and *validity*. Admittedly, all inductive processes are precarious, and it is the function of the 'Methods' to reduce the degree of fallibility involved. On the other hand, all inductive *inference*, and *a fortiori* that involved in direct enumeration, Mill asserts, is certainly 'valid':

> Assuredly, if induction by simple enumeration were an invalid process, no process grounded on it could be valid; just as no reliance could be placed on telescopes, if we could not trust our eyes. But though a valid process, it is a fallible one, and fallible in very different degrees: if therefore we can substitute for the more fallible forms of the process, an operation grounded on the same process in a less fallible form, we shall have effected a very material improvement. And this is what scientific induction does.[67]

The distinction which Mill here makes (it was added to the third edition of the *Logic*, perhaps in response to criticism of the apparent circularity of Mill's earlier argumentation) between fallibility and validity, has not to my knowledge received the attention it deserves on the part of Mill's commentators.[68] On the one hand it considerably softens the apparent inconsistencies of his exposition, and the reader's preoccupation with the 'uniformity of nature'. On the other, it exposes more clearly the 'metaphysical core' of Mill's thinking. In the passage quoted, by 'scientific induction' Mill means of course the employment of his Methods. But what does he mean by 'valid'?

This is a very complex matter which would need more space than I have available to unravel in detail. Certainly this 'validity' is not based on something like the Uniformity of Nature, for that in turn simply expresses what is

[65] Id. (1868), ii. 102 ff. [66] Ibid. 108. [67] Ibid. 101.

[68] Cf. Anschütz (1953) who speaks of Mill's 'uncertain voice regarding the possibilities of simple enumeration'. Mill is (quite correctly) described as employing simple enumeration to demonstrate universal causation, because though being usually 'delusive and insufficient', in this one case it is 'certain'. The author, with a single glance at our passage, then concludes with the brief comment that 'it is on this basis that he (Mill) arrives at the odd [sic] conclusion that simple enumeration is not an "invalid" process but merely a "fallible" one' (pp. 103–4). The conflating of the two notions that I have distinguished is even more blatant in Pap (1962), 156, where we read that 'enumerative induction, he (Mill) argued, *is valid and can be relied on* if the confirmed generalisation has very wide scope and is confirmed by a wide variety of experiences'.

involved in the claim that induction is a valid form of inference. Throughout Mill's book there occur such phrases as, that under certain circumstances, 'we are entitled to' infer inductive conclusions 'from the evidence'.[69] As implied by Mill's discussion of the syllogism, the axiom of uniformity, even if regarded as a major premiss, 'does not contribute to prove' any inductive conclusion, but is only 'a necessary condition of its being proved';[70] as a major premiss it only registers the legitimacy of the inductive inference; basically, inductive inference is from observed instances to the next particular unobserved one. 'All inference is from particulars to particulars: General propositions are merely registers of such inferences already made, and short formulae for making more.'[71] There is no universal premiss formally sanctioning the inference, since, as Mill expressly remarks, there is no contradiction in supposing that all known men have died, and that the next man might live forever; the inductive 'type of ratiocination' is not one that can 'claim, like the syllogism, to be conclusive from the mere form of the expression; nor can it possibly be so'.[72]

Such a position is an inheritance of the Locke–Berkeley–Hume tradition, with its implication that there are no hidden bonds capable of providing a justificatory bridge between the terms of propositions, or rather, between the things or facts asserted in such propositions.[73] In consequence, one ought to expect Mill to subscribe to Hume's doctrine as expressed in the *Treatise*: 'There is no object, which implies the existence of any other if we consider these objects in themselves.'[74] But equally, and again like Hume, Mill wants to retain a place for inductive inference, echoing Hume's interesting remark that both in fact and universally men do draw inductive conclusions.[75] In that case, however, we are still left with Mill's rather stronger claim that induction is a form of valid inference.

In speaking of the legitimacy of inductive inference, we must first and foremost remember that by this phrase Mill again covers two quite different aspects:(*a*) methodological, equivalent to my 'probation', for example the canons of induction; (*b*) inference as such, implying that 'transcendence' or 'projection' is a rational move, that is, 'validation' proper. Here, in order to gain some perspective on the second of these, we may assume that any

[69] For instance, at Mill (1879), i. 358 n. On occasion Mill describes the notion of inductive inference as 'the impulse of the generalizing propensity', ibid. 235; cf. 355. He points out (ibid. 360) that though we have this propensity to generalize, we are not always 'warranted in doing so', but that we need 'a reason to believe' in the uniformity of nature; a reason which is provided of course by the nature of the evidence for universal causation (cf. ibid. 361, where Mill refers again to ii. 101 ff.)

[70] Ibid. i. 356.

[71] Ibid. 221; and cf. the use made of this at id. (1868), ii. 105.

[72] Id. (1879), i. 224, 233.

[73] Ibid. 97, and the whole of sect. I of this chapter.

[74] Hume (1888), 86.

[75] Cf. id. (1902), 34, I have discussed the distinction that has to be drawn between Hume's philosophy of causation and his methodology of causal investigation in Buchdahl (1969), ch. 6, sects. 1–4. Mill's conception of inductive validity corresponds to Hume's foundation of 'necessary connection' on the associative process of the mind, whilst Mill's notion of fallibility which he seeks to reduce by the employment of his Methods corresponds to Hume's deployment of 'general Rules', as elucidated in the *Treatise*, Hume (1888), i. 3. 15, pp. 149, 173–5.

problems surrounding the first have been overcome. Thus we may take it for granted that the inductive situation has already been shorn of all circumstances which usually give meaning to the notion of validity; for instance we may assume that we have already determined the relevant circumstances, eliminated interfering causes, and have in fact arrived at some basic and fundamental putatively lawlike situation. (This corresponds to the sharp distinction in Mill between 'probation' and 'validation'.) But in such a hypothetical case, and given that we do not dispose of principles of inference as for the case of deductive logic, what is left of meaning for Mill's reference to validity, that is, for validation proper?

The answer (and in some respects it reminds us surprisingly of the teaching of Whewell and of Kneale) can only be in terms of 'achievement'. Mill must mean that there are, at least in principle, occasions where the inductive inference presumptively leads to lawlike connections although we are not in the possession of any special signs (let alone a method) indicating whether we have actually reached a requisite law. Now, that there are such lawlike states of affairs, Mill never doubted. There are many passages in the *Logic* implying such a presumption, but in his *Examination of Hamilton* there is a more direct admission of Mill's concealed assumptions, when he asserts that '*there is a connection between one objective truth and another*, which makes it possible for us to know objective truths which have never been observed, in virtue of others which have'. And he makes it clear that it is the province of inductive (rather than deductive) logic to aid us in the 'discovery of truth' of the type here involved.[76]

But what is the status of these 'connections'? On this Mill is seldom explicit. We may gain some hints, however, from his account of causal laws. It is insufficient, he tells us, to characterize causal conjunctions as mere invariably occurring sequences for such sequences constitute only 'conjunctions in some sort accidental',[77] however long continued into past, present, and future. Rather, they must also occur 'unconditionally'; and it is interesting to find Mill adding that 'this is what writers mean when they say that the notion of cause involves the idea of necessity'.[78] Though Mill usually rejects the use of this term when characterizing the lawlikeness of empirical uniformities[79] because of what he calls its 'misleading associations',[80] there is nothing to stop us applying the label of 'nomic necessity' to exactly the characterization Mill has given of lawlikeness. For he holds that laws do not state what does happen but what would happen, given certain conditions: 'All laws of causation, in consequence of their liability to be counteracted, require to be stated in words affirmative of tendencies only, and not of actual results.'[81]

[76] Mill (1889), 477, my italics.

[77] Id. (1879), i. 392. It is interesting to note that Kneale employs an exhaustive division between uniformities due to natural law and those which are merely *accidentally* true, and are thus kinds of 'historical accidents on the cosmic scale', *Analysis*, 10 (1950), 123.

[78] Mill (1879), i. 391.

[79] Cf. id. (1868), ii. 419.

[80] Id. (1873), 168–9.

[81] Id. (1879), i. 518.

Now this reference to the subjective has by other writers, for instance by W. E. Johnson and Kneale, been given precisely as their most important ground for regarding lawlike statements as expressing universals of law and not of fact, and as implying nomic necessity.[82] It has often been held that Mill is here involved in a contradictory position. On the one hand, inference is from particular to particular; on the other, the result of the inference yields 'connections' between 'truths' which have the status of nomic necessity. However, I am not so certain that the charge succeeds. Nomic necessity is neither logical nor metaphysical necessity.[83] Nor in the end is the existence of nomic necessity ('universal causation') used by Mill to justify inference. Rather, it at best merely *explicates* the notion of validity. The claim that inductive inference is 'valid' means, not that it is justified by a special sort of connectedness between the phenomena, but that it leads, on suitable occasions, to nomic connections. Validity does not *support* the fact of nomic connection, nor does the latter provide a justificative ground for the former. That inductive arguments are 'valid' is just Mill's way of saying that inferences to nomic connections are 'reasonable', not that they certify the truth of the conclusion with necessity. Certainly, Mill should not be thought of wanting to justify or deduce validity from nomic necessity; the latter is at best only a means of explicating the notion of validity involved in the central passage referred to. He does not wish to claim that he has a special justification for his ascription of validity, nor—what amounts to the same thing—for the existence of 'connections between truths'.[84]

That there are such connections, Mill never doubted, and we may think of it (to borrow a label from R. P. Anschütz, who has provided a useful commentary on this aspect of Mill) as his 'cosmological doctrine',[85] without however holding, as does Anschütz, that such a position involves Mill in a form of ontological realism which clashes violently with the apparent nominalism of his doctrine that inference is from particular to particular.[86] What we can say is that Mill's

[82] Johnson (1924), 5–6; Kneale (1949), 77. Cf. also Buchdahl (1969), 29 ff.

[83] The position is here much like that in Berkeley's doctrine of universals, to which Mill is heavily indebted. Berkeley likewise argues that we always reason round particulars, but that general conclusions may be drawn where such 'general ideas' do no more than 'represent or stand for all other particular ideas of the same sort'. To the question, what may be meant by the 'same sort', Berkeley would reply that his reference to similarity must not be understood as being *backed* by an additional feature which might in some sense be thought of as *existing* in the way in which exist the particulars, and which might be invoked as a justificatory ground. Cf. Berkeley (1710), Introduction, sect. 12. For a further discussion of this point, cf. Buchdahl (1969), ch. 5, sect. 2, pp. 279 ff., especially 284.

[84] This interpretation is equivalent to Wittgenstein's position as found in Anscombe (1953), sects. 480–3.

[85] Anschütz (1953), 96.

[86] 'Ontological realism' represents a view according to which there is an underlying ontological bond that would *justify* an inductive inference. I do not think that Mill would ever have subscribed to such a view, although his doctrine of 'inductive validity' undoubtedly plays the part of a substitute. This kind of realism should be distinguished from the 'empirical realism' which is Mill's position with respect to empirical laws (colligating conceptions) regarded as 'descriptions', as discussed in the next section.

cosmological assumption is the counterpart to Whewell's transcendentalism, with the notion of necessity straddling the two camps; although Mill's and Whewell's respective views on 'necessity' are of course quite different and the way Mill states his position has the result of emphasizing a notion of inductive proof that will clearly not be acceptable to Whewell. Moreover, the contentiousness of the sense in which the realm of the universal transcends that of the particular is concealed by Mill (when arguing his case for inductive inference) through his frequent employment of the example of unobservable or transcendent scientific entities. But let us follow our two contestants and see how they defend their respective positions.

IV

Whewell never grasped the force of Mill's contention that he had not produced 'proof', nor genuine inductive generalizations, transcendence beyond the particular to the universal. Perhaps he did not see the point, because he might claim that on his own account the inductive process already *begins* with a universal, since he defines induction as the process whereby we 'superinduce' a *general* (universal) conception upon the particular, and that there was thus no need for a special inferential move at all. Mill of course does not recognize the hidden assumption held in reserve by Whewell here, which is the promise of the justificative power of his 'fundamental ideas', with their associated necessitarian axioms; just as Whewell does not grasp the significance of Mill's emphasis on 'proof' which, as we have just seen, boils down to an affirmation of his cosmological thesis, namely that there *are* connections in nature. And so Mill's answer is that colligation of the data, even via a general conception, is only a decription, not an induction. For although the colligating conception may be a universal hypothesis, induction alone can show that the hypothesis applies.

Mill's contention that Whewell's hypotheses are mere descriptions, instead of genuine inductions is, however, argued by rotating round two quite distinct meanings of 'description', interpretations which Mill attempts to use in order to clinch further his arguments in favour of the need for induction. In the present section I will consider the first of these interpretations. This involves a realist approach towards colligating laws and conceptions such as Kepler's law of the ellipse, in contrast with Whewell's position which emphasizes the ideal status of such laws and conceptions. Mill therefore argues that even if it be granted that Whewell 'superinduces' a *general* conception upon the particular data, yet (and here he ignores of course the relevance of the latter's theory of 'ideas') one must ask how far the generality of the conception corresponds to the course of nature. For at this point, Mill's empiricist assumption that the facts of nature present themselves as only 'particular', comes to the fore, causing him to insist on the employment of a special tool which alone would entitle us to define a sense in which we may *affirm the truth* of a *general* state of affairs. And this sense is made good only by the notion of inductive inference.

So Mill's empiricism throws into sharper relief the general point that one can always ask whether one is justified in claiming a general conception to apply. And as we have seen, Mill conceives of justification only in terms of inductive 'validity'. Whewell, of course, in his turn fails to make any intellectual contact with Mill's concealed assumptions, and his answer is as expected. (*a*) The colligating conception applies if it accounts not only for the old data on which it was originally based, but also for newly predicted ones; and applies furthermore, if the conception can subsequently be integrated with others, without loss of simplicity, thus exhibiting consilience through a Table of Inductions. (*b*) That the conception has a chance of applying in principle is guaranteed by the fact that it expresses a 'fundamental idea', and the relevance of such ideas for this purpose is assumed here as a metaphysical datum.

Mill, of course, refuses to consider this last piece of metaphysics. Also he is less impressed than Whewell by the phenomenon of predictive power,[87] though he allows some importance to consilient integration, by what he calls the 'deductive method',[88] whilst still insisting on the need for the independent inductive proof of the truth of the primary explanatory principles, of the *existence* of the explanatory causes.[89]

For the most part, however, Mill develops his objection to Whewell, that 'super-induced conceptions' are not true inductions, via the more circuitous route which employs the argument that such conceptions are 'mere descriptions'. Here, all depends on the meaning of the term 'description' which, as mentioned, is used by Mill in two radically different senses, thus opening the way to considerable misunderstandings between Whewell and himself. In its first signification it refers to the correspondence between the colligating conception (for example, Kepler's formula for the ellipse described by the planets) and physical reality. When discussing 'conceptions', Mill thus writes: 'Every conception which can be made the instrument for connecting a set of facts, might have been originally evolved from those very facts. The conception is a conception *of* something; and that which it is a conception of, is really *in* the facts . . .'[90] This looks like some sort of theory of correspondence, and earlier in the same section he does in fact say that 'description . . . contains the assertion of a resemblance between the phenomenon described and something else'.[91] Moreover, all this simply echoes the point he had already insisted on in his

[87] Cf. Mill (1868), ii. 20–1.

[88] Cf. n. 1 above.

[89] Cf. Mill (1868), ii. 8 ff.

[90] Ibid. 195. This was partly in reply to Whewell's basic contention that all conceptions superinduce something 'new' upon the data of perception which thereby come to be regarded from a 'new point of view'. This (together with the various consolidative and probational procedures described), according to him exhausts the 'logic of induction'. Mill's point is that although the conception may be new, speaking chronologically, it is nevertheless 'in the facts' constitutively, and indeed epistemically, in that it could in principle become an item of perceptual knowledge, contrary to the 'facts' generated by an inductive inference.

[91] Ibid. 190.

dispute with Whewell concerning the status of Kepler's hypothesis of the ellipse: 'If the facts are rightly classed under the conception, it is because there is in the facts themselves something of which the conception is itself a copy.'[92]

The second use of Mill's 'description', as will be shown in the next section, is almost the complete opposite of the first. It denotes what in earlier periods of science we meet under the technical label of 'mathematical hypothesis', a notion more or less akin to the modern idea of a model or analogue, not to say fiction.[93]

In the present section, we will consider Mill's use of 'description' in the first of the two senses noted. Here belongs Mill's contention that colligating conceptions (having the status of 'descriptions') correspond (provided they are 'clear and appropriate') to a 'fact in nature', a fact, moreover, that could have been observed had we the powers of observation that admittedly we normally lack. (For instance, planets 'really do move in elliptical paths'; that fact is not just 'in the mind'.)

This makes it look as though Mill was an 'empirical realist' with respect to colligating conceptions—rather different from Whewell who (as we saw) in this respect is more a Kantian idealist. And this putative realism is here apparently contrasted with what induction alone can yield, namely, the realm of transcendence, of which Mill offers two cases, (*a*) the move from some to all men that are mortal; (*b*) the move from kinematics to kinetics, that is, from motions to invisible forces.[94]

Does this mean that Mill thinks of the universal 'man', and of transcendent forces, as less real? On the question of the universal man Mill was somewhat confused, sometimes holding, as has been shown, to a certain degree of nominalism, at other times acting as though he was a realist. But certainly there is a sense in which Mill would reply that they are not any the less real; only that the claim to their reality requires a 'proof', namely, an inductive inference.

[92] Id. (1878), i. 343.

[93] On 'mathematical hypothesis', cf. Buchdahl (1969), 288; also Blake *et al.* (1960), ch. 2; Palter (1970), sect. III, pp. 152 ff.

[94] Throughout these controversies Mill avails himself of the threefold ambiguity which surrounds his use of the concept of induction (see above, Sect. I) to make his argument more cogent by slipping from the case of 'hypothesis' in the sense of Whewell's colligations of facts, to the case of hypothesis which concerns the 'causes' of these colligated facts (or 'laws of phenomena' as Whewell calls them (1847), ii. 95.) And it is of course much more cogent to discuss the question of 'independent evidence' for the latter case than for the former. Moreover, it is clear from reading Whewell (ibid. 95 f.) that he, as against Mill, never saw any need to modify his general notion of the process of science as that of fitting data to hypotheses. Thus, speaking of 'the laws of phenomena *and* their causes' (my italics), he says: 'In both we must, as far as possible, devise hypotheses which, when we thus test them display those characters of truth of which we have already spoken; an agreement with fact such as will stand the most patient and rigid inquiry; a provision for predicting truly the results of untried cases; a consilience of inductions from varied classes of facts; and a progressive tendency of the scheme to simplicity and unity' (ibid. 106). It is here that Mill opposes his contrary view of the necessity of additional 'methods of proof', as will be explained further hereafter. Cf. e.g., Mill (1879), i. 352, 503, and below, sect. V; also the passage from Mill (1868), ii. 24, quoted below at n. 133.

The above implies a distinction between what is 'described' and what is 'inductively inferred', between 'fact' and 'theory'. Now this is of course a distinction which Whewell again strenuously opposes, in line with his general Kantian approach. Just as for Kant judgements concerning individual matters of observational fact are concept-laden, in the sense of presupposing spatio-temporal and categorical leavening, so for Whewell all scientific facts, at whatever level, are theory-laden to varying degrees.[95] So we may expect Whewell to deny the corresponding distinction between mere 'description' and 'inductive inference':

> But, says Mr. Mill, 'it is a fact surely that the planet does describe an ellipse, and a fact which we could see if we had adequate visual organs and a suitable position'. To this I should reply: 'Let it be so; and it is a fact, surely, that the planet does move periodically, . . . [or that it] is attracted by the sun . . . Perhaps Mr. Mill would remind us that the elliptical form of the orbit is a fact which we could see if we had adequate visual organs and a suitable position: but that force is a thing which we cannot see. But this distinction also will not bear handling. Can we not see a tree blown down by a storm, or a rock blown up by gunpowder? Do we not here see force:—see it, that is, by its effects, the only way in which we need to see it in the case of the planet, for the purposes of our argument? . . . Is not the operation of the sun's force a fact of the same kind, just as much as the elliptical form of the orbit which results from the action: . . .
>
> In truth . . . there is no definite and stable distinction between facts and theories; facts and laws, facts and inductions. Inductions, laws, theories, which are true, *are* facts. Facts involve inductions. It is a fact that the moon is attracted by the earth, just as much as it is a fact that an apple falls from a tree.[96]

Shorn of some of its shadow-boxing, the argument implies that there is no sharp distinction between what can be 'described' (however inferentially) and what has to be 'inferred', as opposed to Mill's contrary opinion that there is a radical difference between what is 'described', and therefore (at least in principle) observed, and what *cannot* be so described, and hence needs inductive inferential support.

Now as just noted, and like some more recent theorists, Whewell denies one of the apparent implications of the 'description' view, namely, that there are any 'pure observations'. One of his chief reasons for this however needs closer scrutiny; it is that *all* observation involves inference. Here, however, one must be careful, for at first sight it seems that this, too, is an insufficient characterization of the difference between Whewell and Mill, since the latter does actually himself seem to ascribe to such a view; a fact that has often been overlooked by those who accuse him of simple-minded empiricism: 'In almost every act of our perceiving faculties, observation and inference are intimately

[95] Indeed, though Kant makes a sharper distinction than Whewell allows between observational and theoretical levels, he does once remark that in 'experimental physics', even the 'principles according to which we perform experiments must themselves always be derived from the knowledge of nature, and hence from theory'. Kant (1910), 199; id. (1965), 6.

[96] Whewell (1860), 250.

blended. What we are said to observe is usually a compound result, of which one-tenth may be observation, and the remaining nine-tenths inference.'[97]

Still, there is a substantive difference here, for Mill evidently holds that we can sort out the 'truly perceived' from the merely 'inferred', this in turn again assuming inductive argument. And this is what Whewell of course denies; there is, for him, no kernel of pure perception. However, in order to defend such a radical view, he avails himself of Kantian arguments, holding that every judgement of experience is an intimate inseparable conglomerate of a 'sensory matter', a 'spatio-temporal form', and a 'categorical conceptualization'. When making this point, he expresses the dichotomy involved as an 'antithesis' between the 'matter' of sensation and the 'form' of 'thought'—'thought' here being represented primordially by the 'ideas'.[98]

Unfortunately, it is not clear whether this moderate Kantian 'idealism' can succeed in overturning Mill's distinction between description and induction. For as was shown at the start,[99] Whewell has dropped the sharp distinction between the three levels of 'nature' which we have noted in Kant, with the result that some of the conclusiveness of the Kantian argument concerning the 'necessity' of conceptual ingredients (which belongs only to Kant's transcendental thesis with respect to 'nature in general') is lost when applied without further ado to the remaining levels. We thus find that Whewell, without noticing it, slides from one distinction (or 'antithesis') to quite a different one, when he argues for the impossibility of the distinction between observational fact and theory by contending for the impossibility of the distinction between 'matter' and 'form' of perception. It is, however, not obvious that the slide from one to the other need be allowed, and it has in fact been opposed by some recent philosophers of science, as for instance by I. Scheffler.[100] Yet, for Whewell at least, the Kantian form of idealism seems to be the only argument sufficiently strong to support conclusively the claim that a sensational core of perception has no independent meaningful reality.

That Whewell's conclusion does not follow automatically from his Kantian premisses becomes clear also if it is realized that Mill—perhaps unconsciously—himself subscribes to a considerable core of these contentions of Kant's that are here relevant, whilst yet denying the consequence claimed for it by Whewell. For Mill certainly admits that even 'in the simplest description of an observation, there is, and must always be, much more asserted than is contained in the perception itself'; one required 'addition' being that in my act of describing it, each particular sensation has also to be 'classed' by me.[101] Now this, I think, is very similar to the Kantian argument; at least it is the argument which Kant employs when supporting his contention that all experiential judging requires a reference to such classification. (Compare Kant's various 'syntheses of

[97] Mill (1868), ii. 184.
[98] Whewell (1847), i. 33 ff.
[99] Above, Sect. II.
[100] Scheffler (1967), 37 ff.
[101] Mill (1868), ii. 187.

apprehension, reproduction and recognition', as elaborated in the *Critique of Pure Reason*.)[102] Only, as I said, Mill (like Scheffler) denies the implication:

We thus see that it is impossible to express in words any result of observation, without performing an act possessing what Dr Whewell considers to be characteristic of induction. There is always something introduced which was not included in the observation itself; some conception common to the phenomenon with other phenomena to which it is compared. An observation cannot be spoken of in language at all without declaring more than that one observation . . . *But this* identification of an object—this *recognition* of it as possessing certain known characteristics—*has never been confounded with induction*. It is an operation which precedes all induction, and supplies it with its materials.[103]

It is thus clear that Mill makes distinctions, where Whewell refuses to draw the line. What is not evident is on what principles a line can be drawn between the two cases, so as to prevent this moderate form of idealism with respect to perception, common to both Kant, Whewell, and Mill, from degenerating into the type of universal idealism espoused by Whewell alone (as well as by some of his modern followers). Which way the issues here divide is not always clear. Certainly, Kant himself believed his moderate idealism compatible with a fairly clear-cut division between descriptive and theoretical language.

On the other hand, it equally clearly does not follow that if we do make the distinction, we are *ipso facto* committed to Mill's form of inductivism. And this is again shown by the procedure of Kant. He, as already noted, makes a sharp distinction between the conceptual ingredients which are a necessary part of the more complex, and higher-level, scientific theories. The need for conceptualization at the level of theory (of what Kant calls 'reason') is thus only *analogous* to that at the level of particulate experience in general (Kant's 'understanding'), but unlike Whewell, who conflates the two cases, Kant argues separately, and twice over, for the subjectivist approach.. Whewell believes he needs only one argument. And in so doing he lays himself open to Mill's strictures. For Whewell, as we have seen, the 'ideas' are the formative elements of elementary perception, as they are also of the different branches of science. Without them, no science could even begin to be formulated. The science is generated by the 'ideas', its conceptual terminology presupposes them.

However, it is clearly far more contentious to say that a given science could not be formulated without its 'ideas', than that common experience, as with Kant, presupposes the categories, for example of substance, cause, or causal interaction. How one resolves this dispute depends perhaps to a certain extent on one's intellectual background and professional employment. Mill had rather less feeling for the relevance of the conceptual aspects of science and hence thinks conceptual disputes of less importance.[104] By implication, despite his remarks about the element of inference in observation, he thinks there is a

[102] Kant (1933), A98–A110, pp. 131–8.
[103] Cf. Mill (1868), ii. 188, my italics.
[104] Cf. his treatment of the notions of 'appropriate' and 'clear conceptions', Mill (1868), i. 200–5.

solid core of observational fact, and hence distinguishes the remainder by insisting upon a further step, namely, inductive inference. For Whewell, the 'fundamental ideas', and their corresponding conceptions, cover all the 'facts' there can be; and with this, no special distinctions are made between 'particular facts' and 'universal conceptions'. For Mill, conceptions, where adequate, are never more than descriptive of potentially observable empirical reality. Hence, in order to advance further, he must needs invoke a special inferential step.

V

Paradoxically, this special step is insisted on by Mill in order to oppose Whewell's Kantian idealism, and is thus meant to support some form of (ontological) realist argument, answering to Mill's 'cosmological doctrine'. The nature of this realism becomes clearer if we consider further the argument between Whewell and Mill on 'description'. In section IV I have emphasized that sense of description, in which it denotes something 'in the facts' of which it is a 'copy'. In this sense, a 'description' represents the facts (if successful) as they are in reality.

Curiously, the opposition between Whewell and Mill is also argued out in terms of a second meaning of description, almost the opposite of the realist interpretation just considered, thus offering fruitful occasion for further misunderstandings. This time the argument is not so much about laws but about the nature of explanatory theories, by Whewell regarded in what we might now call an 'instrumentalist', not to say fictionalist way. According to Whewell, the succession of different astronomical theories, for instance, is never more than a fitting of different hypotheses, continually adjusted to explain new data, due account being taken of the aspects of simplicity and consilience. Mill, this time taking Whewell's idealist approach at its word, objects by making a distinction between a 'mere description' and an 'explanation'. In this new sense, description is anything but a copy of reality; on the contrary, it is more like a mere 'model' or 'fiction' or possible 'representation' of the phenomena, and as such neither true nor false, offering no more than convenient understanding. Being neither true nor false, it is evident that 'descriptions' do not require induction, as contrasted with genuine explanations. The latter must be either true or false, and indeed, there can only be one true explanation.[105]

A prime example which Mill cites is the various astronomical representations of stellar observations in the theories (or 'hypotheses') of Aristotle, Ptolemy, Copernicus, and Tycho. These he contrasts with the different *dynamical* explanations, according to which the heavenly bodies were conceived to be moved, first by a virtue inherent in their nature; then by impact via Cartesian vortices, and finally, by the centripetal force of Newton's gravitational theory.

[105] Cf. id. (1879), i. 346–7, 347 n.

Now, Mill asserts, these cannot all be true; if Newton's gravitational theory is true, that is if gravity exists, then the other assumptions must be false. 'Different descriptions, therefore, may be all true: but not, surely, different explanations . . . [The Aristotelian doctrine of virtue, the Cartesian doctrine of impact, and the Newtonian doctrine of force:] all these are explanations, collected by real induction . . .'.[106]

The interest of this passage lies above all in its affirmation that the contrast between mere 'description' in the sense of analogical fiction on the one hand, and an account which is putatively true (or false) on the other, is provided in terms of the absence or presence of a 'real induction', where induction covers once again the three cases enumerated in Section II, in particular, the two quite different cases of 'transcendence' there mentioned: (*a*) Mill's requirement of the inductive establishment of an existential claim concerning an explanatory cause or hypothetical mechanism; (*b*) the kind of transcendence involved in passing through an inference from observed 'particular' to unobserved 'universal'. These are of course precisely the arguments advanced against Whewell's 'descriptivist' account in the first sense of description (*qua* 'copy of reality') of our previous section, and it explains, incidentally, how Mill could have been sliding so easily from one use to the other. At any rate, Whewell, so he objects, has made no provision for either of these logical functions.

Whewell of course does not accept this argument here any more than in the previous case. He prefers to say that the different dynamical theories are just as much hypotheses as were the first set of 'descriptions' afore-mentioned; and that they succeeded each other through modification of each of their predecessors. For instance, the vortex theory can be stripped of its superfluous features, so that the essential element in it, as in Newton's theory, is the inverse-square acceleration of the planets towards the central body.[107]

But once again Mill replies:

> If the [vortex] hypothesis had been *merely a mode of representing* the facts, not an attempt to account for them; if, in short, it had been only a description, it would no doubt, have been reconcilable with the Newtonian theory. The vortices, however, were not a *mere aid to conceiving* the motions of the planets, but a *supposed physical agent*, actively impelling them; a material fact, which might be true or not true, but could not be both true and not true. . . . Dr Whewell probably means that since the phrases, centripetal and projectile force, do not declare the nature but only the direction of the forces, the Newtonian theory does not absolutely contradict any hypothesis which may be framed respecting the mode of their production. The Newtonian theory, regarded as a mere *description* of the planetary motions, does not; but the Newtonian theory as an *explanation* of them does.[108]

This passage is of interest because it suggests a further sense of 'description', subsidiary to the other two so far mentioned, and which always remain primary. In this third sense, Mill makes a contrast between kinematical and

[106] Mill (1879), i. 347. [107] Whewell (1860), 251–2.
[108] Mill (1879), i. 348 n; cf. also ii. 23, first three italics mine.

kinetical theory; between accounts which restrict themselves to a 'description' of motions, and those which involve forces introduced in order to 'explain' these motions. Whilst phrases such as 'mere mode of representation', and 'mere aid to conceiving' allude to the 'analogue' sense of description, 'mere description of planetary motion' as contrasted with their 'explanation' by means of a 'supposed physical agent', involves reference to two contrasted types of theory that we meet in physical science. The earlier contrast concerns a certain philosophical position, which for instance was taken occasionally by Newton, when he insists that his theoretical terms (including 'forces') are used by him 'not physically, but mathematically', the latter making abstraction from the 'true physical sense'.[109] The second contrast concerns the attempt of some physicists and philosophers of physics to exclude consideration of 'transcendent entities' altogether. (Compare, for instance, purely phenomenological thermodynamics which involves consideration of empirically defined variables of state only, with statistical thermodynamics which makes reference to 'underlying' molecules.)

By conflating the two contrasts, Mill makes it easy for himself to answer Whewell; for whilst the difference between 'mere method of representation' and genuine 'explanation' is one that is extremely difficult to define, and which moreover appears to have little relevance to the employment of induction, the second contrast can be dealt with in a straightforward way, if one insists (as Mill does) that the hypothetical employments of explanatory agents involves 'induction'. It is however no wonder that Whewell and his contemporaries, and indeed most subsequent readers of Mill have been confused by these vacillations in his argument.

To see all this more clearly, let us here interpolate a distinction between a number of different theory-types. The explanatory hypotheses ('explanans') of the first type, which we will call 'nomological', of which an example would be phenomenological thermodynamics, employ concepts only which either are employed in the formulation of the laws governing the phenomena of the system to be explained ('explanandum'), or are straightforward functions of such concepts. Contrasted with this type there are two others: the teleological type, which does not concern us here, and the type involving reference, in one way or another, to causes, and which we may call the 'aetiological' type.[110] This consists of those theories which involve additional concepts, not employed in the explanandum. (It is of course an open question whether such a clear-cut distinction between explanans and explanandum can be maintained; on the whole, Mill would have thought that there was a fairly sharp dividing line, whilst Whewell would have denied it, if either had ever asked this question

[109] Newton (1962), i. 5–6.

[110] Above, n. 94, we noted the distinction between 'laws of phenomena' and 'causes'. Now aetiological theories *par excellence* deal with 'causes'. What nomological theories do is to model the case of causal enquiry on the non-causal case of the 'laws of phenomena'. Mill, as our text shows, may have wanted to say that such reductions did not yield genuine 'explanations', but only 'representations' or 'descriptions'.

explicitly.) The aetiological group is divided further into 'infrastructural' and 'interstructural', depending on whether the 'cause' generative of the observed phenomena is given by the components of the system (as in atomistic theories) or whether it consists of certain 'interphenomena' which form links between the phenomena (for example when we explain optical phenomena by reference to photons).

There is, however, a second contrast which cuts right across this classification. We may distinguish between a purely theoretical calculus, on one side, in which the explanans does not involve any 'interpretation', and is held to be a purely formal mode of expression where we restrict ourselves to mere logical or mathematical functions ('formalist theories'); on the other side, the same explanans fully interpreted ('interpreted theories'). The interpretation of the formalism will be given in terms of 'correspondence rules'.[111] Moreover, the distinction between nomological and aetiological theories frequently overlaps with the distinction between those theories in which all the concepts have correspondence rules, and those in which some concepts lack such rules.[112] Finally, even where an interpretation is forthcoming, this may be only in terms of a 'model' rather than of some putative underlying 'reality'. For example, take an infrastructural theory like the kinetic theory of gases. Many nineteenth-century positivists, for example Mach and Ostwald, held that the interpretation of those theoretical concepts for which no correspondence rules existed (for example, the mass *m* of the 'molecule of the gas') should not be given a realist interpretation but at best only an interpretation in terms of 'model', for example a set of observable particles subject to the usual laws of Newtonian dynamics. Moreover, they usually preferred nomological theories since, in our terminology, all the concepts in such theories could easily be given correspondence rules, yielding a realist interpretation.

We can now grasp more clearly the nature of Mill's methodological preferences and of his insistence on the need for induction. When he speaks of 'descriptive' theories, though this may sometimes be a reference to nomological theory-types, his main concern is with those aetiological theories whose interpretation he conceives as having been given only in terms of models. And he argues, against Whewell, that without induction we must for ever remain at the level of 'mere descriptions', that is, either of nomological theories, or of non-realistically interpreted aetiological theories. On Whewell's account, so he charges, we perpetually remain at the level of 'description' in the sense of 'model' or mere 'mode of representation'. For the same reason, we remain perpetually at the level of mere hypothesis-fitting. Only an inductive inference can thus enable us to 'transcend' the phenomena, and reach a realist realm of explanatory causes.

It will be clear that the distinctions involved are extremely complex, and even nowadays still very contentious. And when we add to this the constant changes in Mill's use of the term 'description', it is no wonder that our two

[111] Cf. Nagel (1961), pp. 91–7.
[112] This overlap has also been pointed out by Nagel (1961), 127–8.

antagonists did not arrive at any agreement. At any rate, where Mill began his criticism of Whewell by complaining that the latter had remained entirely at the level of 'description', in the sense of 'observational reality', he ends by arguing that he has made insufficient contact with the 'underlying', or theoretical reality of things, replacing it instead by 'descriptions' which are no more than mere analogues, and can have no claim to represent putative 'fact'. In both cases, we have a contrast between some form of 'empirical realism' and Kantian 'idealism'—although Kant himself would perhaps have described the contrast as one between 'transcendental realism' and 'empirical realism', for reasons that cannot be considered here in any detail.

VI

Mill's insistence on the importance of inductive inference thus turns out to be tantamount to some form of empirical realist interpretation of theory; just as in the earlier set of arguments, he had insisted on a 'realist' interpretation of Whewell's colligating conceptions—equivalent to the first sense of 'description'. We also have seen that with this went a sharper distinction than Whewell would permit between 'fact' and 'theory'. In Whewell, where in some sense all is theory, the conviction of a need for special 'proofs' of the adequacy of one's theory easily evaporates; yet on the other hand, Mill, who so much emphasizes the need for such special proofs, cannot in the end be said to have provided more than a formula, rather than an adequate justification of his programme of inductive inference.

This has some curious consequences in practice. Sometimes, the greater stringency of Mill's programme, his insistence on an independent support for hypotheses, and his parallel contention that inductive proof is required to demonstrate the *existence* of physical agents, makes him more careful about the results of contemporary theorizing. Thus he is less sanguine than Whewell in accepting the wave-theory of light, current in his time, on the ground that no independent inductive proof has been supplied concerning the existence of an ether as the carrier of optical 'undulations'; hypotheses in the form of wave-equations are insufficient to 'prove' the existence of a 'carrier' of actual waves as against the competing theory of the emission of light corpuscles: 'Who knows but that some third hypothesis, including all these phenomena, may in time leave the undulatory theory as far behind as that has left the theory of Newton and his successors?'[113] On the other hand (we might observe) this third theory, wave-mechanics, can hardly be said to have been arrived at on the basis of some sort of 'independent inductive evidence' for a wave-particle theory of light!

What is more important, by resting the centre of gravity of his logic on

[113] Mill (1868), ii. 22.

inductive inference Mill becomes less sensitive towards those aspects emphasized by Whewell's notion of the 'formation of ideas'. Believing himself in the possession of a special logic of induction, he thinks he has no need for those supplementary criteria, so much emphasized by Whewell, for example the 'explication of conceptions' and the 'simplification of theories'.[114] Where in the previous instance Mill was shown to be the sceptic (with respect to optical theory), in another case his logical approach has the result of making him more complacent, as for instance towards gravitational theory. He is less concerned with such conceptual questions as the correct definition of matter, has less sympathy with such controversies as whether matter can attract matter at a distance.[115] Perhaps this is in part due to the fact that the idea of an inductive proof seems to certify the possibility of closure of a chapter of scientific theorizing: since inductive inferences are valid inferences, and may thus lead, at least in principle, to the establishment of connections between truths. Nevertheless, as we have also seen, this is accompanied by the notion of the fallibility of inductive conclusions to an indefinite degree. Even the law of causation, as already noted above, possesses a 'certainty' which is 'absolute' merely 'for all practical purposes', and which does not hold 'in circumstances unknown to us, and beyond the possible range of our experience', such as 'in distant parts of the stellar regions'.[116]

Whewell's theory displays a similar lack of definiteness concerning the relation between validation and probation. Not allowing for an independent inductive move, the fitting of data to hypotheses, except for practical or psychological purposes, never comes to an end, nor can we be certain that our ideas have ever been sufficiently 'clarified' to the point where no conceptual questions need be asked. As already noted (section I), we cannot be certain that on any given occasion we have reached the ultimate and correct 'mode of interpretation' of our ideas and their corresponding axioms. Yet clearly it is one of the functions of the notion of 'fundamental ideas' to hold out the hope of arriving at least in principle at an irrefragable end to the scientific process—though only in principle! Let us in conclusion take our observations on Whewell's oscillations a little further in regard to this topic. Ideas come to be explicated in the course of a process of historical development, during which they become progressively clarified and more 'distinct'. In this way, adequate ideas concerning force, polarity, chemical element, and the like take a considerable time before they can become 'distinct' conceptions.[117] In consequence, Whewell here, too, allows that although if we *were* to come to perceive these ideas (for example, if we could conceive the composition of bodies distinctly),[118] we

[114] Whewell (1847), ii. 12–16, 68, 73–4, 78.

[115] Mill (1879), i. 275–6; ibid. ii. 319–20. I have dealt with Mill's reaction to the controversies concerning action-at-a-distance in Buchdahl (1970), sect. III.

[116] Mill (1868), ii. 100, 108.

[117] Cf. Whewell (1860), 250–61.

[118] Whewell (1860), 340.

might be able to perceive the *necessity* of the corresponding truth (for instance 'that the modes of [chemical] composition should be definite'), yet it is only a *supposition* that we have a distinct conception of chemical element at the present time. This brings us back to the problem of 'necessary truth'.[119]

Like Kneale later, Whewell has a notion of necessary truth whose affirmation at any particular moment is problematic. A discussion of the question of whether such a position is untenable, not to say self-contradictory, would transcend the confines of this essay. Whewell's views in any case are not clear-cut, and are constantly modified in the course of his many writings on this subject.[120] He certainly claims that when we have learnt necessary truths, 'we see that they could not be otherwise';[121] the negation of such truths 'is not only false, but impossible', and we cannot here 'even by an effort of imagination, or in a supposition, conceive the reverse of that which is asserted'.[122] Nevertheless, we must read these claims in the light of Whewell's subsequent, and perhaps modified, position. Thus, once more in connection with the laws of chemical composition, regarded as putatively necessary truths, he adds that they must be considered to be 'assumed as a necessary consequence of the ideas of Kind and of Composition'.[123] And with this, we are back to the question of whether these ideas have yet reached a sufficient state of 'distinctness'. At any rate, what prevented Whewell here from pursuing further the difficulty of his position, which he never faces, were evidently the following two slides: (*a*) From regarding the laws of composition as a necessary consequence of his 'ideas', he ends up by viewing the latter themselves as necessary truths; (*b*) connected with this, from attributing categorical necessity to the 'ideas', he slides into regarding these as regulative principles, and as conceptual forms, the correct interpretation of which is always capable of being turned into a problem. Thus, if we found bodies not obeying the laws of composition,

> we should not at all doubt of the principle just stated, but conclude at once that the bodies were *not* of the same kind;—that the marks by which the kinds are distinguished had been wrongly stated. And thus we have it, not as an experiential fact, but as a necessary principle of chemical philosophy that the elementary composition of a body determines its kind and properties.[124]

In sum, Whewell, like Mill, finds it very difficult to reconcile the validational principles of his logic, suggesting closure, with the openness and fallibilism of his methodology. But where Mill regards Whewell's regulative principles as mere creatures of 'past history and habits of our own minds',[125] Whewell is

[119] Cf. above, Sect. II.

[120] On Whewell's theory of necessary truth, cf. Butts (1968), 5–17.

[121] Whewell (1847), i. 19. [122] Ibid. 55. [123] Ibid. ii. 662.

[124] Ibid. It is true that in his philosophy of mathematics Whewell espouses more explicitly an intuitional position with regard to necessity. But whilst he allows considerations of this to infiltrate into and buttress his general position, owing to the absence of sharp distinctions between mathematics and physics, we may ignore this aspect here. In the present context, Whewell's position approximates more to what has sometimes been labelled 'conventionalism'.

[125] Mill (1879), i. 275.

more aware of the constructive procrustean effect of predisposing ideas, a forerunner in some ways of our own latter-day philosophy of 'paradigm' situations in science.[126] In practice, as contrasted with his principles, it must be admitted that the theory of ideas is used by Whewell to support the promise that, at least in principle, and after sufficient 'clarification', we may arrive at a final state of stability; and as we have noted, Whewell thought that there were instances of scientific enquiry where we had reached such a state of perfection.

Undoubtedly Whewell caught much of the actual methodology of science: the need for conceptual clarification; the endless process of testing; the advance via hypothesis; the increasing enlargement of theories via consilience; and last but not least, the notion of the regulative employment of 'ideas', a forerunner of the theory of the 'paradigm' of our own day. Whewell thus steers us away from the preoccupation with the model of a logic of inference; and even the degree of 'deductivism' in his methodology, exemplified by his doctrine of hypothesis-testing, is modified through his emphasis on the very much competing operations of 'clarification of conceptions', reminding us thereby that much of scientific practice consists in informal criticism of its operative notions.

As against this, Whewell's conception of 'ideas' is rather limited and static; and he also believed that there could be a *final* clarification of ideas, so that the axioms associated with these ideas, for example, those of statics and dynamics, would be genuine examples of 'necessary truths', not likely to be impugned by future research.

Indeed, as explained briefly at the start, the notion of 'fundamental ideas', in Whewell sometimes plays precisely the part invoked by Mill in favour of independent inductive proofs for the existence of the 'causes' invoked to 'explain' the laws of phenomena. Such proofs may clearly be regarded as 'constraints' on the choice of hypotheses concerning causes. For Whewell, by comparison, the provision of such constraints is the task of the 'ideas'. Thus he says:

> What these characters [of the causes of the phenomena] are, has appeared in the survey which we have made of the Fundamental Ideas of the various sciences.[127]

And later

> The principal Maxim in such cases of induction [of causes] is the obvious one:—that we must be careful to possess and to apply, with perfect clearness and precision, the Fundamental Idea on which the induction depends.[128]

Now although the status of these 'ideas' in Whewell is controversial, and though he does not think that we are born with them, but on the contrary, as has been shown, that only a lengthy study of the history of the sciences will elucidate them, he nevertheless ultimately holds that they are not entities which have the status of those elements which are given by 'sensation', but that instead they

[126] Cf. Kuhn (1962). [127] Whewell (1847), ii. 99. [128] Ibid. 431.

are—as in Kant—'the work of the mind itself'.[129] So, as in Kant, instead of Mill's inductive constraints, we have here constraints of a 'transcendental' kind, in the sense that without such 'ideas' ultimately no scientific knowledge would be possible.

In his discussion of Newton's Third Rule of Reasoning, which authorizes us to affirm that 'qualities which are observed without exception [may] be held to be universal', Whewell makes it indeed quite clear that he regards his theory of ideas as fulfilling the task which for Mill is supplied by inductive inference. The mode of reasoning implied in this rule is declared to be 'far from conclusive': 'The assertion of the universality of any property of bodies must be grounded upon the reason of the case, and not upon any arbitrary maxim.'[130] In the discussion which follows, Whewell again insists that we are entitled to adopt lawlike propositions only if we are 'authorised, to look upon those as not only universally, but necessarily true'.[131] He then contrasts the generalization that all bodies attract with that which argues that all bodies are extended, which, unlike the gravitational law, cannot be 'conceived' to be otherwise. And he concludes: 'It cannot, therefore, be deemed philosophical to cut these Gordian knots by peremptory maxims, which encourage us to decide without rendering a reason.'[132] Evidently, this reference to 'reason' is to the theory of ideas, and it is the latter alone, and not inductive inference, which, according to this view, sanctions the ultimate adoption of a generalization as lawlike. It is thus clear that Mill's inductive validation, with its cosmological assumption of 'connection between truths', is the exact equivalent of Whewell's theory of ideas.

In conclusion, all this draws attention to an interesting psychological phenomenon. Because both Whewell's position with regard to the status of ideas and their associated necessary axioms, and Mill's conception of validity and the associated 'cosmological doctrine', involve in their nature metaphysical positions, it was easy for either to ignore its relevance in the work of the other. Just as Whewell would charge as arbitrary and irrelevant a position which involved the formulation of inductive rules like those of Newton, which are equivalent to a specific formulation of inductive inference, so Mill will not recognize the relevance of the function of fundamental ideas for the process of science, and instead will charge that Whewell's development of hypotheses involves an element of arbitrariness, ultimately resting on sand. And so he criticizes Whewell's method as one which consists in

> trying hypothesis after hypothesis until one is found which fits the phenomena; which one, when found is to be assumed as true, with no other reservation than that if on re-examination it should appear to assume more than is needful for explaining the phenomena, the superfluous part of the assumption should be cut off. And this without the slightest distinction between the cases in which it may be known beforehand that two different hypotheses cannot lead to the same result, and those in which, for aught

[129] Cf. ibid. i. 25. [130] Ibid. ii. 289. [131] Ibid. 290. [132] Ibid.

we can ever know, the range of suppositions, all equally consistent with the phenomena, may be infinite.[133]

Mill's positive point once again is complex, and looks in several directions at once. He may be referring to the idea of independent testing; or to the need for excluding alternative hypotheses, allowing of a 'crucial experiment' between two such competing hypotheses; more generally, to the need for ascertaining the prior antecedent probability of a hypothesis, though he nowhere operates explicitly with this notion. More in line with Mill's general argumentation, what does his criticism imply?

He requires some independent means of assessing the inductive probability of a hypothesis; in the absence of this, the method of unlimited testing has no chance of indicating whether we are approaching anywhere near truth, or even finite probability. For this reason, he insists on the need for independent proof; and his technical paradigms for this are threefold, (*a*) an application of the Canons of Induction; (*b*) a more specific application of the method of 'induction of causes', whereby we seek to determine the actual physical agent responsible for a set of phenomena; (*c*) his general insistence that only enumerative induction, regarded as a form of inference, can even begin to guarantee the universal applicability of a hypothesis that has been found 'descriptively' to fit the data.

But it cannot be said that Mill ever made very clear the nature of his opposition, short of the irrelevancy of the point concerning the need for 'independent evidence' for hypotheses, and short of his own metaphysical position, evidently replacing Whewell's doctrine of ideas, namely his assumption of the fundamental *validity* of inductive inference.

133 Mill (1879), ii. 24.

13. *Whewell's Problematic Heritage*

by John Wettersten and Joseph Agassi

Introduction: The Suppression of Whewell's Achievement

William Whewell is a very important writer whose contribution to the philosophy of science is enormous. Ignoring him presents a gap in our understanding of the history of the field. Yet he is ignored, and though he was famous and influential in both philosophical and scientific circles, he was maligned to the degree that he had to be rediscovered.

The philosophy of science was divided between the Baconians or empiricists and the Cartesians, intellectualists, or apriorists. The intellectualists declared knowledge inborn; they thus could not account for the growth of science. The empiricists accounted for the growth of science by declaring factual history to precede theory and by declaring theory to be the outgrowth of empirical factual knowledge achieved with no aid from the imagination, or with minimal such aid. The hostility of empiricists to bold conjectures and speculations was taken for granted—in many scientific and philosophical circles it still is. William Whewell claimed that without bold imagination science cannot proceed. And in this way Whewell deviated from both the intellectualists who declared scientific theories inborn and the classical empiricists who declared scientific theory born not out of the imagination but out of facts. Yet he was an empiricist: most of our bold conjectures, he urged, are refuted by facts and then rejected; some, however, due to luck and ingenuity, are verified by facts and enter the body of science.

In brief, Whewell's ideas were declared defeatist, since he made science dependent upon ingenuity and luck. And so, though his books were influential in scientific circles, and (together with the works of Faraday and Maxwell) led to a scientific tradition of honouring bold conjectures, because philosophers ignored him and scientists did not write histories of philosophy, he was forgotten.

The act of consigning the works of Whewell to oblivion began at once. One of the very first reactions to Whewell's work came as a review from the pen of his friend Sir John Herschel. However, it does not have the style of a highly critical review. For Herschel seeks to portray his task as one of great seriousness and one which requires that he grant as much credit as possible, that is, as much as the facts allow, to the author whose books—*The Philosophy of the*

Inductive Sciences and *The History of the Inductive Sciences*—were under review. He does this by pointing with sincere approval at Whewell's scope and the accuracy of his descriptions of science—even while maintaining that it was too ambitious to try to bring the history of science under one scheme: Whewell had tried to do too much, he should have been content to relate the facts.

Herschel's critique of Whewell is not, of course, critique but a reiteration of a disagreement. Classical Baconian empiricists demand that we amass vast collections of facts from which to allow theories to evolve, whereas Whewell presented his bold conjecture about the growth of science and presented the historical material as the empirical verification of his conjecture. Thus, on Herschel's traditional, establishment view Whewell's portrayal of the facts, but not his history, should be highly praised. Herschel claims that the facts had already been known yet he praises Whewell's portrayal of them as well. Now, the most significant aspect of Whewell's theory concerns the roles of hypotheses and ideas. Yet Herschel deemed inadequate Whewell's theory of the development of the historical process of the growth of science and of the role of facts and of hypotheses there. All the judicious praise which Herschel lavished on Whewell's collection of historical information was only part of a general dismissal of Whewell's chief ideas.

This view of Whewell, as having enormous erudition but as having offered an inadequate and so a worthless theory of science, is a mere application of Herschel's traditional view of science to the case of Whewell. As Herschel valued knowledge and sophistication, he could not but praise Whewell for his knowledge and sophistication. As Herschel followed Bacon and identified science with theories inductively emerging from facts, by contrast to Whewell's view of the growth of science as the outcome of verifications of bold conjectures, Herschel could not but dismiss Whewell's central thesis, especially his view that the sources of knowledge are not observations alone but ideas and observations. A Baconian cannot but deem this to be on the wrong side of the divide in philosophy: Whewell stands on the wrong side, with Hegel and Schelling, while Herschel stands on the right side, with the established tradition of science, the hallmark of which was the view that there is only one source of knowledge: experience or the observation of facts.

Herschel's view of science has been rejected recently, on the authority of Einstein, by Herschel's legitimate successors, such as Rudolph Carnap. Herschel's view of Whewell's research, though derivative of his defunct view of science, is the dominant one to this day. Although Herschel's review is a major cause of the traditional neglect of Whewell's theory, the dominant view received its greatest support not directly from the authority of Herschel, who in this regard is all but forgotten, but indirectly so: the tradition explicitly rests on the work of John Stuart Mill, who followed Herschel. Mill emphasized, quite in accord with Herschel's analysis, that Whewell had provided an excellent description of the methods of science, a description from which he had learned and which he had even taken over in his own work; in the final analysis, however, Mill felt that he could not be charitable: Whewell's theory of science must be judged at best quite incomplete. It is merely a theory that science is a series of conjectures and refutations, without any theory proper concerning the heart of the matter—concerning scientific proof. Mill proposed to avoid all quarrel with Whewell's historical studies as well as with his methodological observations. These, he said, can and should be endorsed, incomplete as he deemed them. Rather, he wished to offer a logical theory, a theory of proof, which should overcome the incompleteness of Whewell's work; one which would complete Whewell's schematic descriptions. This theory is that all proof is inductive, and it is reducible to Mill's celebrated four methods; it explains not merely how scientific knowledge is obtained, but also how it is established. And it does so by acknowledging the role of hypothesis in historical fact but not in philosophical or scientific principle: in principle all hypothesis is redundant. The question debated between Whewell on the one side and Mill supported by Herschel on the other, was about how many unavoidable or ineliminable sources of knowledge there are.

Let us say at once that we find it impossible to be charitable to Mill: it is very clearly quite impossible to endorse the claim that Mill's theory of method is in any way a completion of Whewell's theory of method: the two theories obviously contradict each other; the one cannot, thus, include the other. Yet only by viewing Whewell's reference to luck as a lacuna can one say that the lacuna was perhaps filled by Mill. What Whewell said is not that luck serves where method is not present, but that method is never present so that luck is always needed. Hence it was a plain distortion of facts that permitted Mill to object to Whewell's theory of method; yet only on the strength of this view could Mill object, as he did, to Whewell's view of science as rooted in both lucky conjectures (ideas) and observed facts: his objection rested on his claim that science is merely safe deductions from observed facts. To make this judgement transparent, let us, the authors of this essay, confess that we, too, reject Whewell's theory of verification. Yet we reject all theories of verification and find the strength of Whewell in his discovery of the indispensability of conjectures for science. Mill, however, was concerned with verifications; he rejected Whewell's theory of verification, and developed the traditional one which eliminates conjectures from science as never essential even if at times they admittedly provide useful short cuts; we have no objection to the claim that Whewell thought conjecture essential and Mill did not; we object to Mill's claim that Whewell failed to eliminate conjecture but Mill succeeded and thus completed the job done by Whewell.

Whewell's philosophy is not clear to us. His view of scientific truths was that they are necessary. Were they analytic or logically necessary, on transcendental grounds, then he would count as a precursor to Duhem and Poincaré, the great conventionalists. Were they synthetic yet necessary, on transcendental grounds, then he would count as a successor to Kant, as was his friend William Rowan Hamilton. Intuitively, we can conjecture that he saw the necessity of scientific truths in their integration into hierarchies of hypotheses culminating with very simple and very general laws; this makes him a precursor of Charles Saunders Peirce, perhaps, except that we are not clear in our own minds about Peirce's views either. We shall return to him later. Here we are content to express our failure to comprehend parts of Whewell's work which we shall not discuss here.

This is not to say that all options are exhausted. In particular, it may be noticed that the—alleged or real—convincing power of scientific ideas due to their high level of integration into a coherent system is an obviously metascientific claim, or, as Menachem Fisch suggests, an idea about ideas. But here we need not go into all of this, especially since very few thinkers were clear about the difference between the object-level and the meta-level; nor is the matter clear of all difficulty to date.

For us, Whewell presents a breakthrough in his discovery (*a*) of the role of bold conjecture in science and (*b*) of the procedures of testing them scientifically. It is the discovery more of a problem than of a solution. In part the problem was solved by Einstein and Popper, in their theory of science as series of conjectures approximating the absolute truth in a series of stages. Yet the problem is not fully solved even today; but solved or not, it cannot be overlooked. Yet nineteenth-century thinkers feared that admitting an unsolved problem about science may amount to a betrayal of the life of reason: they systematically agreed to recognize a problem officially only upon its having found an adequate solution, never before.

This is why Mill's pathetic theory of verification won and was taught in philosophy classes until very recently all over the western world. It was on the strength of this theory that Whewell's theory of the two sources of knowledge, ideas and observations, was neglected and his theory of the development of science was reformulated in traditional terms—so far as that was possible—in order to avoid the profound and difficult philosophical problems which it raised. Whewell's observations concerning the historical development of science on the basis of both hypotheses and facts were nevertheless taken to be among the best extant, if not the very best, except that they had to be further developed so as to prove the two sources of knowledge, hypotheses and facts (that is, ideas on the one hand and facts on the other), to be really one: observed facts. To this end, Whewell's historical, methodological, and even psychological observations had to be rejected by the very act of reconciling them with the traditionally established view as expounded and defended by Herschel and Mill. They both offered a technique of how to absorb Whewell's theory into the traditional one. This technique was to take (much of) Whewell's descriptions as true but not as ultimate. But how should the Whewellian heresies, expressed in historical studies, methodologial analyses, and psychological observations, be reduced to acceptable ones? The task required that one show explicitly what is implicitly (dogmatically) taken for granted: that in cases where it appeared that one was employing hypotheses not derived from observations, one could, with the aid of a more detailed analysis, reveal the empirical source of these hypotheses: all genuinely scientific hypotheses are in essence theories derived from observations.

Now the idea that all genuinely scientific theories are reducible to observations, perhaps in stages, is already traditional. But tradition only refers to true theories, of course. Whewell radically altered the situation when he agreed that bold conjectures play enormous roles in science even if they end up refuted and rejected. So Mill now added another technique: show that even these false-but-useful conjectures are not really conjectures but based on induction. This already revolutionized the establishment view radically, and not for the better: the change made a pack of confusions out of the traditional approach. Any attempt to straighten it up led to a rediscovery of the ideas of Whewell, at times with little variation.

In the following pages we shall outline a tracing of a small part of the clandestine history of Whewell's ideas, partly as alive in the programmes of the philosophical establishment to overcome them, partly as alive in the writings of new generations of authors who attempted to straighten out the philosophical establishment's mess.

The Whewell literature began a couple of decades ago: between his death and our age there are very few scanty references to him and one forgotten dissertation. Now most of the contemporary literature on Whewell is either Millian or Whewellian. The Millians are still attempting to reconcile what they cannot reject of Whewell with the established philosophy—except that this philosophy is now excessively confused and no longer established: ever since a precedent was set and Sir Karl Popper was elected Fellow of the Royal Society of London as a philosopher of science, the established version of Popper, which is a variant of Whewell, is the new voice of the establishment. As to the Whewellians, they are attempting to explicate him, that is, to make his theory clearer, to modify it so as to answer some of the Popperian objections, or to make it better cohere with twentieth-century prejudices, or to incorporate it into some forgotten Millian archaism as if it were the greatest innovation. We frankly reject all variants of Whewell's philosophy; yet we accord it a place of honour—in the history of philosophy in particular. We shall assiduously avoid explicating his view or discussing any of its fine points. All we do is trace its underground history.

1. Four Whewellian Problems

We propose to look at the background material a little more closely, so as to see the problem-situation which the Herschel–Mill response to Whewell created for the philosophical establishment. We may then use this analysis to portray the context in which later thinkers, including Ernst Mach, Charles Saunders Peirce, Claude Bernard, and Pierre Duhem, operated and to see in this analysis the roots of contemporary problems in the field of methodology and even in the field of psychology. Our task here, however, is primarily to point to the problems in the Whewell–Mill controversy which led to the later developments, and also to illustrate attempts to solve them.

There are four central problems which Whewell raised for the traditional view of science. One of these is a product of his central study in the philosophy of science, and two are of his study in the history of science. The fourth is due to some psychological observations he used to support his theory. All four problems are, of course, intimately connected, but we may still deal with each in turn; we begin with a summary of his philosophy and the four problems which it raises.

Whewell's philosophy is founded on the idea which he took over from Kant—that scientific ideas are not derived from scientific experience but are, on the contrary, conditions for such experience; they are imposed on observations so as to make scientific experience at all possible. Whewell made it amply clear that he considered his own discussions of space and time mere paraphrases of Kant's discussions of them which may easily be found in his celebrated *Critique of Pure Reason*. Whewell further used Kant's transcendental argument to defend Kant's view: he argued that science incorporates statements of necessary relations; since these statements are known to be true, and since no inference from observations alone to theory could provide us with such knowledge, the content of these statements cannot be reduced to the mere combinations of concepts in accord with the rules of association, resemblance, and contiguity; nor can the proof of the truth of scientific statements rest on a derivation of these concepts from observations. New theories of the content and of the proof of the truth of scientific theories are needed, he declared.

Whewell had a severe criticism of Kant's theory. The weakness of Kant's theory is that it deems all certain ideas, all ideas of science, to be a priori valid, to be merely imposed on, and not in any way learned from, scientific experience. Such a theory does not account for, or even allow for, the growth of scientific knowledge, let alone describe the role of experience in this growth. Moreover, Kant's theory of method was quite traditional, and it did not sit well with his theory of knowledge. Whewell rejected, therefore, the theory that synthetic forms of intuition and synthetic principles of reason and synthetic principles of understanding are all given a priori and merely invite, in order to be justified, some transcendental analysis of the way humans reason about the world, understand it, and perceive it. It is also particularly difficult to see why they have to be discovered empirically yet proven transcendentally. Rather, Whewell deemed the very development and growth of ways of understanding and perceiving the world a task for science; likewise, the development and growth of ways of understanding and perceiving science he deemed the task of the philosophy of science.

The problem which thereby arises seems quite difficult. For it is the task of explaining how empirical research leads to the growth of necessary knowledge, no less. This is a dilemma: ideas which are necessary must be proved by experience alone, or else by some method other than mere inference from observations; this means by definition, that we have only a posteriori or a priori methods of discovery coupled with proof. And both are inadequate. We may combine the a priori with the a posteriori methods, but the defects of each solution may show up in the combination.

Whewell sought to combine the a priori method with the a posteriori method and to block the reappearance of the defects of each ingredient in this combination by first introducing some fallibilism into Kantianism (if we may use an anachronistic label); he assumed, that is, that hypotheses which we employ in experiencing the world are not *given* a priori (in Kant's sense of a priori valid) since they are probably false, and even unavoidably so. Rather than given, they are conceived: humans create ideas, they make up guesses or conjectures which they use to order experience and to understand the world. Thus Whewell's theory, contrary to all Kantian theories of a priori valid knowledge, could account for growth. Just as Kant's synthetic a priori knowledge, so the hypotheses which we invent, possess content in excess of that provided by experience alone. Whewell showed this by analysing many of the conjectures which had played important roles in the course of the growth of science. They provide at least some ordering of our experiences—an ordering which no amount of mere observations could provide. This is a dualistic theory, the theory that science employs abstract conjectures and observations of facts; it is the fundamental thesis of Whewell's philosophy (the fundamental antithesis, as he called it). It sets the context in which all else is developed. It leads to difficult problems, some of which cannot be solved within his philosophy. Perhaps no philosophy can offer a solution for them, as he himself proposed: perhaps these lie beyond the limits of human knowledge. Nevertheless, it is the only possible view: anyone who has studied the alternatives to it can recognize its necessity through the fact that contradictions ensue from any attempt to reconcile the known historical facts with either of the other two views, the Cartesian intellectualist view or the Baconian empiricist view: mere ideas and likewise mere observations are not and cannot be the sole source of scientific knowledge.

This, let us notice, is a mere shadow of Kant's transcendental proof. It is no doubt a proof, and the proof is perhaps transcendental, but then at most merely of the existence of a dual source of knowledge, merely a proof of a small part of Kant's system.

The basic thesis that there are two sources of knowledge—in theories and in observations—enables one to explain the content of scientific theories which is not reducible to observations and their combinations. It does not explain, however, the necessity of scientific theories. For the history of science is constituted mainly of conjectures which were usually refuted and then improved upon. How, then, were some conjectures rendered necessary truths? How can we explain the excess content of true theoretical science? How can we, that is, explain our having genuine knowledge, not merely of the content of our conjectures, but also of the status of necessary truth which some of these do acquire?

Whewell's solution to this problem is at the heart of his philosophy and its most criticized portion. For Whewell sought to avoid both the view that the necessity of ideas could be explained merely through the proof that they are a priori valid (as Kant would have it) and the proof that this necessity is achieved when theory is reduced to a mere conjunction of observations (as Herschel and Mill would have it). The tradition has sought to show that any claim for the status of necessity for some ideas must be either one or the other sort—even while conceding that neither one nor the other sort has ever yielded successful results; for any alternative to these two must be a confused combination of both. Whewell is often accused, even today, of not having rightly understood Kant, or of mixing Baconian, Kantian, and even Platonic ingredients in some obviously unacceptable way.

Whewell's theory of science as a series of conjectures, refutations, modifications, and occasional verifications seems to us false, but let us declare our own view at once: Whewell's theory is false but sufficiently clear; it was and still is rejected by the philosophical establishment as either confused or failed Kantianism or failed empiricism. We find the source of this philosophically received opinion to be the intellectual dishonesty of Herschel and his followers who refused to admit, yet did endorse, Whewell's historical criticism of their official doctrine. That they endorsed it is proved by their repeated (lame) attempts to mend their fences. That they did not admit his criticism is betrayed by their condescending dismissal of it.

The first of the three central problems which Whewell imposed on the philosophers of the establishment, from Herschel on, was this: how could the content and methods be explained without appeal to the special and independent role of ideas? This problem was, of course, not new. It was central to the works of John Locke and David Hume, to mention the giants of the tradition. But Whewell had given the dualist view a new and powerful formulation which was not so Kantian and he backed it with massive historical evidence. And as it was not quite Kantian, it was no prey to the objections which destroy Kantianism; above all, it accounted for scientific growth. Whewell's challenge was dismissed by the mere dogma that only the Kantian and the empiricist justification of science were available but no mixture was possible.

The development and justification of Whewell's theory of science depended above all on Whewell's history of science. He sought there to illustrate the process of learning from experience as well as to show the attainment of specific scientific theories which were learned from experience and which competent scientists regularly consider necessarily true. This theory clashed head on with the standard theories of method in that it took all science to begin with hypotheses and the central method of science to be the testing and improvement of hypotheses, rather than the accumulation of facts and the making of inductive inferences to theories on the basis of these facts alone. Whewell, it may be noted, was not the first to have seen that hypotheses play a significant role in science. In particular, Herschel makes a point of observing that scientists had long noticed the significant role which hypotheses play in the growth of science. Yet Herschel's own work—his celebrated *Preliminary Discourse to the Study of Natural Philosophy*—combines two descriptions of the growth of science: in one description hypotheses are declared insignificant, in the other science is presented as a Baconian collection of facts from which theories evolve. Clearly Herschel gave priority to the latter (as he did in his review of Whewell) so that, against his own emphatically declared intentions, he played down the role of hypotheses and declared them insignificant. Thus, he both warned us sharply against the use of hypotheses and he censured those who spoke ill of them. By contrast, Whewell provided a fairly comprehensive account of the growth of science without reference to any Baconian procedure, and the role he ascribed to hypotheses in this growth is very clear. He further claimed, quite explicitly and quite contrary to Herschel, that no Baconian account of the history of science is possible.

One of the reasons why Whewell's challenge was so powerful was just because of its methodological comprehensiveness, which by itself rendered Herschel's work pallid in comparison. He set himself the task of employing his philosophy of science as the explanation of the history of science. He developed his thesis that there are two sources of knowledge as a methodological theory and showed that the methods which were described in this theory were universally applied in the history of science; his thesis was thus shown capable of accounting for the whole history of recognized scientific research.

The second challenge which Whewell offered those who rejected out of hand his excellent philosophy of science was to account for historical research. Whewell had based his philosophy of science on his extensive historical research and maintained that his theory and his theory alone could account for the details of the history of science. The conservative establishment represented by Herschel and Mill had maintained that his description might be true but was not ultimate: the process he described was correct but partial. It could and must be reduced further by a Baconian process in which the source of ideas is explained away and reduced to experience—the employment of an idea in science is justified by the claim that the idea had been justified by the mere use of past inductions from observations alone.

At the historical distance of over a century we can only gasp at the cavalier attitude of Herschel. His own historical sketch, even when viewed as a mere sketch, is easily seen as very unsatisfactory and apologetic. Whewell's history—dedicated to Herschel!—still challenges us and alerts us to highly important and interesting historical details. Yet all that Herschel could say of it was that Whewell found many important historical facts from which theory now should be extracted. Which theory? The theory that theories are extracted from facts.

Yet in every chapter of his history Whewell offered empirical refutations of this theory! Why, in particular, are we to extract the theory that theory comes after facts when we had that very theory before Whewell marshalled his facts? No answer. The establishment has the last word. We find this high handed and disingenuous.

The first effort to reduce the Whewellian description of the procedures of science to a Baconian one turns out not to be so much a completed theory as a project for developing and completing a theory which will, in retrospect, justify the establishment view. It was contributed by John Stuart Mill. Mill had two central tasks. The first was to show that mere observation independent of theory had played a significant role in science and in its history. The second was to show that the important role which hypotheses had played in science did not conflict with his own Baconian view of hypotheses as redundant. The first task he sought to perform by appealing to an analysis of the story of the discoveries of Kepler. He claimed that Kepler's discovery of the planetary orbits was a product of Tycho Brahe's observation of the positions of the planets which was done with no knowledge of Kepler's laws. Kepler's major accomplishment lay in the tedious observations Tycho Brahe had made and in the tedious calculations Kepler had performed—not in any hypothesis. This very alternative was, we think, effectively criticized by Whewell: as he has argued, Brahe used many theories in his construction of instruments; he discovered time-series, and he observed his data as deviations from known rules—as what Whewell called 'residual phenomena', now better known as anomalies. Thus, neither Mill's chief instance, Tycho Brahe's case, nor any other instance, was to any avail. Mill could have no success in seeking to demonstrate the role of fact-gathering unaided by ideas as long as he could not respond to Whewell's criticisms. Mill was usually a fine critical thinker. This was his great *faux pas*.

Whewell's defence of his dualism was not limited to his historical studies. He offered a competing theory of perception (in science) which was in part based on the physiological conclusions of Charles Bell. Like Kant before him, Whewell sought to explain our ability to perceive, especially positions in space. He deemed perception to be an activity which involved the procedure of guessing and making tests. This hypothesis was supported by Bell's theory of the role of eye movements in the perception of space. Yet Bell's theory had more details, including some on the accuracy and rapidity with which perception occurred; and these required explanation. And Whewell offered an explanation. One of the items which it contains became central to all psychological theories of perception of the late nineteenth century: it was the theory of unconscious inference.

Whewell offered thus an idea which the empiricists could use; they used it freely, and they thereby undermined the very psychology which traditional empiricists were attempting to employ in order to reduce theory to observation. Thus, even if the establishment successfully responded to Whewell's challenge, at the very least they had to admit that he had forced them to modify their views. Instead, they pretended to have improved their empiricism so that definitely there was no more room for anything like Whewell's challenge. This, again, is obviously apologetic and quite disingenuous. It really would be condescending of us to pretend that we condone it in any way.

Whewell introduced his theory of unconscious inference in order to explain the quickness and exactness with which we perceive the world in accord with existing theories—with which we impose our concepts on our percepts, to use Kant's idiom. John Stuart Mill took over the idea that we perform unconscious inferences, though he rejected its Kantian garb, maintaining that the unconscious inferences merely represented the application of knowledge empirically acquired in the forgotten past. This move was generally endorsed, and it opened up a new challenge for psychology, to study and develop empirically our knowledge of unconscious inferences so as to be able to explain perception, especially in science, by reducing it to sensations plus unconscious inferences from them. The battle over reduction thus started within the field of empirical psychology, where, until then, the association of ideas had looked less problematic. As it happens, victory was achieved by the followers of Whewell long after the contestants had died. The battle had been started already by the contestants, and immediately Herschel and Whewell argued about unconscious inference with regard to mathematics. Herschel and Mill are deemed to have won, yet this problem is still on the agenda of the establishment philosophers of mathematics: the psychology of mathematics is still less advanced than that of experience, at least with regard to unconscious inference.

Though it is still not our intention here to analyse further the development in the theory of mathematics and logic, we would like to note that the most powerful response to the establishment views of these disciplines was presented by George Boole. With explicit reference to the work of Whewell he offered a deductive theory of these disciplines, a theory which was intended to portray the fundamental idea which unified them. In the last chapter of his *Laws of Thought* he provided a Whewellian interpretation of his own work. He rejected Mill's theory of science without mentioning him: he considered and rejected the view of Kepler's discovery of his laws unaided by hypothesis. Mentioning Whewell approvingly he spoke of its importance for the growth of the activity of the mind.

Mill's answer to Whewell amounted to a tacit admission of failure, and so it was not characteristic of his usually candid and brave spirit. His answer was a mere programme, and one taken over rather lamely from Herschel with no improvement. It was actually worse: a reaffirmation of the faith that the proper use of ideas is the mere use of previous inductions. Due to the swiftness of unconscious inference, he said, we do not notice that we are merely appealing to past experience when properly we refer to ideas, since ideas properly used are mere ideas based on past experience. Yet this view in itself was not based on past experience; it was frankly sheer speculation, and one whose role was apologetic: a mere programme for overcoming Whewell's challenge with some Baconian alternative to Whewell's impressive account of science.

The question of how this Baconian programme could be executed was the third one which Whewell presented to his opponents. It turned out to be quite too formidable.

There was a fourth problem which arose primarily due to Whewell's sophisticated historical studies but which is worth handling separately. It concerns the reconciliation of logic with the history of science. This should, prima facie, be no problem at all since, with the possible exception of mathematics, science is the most successful application of logic we have. Yet it was Whewell's failure to effect this reconciliation which led to his oft-quoted quip that if his theory conflicted with logic so much more the worse for logic, made in response to De Morgan's critique and rejection. Whewell had, apparently, sought De Morgan's support after he saw his theory losing ground. The quip—in a private letter—seems to us to be merely the outburst of a frustrated individual. But the conflict was real and it even led Whewell to a Hegelian view of scientific truth: he deemed truth to change and sought to reconcile this with logic by claiming that, although two theories at the same level could not be contradictory and true, two theories at different levels and perhaps at different times could be both contradictory and true.

This corollary was not, we think, desired by Whewell. But he saw no way to avoid it if he was to account honestly for the facts. For he knew quite well that, as new theories in the history of science superseded older theories, they did not merely incorporate them as special cases, but changed both their fundamental ideas and the precise factual information. The new theories, in brief, contradicted the old ones. Whewell could not ignore this. His only important contribution to empirical physics was a variant of a crucial experiment between Newton's theory of gravity and Galileo's. And, of course, Newton's theory won. These are hard facts—logical and empirical. These facts are hard to explain on any theory. For no theorist dared deny that Galileo's and Kepler's theories were true, nor that Newton's theory had superseded them and was also true. But Newton's theory contradicted and corrected those which it superseded. Whewell was the only theorist of his generation to meet the problem head on and try to solve it. He proposed that truth changes. His theory of the consilience of inductions was designed to explain how two theories at one level could be unified by a theory at a higher level even when this theory contradicted those theories which it unified.

The problem Whewell posed then, was how can one avoid his unacceptable theory that truth changes and account for the history of science? How can science be true in the face of the fact that new scientific theories contradicted older ones? One would have thought that Whewell's theory of the transformation of hypotheses and that of levels of truth would have been the subject of sharp criticism. For here Herschel and Mill could have shown in detail what they otherwise had merely suggested by something like guilt association: that is, they could argue with some cogency that Whewell was Hegelian. But they never did this. They criticized his view that truth changes as a violation of the principles of logic, and here their criticism was valid and could be effective. Yet they never discussed the reasons for Whewell's construction of this idea. The reason is clear: they too could not account for the facts.

The strategy which the entire tradition has used in order to avoid looking at the facts, from Mill to the present, is to distinguish and separate the context of discovery from the context of justification so-called. This distinction, at times attributed to Reichenbach, has been used by him and by others rather disingenuously. For it has been used to justify focusing on the problem of justification as a mere problem of abstract inductive logic which invites no concrete examples or which invites as examples at most only the latest established scientific theories. If any serious attempt had ever been made to apply this view to the history of science one would have seen that Kepler's theory is part of the context of discovery of Newton's and Newton's of Einstein's, yet they contradict each other.

The accepted policy still follows Herschel, Mill, Reichenbach, and others, and it is that of studiously ignoring the fact that new theories contradict their predecessors. In this the tradition has been enormously successful. In 1905 Pierre Duhem argued, and in 1963 Karl Popper argued too, and in detail, that Kepler's theory is contradicted by Newton's. Neither made any reference to Whewell. This was not a mere anachronism, for the knowledge of the contradiction had been obliterated by careful tradition. It is only now that tradition is slowly altering as writers begin to timidly concede that this sort of contradiction should not be ignored.

Whewell's theory was quickly taken off the agenda. There were three central justifications for this move. His theory of the growth of necessary ideas broke with Kantianism and as such was deemed confused. His theory of the growth of science thus allegedly provided no theory of the proof of scientific theories. His theory of the history of science was thus declared over-ambitious, seeking to bring all science under the scope of one scheme, and a scheme which was defective. This rejection of his theory was combined with an acceptance of his methodological observations and, to a degree, his descriptions of science, but under the proviso that they needed a face-lift so as to bring them into accord with official quasi-Baconian views of empirical verification. Herschel and Mill had proposed a programme for providing a response to Whewell but they could not begin to carry it out. It was taken up in numerous fields, from logic and mathematics, to methodology, to psychology, to history, and more. In spite of all effort, results were insignificant in themselves and so represented no progress towards the realization of the Herschel–Mill programme of coping with the results of the Whewellian opposition. Yet efforts were repeated for a century or so to invalidate them or to incorporate them into the old crumbling framework. Success in this endeavour is useless, anyway, since the opposition provided ever more new results which could be substituted for the old ones.

Since all this was done in the name of Science and with the support of the establishment, the programme could remain dominant to the present with no end in sight. Fortunately it is now dead.

It is not for historians to complain, but they need not desist from judgement, especially when the unfairness in question is so very conspicuous (the Whig interpretation of history is commendable, though not necessarily obligatory). The charges against Whewell were, to repeat, question begging: the charge laid at his door amounted to a censure for his dissent from official doctrine. His critique was dismissed with a programme designed to find some answer to it. His success was declared too little and too much, and with the same excuse: it did not tally with official doctrine. The presentation of a programme as an answer to an opponent as if it were an effective answer which has the opponent defeated, is a bit premature, to say the very least.

The programme was, of course, a mere continuation of the traditional plan or theory of the British empiricist tradition, going back to Francis Bacon; it aimed to show how all science is derivable from observation. Yet the new programme also posed a new problem-situation. It did so because there existed after Whewell's work a test for success; Whewell's work created two sets of desiderata for the establishment's programme to fulfil. First, a successful execution of the programme had to explain away (*a*) the content of the scientific *theories of science*; (*b*) the role of theory in perception; (*c*) the role of hypotheses and conjectures in science, especially of false conjectures; and (*d*) especially those false conjectures, such as Kepler's, that the scientific community deemed established, yet which contradicted other scientific theories. Secondly, a successful theory of science had to be at least as good in details as (*a*) Whewell's theory of science and (*b*) his history of science. Whewell thus provided detailed desiderata for those who still wished to maintain the traditional view in the face of his studies of the nature and the development of science. This, we contend, is how Whewell won, even though both he and his writings still suffer official neglect.

The most important lesson from all this is that valid criticism should not be dismissed on the ground that the critic's own views are erroneous.

2. The Aftermath: Three Examples, Mach, Peirce, and Duhem

The impact of Whewell's philosophy has been on the one hand deep and far-reaching and on the other hand often difficult to trace due to the official silence of the establishment and the pretence that his work had been superseded by that of Mill. We wish here merely to illustrate the depth and importance which Whewell's philosophy has had. We have chosen three thinkers who are all of importance, who worked within somewhat different traditions, who exercised considerable influence within these traditions, and whose basic problems are all versions of those fundamental problems which Whewell's philosophy had created. We note thereby that there is no one way in which these problems are formulated, for one can seek to improve on Whewell's results in various ways. Consider, for example, the problem raised by the discovery of a contradiction between successive scientific theories. One can seek to develop a new theory of how scientific truth changes, as Peirce did; one can attempt to overcome the same difficulty by attempting to reconcile all the theories of science, as Mach did; or one can seek to explain science and its history without any reference to the truth value of scientific theories, as Duhem did. Yet all these approaches are still attempts to take account of Whewell's historical studies with a new and, for some reason, more acceptable, methodology. We leave completely aside the influence of Whewell on the followers of Mill, even though his discoveries have continually served as one of the most important sources of problems for their tradition: we deem the results this tradition has produced to be of little or no interest. We also leave aside the influence Whewell had in psychology, which should be presented elsewhere; and leave aside some important direct influences. We have mentioned in passing Boole's and thereby Whewell's influence on the history of mathematics and logic, but we shall not elaborate. The same holds true for other examples, such as Robert Merton's use of Whewell's ideas to develop both a sociology of science and a theory of method for sociology, and such as Thomas Chrowder Chamberlin's introduction of Whewell's view of the need to conjecture freely, into the traditions of geology. We restrict ourselves to methodology but we do not discuss the most startling result even in this field, the theory of Karl Popper, since this result was indirect, arising as it did (as has been explained elsewhere) out of psychological research, and out of Whewell's influence on the Würzburg school of psychology, via its founder Oswald Kulpe: Popper is in the cast of that school.

We begin, then, with the philosophy of science of Ernst Mach as an answer to Whewell's challenge. The analysis of the work of Mach in terms of Whewell's problems is prima facie quite simple. Mach's *Die Mechanik* corresponds to Whewell's *The Mechanical Euclid* and *The History of the Inductive Sciences*; his *Erkenntnis und Irrtum* corresponds to Whewell's *Philosophy of the Inductive Sciences*, and his *Die Analyse der Empfindungen* addresses the task of the reduction of ideas to sensations and their combinations, the defence of sophisticated associationist psychology and thereby to demonstrating Whewell's dualism as monistic at bottom. These books are thus quite clearly designed to overcome just those problems which Whewell's theory puts on the agenda of the establishment in philosophy. The history is designed to show that the development of science does not lead to the establishment of rock-bottom dualism but rather that all scientific ideas may be removed by further analysis and that further progress in this direction is still possible, enabling us to remove those remnants of such ideas which are still found in science. Newtonian mechanics presented Mach with the central and most difficult case, and he repeatedly tried to eliminate Newton's theory of force without destroying

Newtonian mechanics. In *Erkenntnis und Irrtum* Mach seeks to explain away Whewell's claim that the activity of the mind produces ideas which, after repeated refinement due to repeated tests, finally correspond to the world. The very research methods which Whewell described he wished to have reinterpreted as mere psychological processes leading researchers to a better orientation. They would thus be seen not as real research methods, but only as preliminary to real (Baconian) research. Mach's *Die Analyse der Empfindungen* is an attempt to give a theoretical basis for explaining away any metaphysical content of scientific theories (such as Newtonian forces). In all three of Mach's books just mentioned there is one theme: scientific theories are neither true nor false descriptions of the world. This view seems obviously designed to remove just those problems which led both to Whewell's quasi-crypto-Hegelianism (as explained above) and to his view that there was an ultimate mystery found in the fundamental duality of the sources of scientific knowledge. Ultimate analyses were possible, Mach thought, which did not prohibit the existence of knowledge as described in the purist empiricist tradition.

We propose that the logical and methodological parallels we draw are not accidental: they are too extensive and too deep for that. Also, it is easy to show that Mach not only knew Whewell's work but that he had studied it. He deemed his central task as the overcoming of Kant's view of the thing in itself. He pointed to the works of Fries, Apelt, and Whewell as having failed, as having been led to fantastic results due to the inability to be free of Kantian errors. He would do better. He viewed the capacity to avoid all vestiges of Kantianism as the prime desideratum of his own philosophy.

So far as we know Mach made no analysis of his debt to Whewell, or to anyone else for that matter. He mentions Spinoza but does not explain this, and Kant, of course, whose views he wished to overcome, and Hume and Berkeley with whom he expresses affinity. But he gives precedence to comparisons rather than to historical analysis. Indeed, in a footnote to *Die Analyse der Empfindungen* he attributes the development of his view to two things: an intuition he had as a youth—an intuition of the world as made of sensations—and his strong reaction to Kant. We thus cannot expect to learn very much from Mach concerning his intellectual debts. *A fortiori* we do not have direct evidence that Mach modelled his history after Whewell's work and in response to it, but we think that he did. We will proceed then with a discussion of his rather late (1905) *Erkenntnis und Irrtum*, where the use of Whewell as a foil appears explicitly (in the other texts Whewell is nowhere mentioned). It is clear that he presented the successful resolution of Whewell's problems as one of the prime tasks for his own theory.

There are approximately ten references to Whewell in *Erkenntnis und Irrtum*. Three of these are of interest. In the foreword[1] he mentions the difficulties which the Kantian theories of Apelt and Whewell fall prey to (he does not

[1] Mach (1917), p. vi.

explain them) and refers to Beneke whom he later quotes; Beneke refers to Whewell and claims to have overcome his criticism. Mach endorses the view quoted. We mention this only to show that Mach was himself quite clear concerning the need to overcome Whewell's Kantian criticism and had examined earlier attempts to do so.

There are two references to Whewell which indicate in detail how this was to be done, in accord with the above general sketch of the parallelism between the two. The first concerns what we regard as the central passage of the book, that is, the discussion of the controversy between Mill and Whewell which occupies the latter half of 'Deduktion und Induktion in psychologische Beleuchtung'.[2] Mach here seeks to resolve this dispute. He explicitly accepts Whewell's view, as opposed to Mill's, concerning the role of hypotheses and of intuition in the development of science. He does not accept Mill's account of Kepler's debt to Brahe as a refutation. Nevertheless, he wishes to explain away any use of ideas in research: there is no mystery in the capability of an idea to describe the world, since ideas are inessential. He admits that the description of the methods of science provided by Whewell is true; but it is, he says, a description of a psychological phenomenon and not of the logic of science. Ideas, then are merely provisional, parts of the path to truth which proceed until science is complete but then are totally eliminable. The use of hypotheses can be psychologically explained as due to the need of the organism to find its way in the world and they must be provisional because they are merely psychological (he refers once more in his last and summary chapter to Mill and Whewell in offering his own alternative).[3] The purpose of the material preceding this central section has been to build up a psychological explanation and the purpose of the material following it is to explain how the psychological phenomena as hypothesis-making may be developed into sound explanations, that is, into mathematical relations between sensations.

Mach's third reference which interests us is to Whewell's interpretation of the principle of conservation of momentum. Whewell claimed that this principle presumes a principle of causality which is, in turn, a basic principle for science.

Basic principles are conditions which make science possible. The Kantian view is to prove the validity of the conditions which make science possible in a transcendental manner. Mach's contrary move was to show that they are superfluous; or, as he could not do that, to show that they are only psychologically necessary, not methodologically or epistemologically so. And as Whewell proved that the principle of conservation of momentum rests on the principle of causality, Mach argued that causality is redundant.

Mach claimed, contrary to Whewell, that as humans are primarily logical, rather than psychological, then the principle of conservation of momentum would have been found much earlier:[4] Whewell has been misled by taking

[2] Ibid. 313–20. [3] Ibid. 450. [4] Ibid. 138–9.

psychological phenomena also to be logical or epistemological. He referred back to his discussion of the development of mechanics and contrasted his view with Whewell's.[5] One purpose, then, of *Die Mechanik* has been to overcome Whewell's history with an alternative. Kant and Whewell had argued that if Newton's theory were taken as fundamental in any sense then Kantianism of some sort is thereby admitted. Mach's wish to reject Kantianism, then, forced him to reject the view of Newtonianism as fundamental. He attempted this by venturing to show that Newton's theory can be further analysed, that some abstract principles not referring to Newton's metaphysical entities can be developed in the same way as Newton's theory had been. This would render it a mere stage to pass through due to our psychological nature. This idea, it may be noted, is usually read as an expression of conventionalism, in line with the ordinary interpretation of Mach as a conventionalist. Yet here it is read as a critique of Whewell. This reading resolves much of the apparent conflict between the young Einstein and Mach. And indeed, Einstein sought then to carry out the programme overcoming Newton's theory with an approach that was simultaneously more analytic, seeking to find more abstract relations among more fundamental entities. And Einstein expressed his indebtedness to Mach in this very context, generously taking his deviation from Mach (his rejection of Mach's identification of the fundamental entities with sensations) to be rather minor.

Before concluding our discussion of Mach it is useful to say a few words about *Die Analyse der Empfindungen*. On our view the purpose of this text is to provide the proper foundations for methodological studies—while thoroughly overcoming Kantianism. To this end he sets himself the project of exhibiting, or, if not exhibiting, at least pointing to, those elements which may be deemed fundamental. By doing so he thought he could explain the relation between mind and body, between the psychical and the physical. This reduction had above all to meet the epistemological aim of closing any possible mysterious gap between subject and object. This is the very same gap which, according to Whewell, is the Fundamental Antithesis in philosophy, the dual source of science, thought, and facts, and which can receive no solution without thereby explaining knowledge away. Mach's view of sensations as fundamental facts came to solve the problem which Whewell thought was insoluble—to solve it not philosophically but scientifically and to show thereby that knowledge is possible. Thus the superstitious conclusion which this dichotomy has led to, he said, may be overcome, and contrary to the opinion of some of his critics, the world is not dissolved into a chaos but may still allow for knowledge. We do not wish to appraise this claim but only to point out that on Mach's view he solves just that problem which Whewell deemed as central and unsolvable. He did not, we should repeat, mention Whewell in this book: we put forward our view as mere conjecture.

[5] Mach (1917), 173–4.

We see, then, Mach's philosophy as a concerted attempt to solve or explain away all those problems for the traditional establishment view which we deem Whewell to have left, and which had remained unsolved since the Whewell–Mill debate, and to do so in a way which met, or sought to meet, Whewell's standards in history and philosophy. The fact that he failed and that his view led to new problems is not to be used against him; perhaps not even that he muffled the importance of Whewell for his own work. After all, his work was one of the most powerful, sophisticated, and fruitful attempts of that fermenting period.

Our second example of a thinker, who attempted to respond to Whewell's problems while taking into account his insights and progress rather than suppressing them, is Charles Saunders Peirce. Peirce's philosophy was enormously ambitious: in the wake of Whewell he sought a theory of the sciences which would itself be scientific. Yet he compared his work not to Whewell's but to Aristotle's, whose most basic ideas he deemed to be obsolete yet for which he had no adequate replacement as yet. His philosophy, as well as his manuscripts, lay unfinished, and most of his output was published posthumously. His ideas have nevertheless exerted a large influence, above all, of course, in America. Ironically, in posthumously published letters Peirce praises Whewell for his having gone about his philosophy scientifically; yet this, too, has now an antiquarian value: at the time no one looked at Peirce as at a disciple of Whewell in any sense whatever.

The centrality of Whewell's results for those of Peirce is fairly easy to document. For example, 'If you carefully consider the question of pragmatism you will see that it is nothing else than the question of the logic of abduction'.[6] Or the claim that there are three methods of reasoning: deduction, induction, and reduction.[7] That the third method is nothing other than Whewell's, Peirce himself noted before he proceeded to reject Mill's view and to explain his method with the very example of Kepler's debt to Brahe—the key example in the Whewell–Mill debate. The interpretation of pragmatism as the study of the logic of abduction, that is, of Whewell's methodology, is bolstered by the consistently high regard which Peirce expressed for Whewell, by his rather negative comments about Mill's superficial philosophy, and by the debt that his own forays into the history of science owed to Whewell. A further examination will show that Peirce sought an alternative to Whewell's answer, but it will have to be an anlaysis of the logic of the situation, not a more direct testimony. For Peirce did not describe the way in which his programme was meant to meet the problem situation as he then saw it.

Peirce's central problems were foundational. He took for granted that there are three methods, deduction, induction, and abduction, none of which is complete or properly founded. There was then no systematic analysis of any of them. Peirce proposed to provide this. Like Mill and Mach, Peirce was concerned to overcome the incompleteness of Whewell's work which was, of

[6] Peirce (1931–58), vi. 121.
[7] Ibid. i. 28–9.

course, the unexplained fundamental antithesis, the thesis that there are two sources of knowledge: thought and observation. In order to do this he sought to develop a psychological theory of how ideas develop. This theory should reduce ideas to experience. The difference, however, between Mill and Peirce is that Peirce rejected sensationalism; he saw experience as a process which requires new analysis. He studied this process through the application of his principle of continuity—the major principle of his philosophy. The continuity between thought and sensation should somehow bridge the gap between them. Indeed, already Kant had said that experience is the marriage of conception with sensation; yet in Kant we have three separate entities, sensations, conceptions or ideas, and (scientific) experience. Peirce wanted a continuity between the three as a way to reduce all ideas to experience. This is very vague: we do not pretend to understand Peirce's theory of learning from experience; we only claim that his intentions were clear. He sought to defend something like an associationist psychology even while denying existence to the elements of association as traditionally conceived. They needed to be redescribed and identified, and knowledge derived from the contents of experience had to be analysed. More we cannot say.

Peirce also attempted to develop a new theory of truth. Why? We suggest that it was rendered necessary by precisely those studies of the history of science in which Whewell had shown inconsistencies within the body of science; that, for example, Newton's theory contradicts Kepler's. This had led, we have claimed above, to Whewell's quasi-crypto-Hegelianism. Among all those who seriously responded to Whewell, Peirce was the only one to follow Whewell's Hegelian traits rather than seek to avoid them. But he attempted to make it reductionist and empiricist by proposing a theory of truth as, on the one hand, satisfaction at any space-time section and, on the other hand, the mere ideal of science, the mere end-point of all scientific investigation. Now this is puzzling. To say that the truth is but the end-point is to deny that today's scientific truth need be true; it is to allow scientific truth to be falsehood; and so it is to allow some inconsistency within the body of science. This is Karl Popper's solution. Yet Peirce's theory differs from Popper's in that it allows us the conclusion that all scientific truths are true. Yet scientific truths cannot be both true and false, unless we postulate two kinds of truth, the truth of the idea of science alone as different from the truth of all extant scientific truths, today's as well as yesterday's. This is Pierre Duhem's solution: he deemed only the ideal truth absolute, but the historical truths relative and pragmatic. Did Peirce agree with Duhem and disagree with Popper, perhaps? We confess, we do not know. This, however, we can and will say: all three, Peirce, Duhem, and Popper, offered the view of the ideal, of the end-point of science, as the absolute truth; and this solves the mystery of how facts correspond so marvellously to ideas. For it is claimed that the marvellous fit is but the ideal, that facts and ideas are only hoped to fully fit with each other. But Peirce also wanted, like Whewell but unlike Duhem and Popper, truth here-and-now. Turning to Whewell's observation that thought and object cannot be sharply separated (that is, there are no sense data utterly free of theoretic bias), he reversed it and allowed no object outside thought, since that would violate his (to us incomprehensible) principle of continuity and be thus meaningless. This might enable Peirce, perhaps, to have his truth here-and-now as a variant of Hegelian idealism; but we have no intent here to go as far as to say whether Peirce was Hegelian all the way or not, since we wish to by-pass his theory of meaning. We can say, however, what that theory was meant to accomplish: all metaphysical problems as well as the problem of knowledge should be explained away with its aid—by their reduction to merely empirical questions: ideas should only be admitted whose empirical consequences are identified, and the empirical significance of a proposition should be made equivalent to its meaning. If all this is accepted, ultimately all knowledge is then to be shown as derivable solely from experience; this will account, though only in retrospect, for the growth of knowledge and for the methods used in research. And this way a better understanding of what we already know of the growth of science thus far can be achieved. Peirce thus tried to meet Whewell's challenge by accounting for the history of science and by developing a methodology as sophisticated as Whewell's and as open-ended, yet one which, in principle or ideally, at the end-point, will have reduced all ideas to experience through a new psychology, a new theory of meaning and truth, and a new theory of the scientific endeavour. Peirce himself viewed his theory as a version of Hegelianism even though he judged Hegel's contempt for facts quite harshly.

Let us note in passing that Peirce devoted great energy to developing a logic within the framework of his triadic system of deduction, induction, and abduction. This, we suggest, had its roots in De Morgan's summary rejection of Whewell's methodology as contrary to logic: Peirce set himself the task of exhibiting a new logic in the context of his new methodology, of unifying Whewell's methodology and the new logic which De Morgan, Boole, and others had begun to forge.

Neither the intrinsic value of Peirce's work nor its influence on posterity are easy to assess; even the meaning of some of his central ideas or concepts is controversial. Yet, we propose, efforts to understand and benefit from him have been a constant source of rediscovery of Whewell's ideas, which facilitated the revival of his writings which we are now witnessing, as well as the growth of interest of both historians and philosophers of science in his ideas and their possible usefulness in the solution of problems of our day. Even this, the topicality of Whewell's ideas, we propose, is no mystery; it is due to the fact that standard problems of today's establishment of philosophy go back to Whewell, via Mach, Peirce, and Duhem more than any other channel.

And so we have reached our third and final example, the work of Pierre Duhem. It is hard to introduce Duhem here as he was in no way a member of the establishment of philosophy, as Mach and as Peirce aspired to be: he was an orthodox Roman Catholic who wished to revive the instrumentalism of St Robert, Cardinal Bellarmine and combat the realism of Galileo which had won the unqualified approval of the establishment of science. Being an instrumentalist, Duhem was, a priori, receptive to the method of hypothesis; not being a realist, he naturally did not have to solve the problems which Whewell faced and those which Whewell had imposed on the establishment. Moreover, he claimed explicitly that he drew enormous encouragement from the attitude exhibited by Hermann von Helmholtz towards Maxwell's theory of electromagnetism. Helmholtz was a realist and a Newtonian at that. He declared that he was willing to endorse Maxwell's theory, since it was excellent, on the condition that it could be reconciled with Newton's system. Otherwise, he said, he was willing to endorse Maxwell's equations only as a mere instrument.

This was not in the least a new attitude: the establishment view was always realist and always saw the view that a good theory is a good instrument as a consolation prize. The novelty was in the fact that Maxwell's theory was not just a minor idea or a transient one, yet it conflicted with Newtonianism which was official. And Duhem's leading idea was that we should view not only Maxwell's theory but also Newton's as a mere instrument and thereby give up realism altogether and thus hope to repair the rift between the scientific and the religious establishment caused by Galileo when he created the conflict between faith and science.

Yet Duhem had no hope of raising any interest in revising Bellarmine's view of science as it stood, since that view was too primitive. Rather, he fortified it and modified it. The only source for his methodological innovation which he named, and this too only in passing, is the classical and best-selling *Introduction to the Study of Experimental Medicine* of Claude Bernard. Now this book includes many exciting ideas which do not belong to methodology in general, much less to the methodology of the physical sciences. But it also has the important idea that experiment is the rigorous test of a hypothesis. Whether Bernard got this idea from Whewell, directly or indirectly, is a question we should leave open. That Duhem's presentation of this point is much sharper and clearer is uncontestable. Yet it is impossible to say whether Duhem was familiar with the works of Whewell—methodological or historical—or whether he was influenced by Whewell's ideas which had in part, at least, become common property, or whether he reinvented these ideas. Some observations, however, may sharpen the problem.

First, there are a number of coincidences here, such as Duhem's delight in his discovery of Mach's works and his affinity with him, Mach's discovery of similar writers, such as Kirchhoff, Jerusalem, and Stallo (and his silly priority contests), and more. Much of the coincidence is the revival of instrumentalism in the face of the crisis in physics—which is a recurrent phenomenon. It is the significant detailed argument that sorts out diverse influences, but detailed arguments alter, or may be reinvented, when they are transferred from one intellectual framework to another. The contradiction between Kepler and Newton, which Whewell discovered, may have been rediscovered by Duhem.

And he used it to prove that not all scientific truth is true (which we must admit anyway) and to argue that we had better declare scientific truth neither true nor false so as to be saved the embarrassment of viewing it as false. This embarrassment, of course, gave way to Einstein's theory of scientific truth as approximately true. Duhem already had a similar idea since, unlike Bellarmine, he could not suggest that scientific theories are merely useful fictions without giving a hard look at the series of such fictions and finding some correlation between them.

It is here that Duhem appeared at his best, as a historian of science, and we can hardly avoid seeing him as a successor to Whewell—both as a historian and as a philosopher of science demanding of himself a historical approach so as to explain the growth of science.

Duhem was clearly familiar with the difficulties which Whewell presented to the establishment of philosophy, since he used them with a vengeance: it was easier for Duhem than for Whewell, since he attacked scientific realism and so did not have to trouble himself over the difficulties which Whewell felt obliged to solve. He denied not only empiricism, but any claim for truth of any scientific truth. Yet he faced other problems which need not concern us here. What we wish to stress is rather his Whewellian view of empirical tests of hypotheses.

The starting point of any theory of tests must be that of Francis Bacon. All hypotheses, he said, when tested by empirical means, are confirmed by experience to the satisfaction of their propounders. This is why schools exist: every school is prejudiced in favour of its own hypothesis and is confirmed by experience in its own belief. This, said Bacon, could only be overcome by the resolution never to hold a hypothesis. This, also, is why followers of Bacon insisted on their claim that they held no hypothesis, that all their theories had been extracted from experience at one time or another. To make things messier, Bacon already issued permission to the intellect to entertain a hypothesis. The same ambivalence is seen clearly in the work of Sir John Herschel, Whewell's immediate predecessor.

Whewell is the watershed in the history of methodology because he refuted Bacon's doctrine of prejudice: no matter how strongly prejudiced the scientific community was in favour of Newtonian optics, and no matter how well confirmed by experience it was (erroneously) judged, it was given up under the pressure of experience. Hence we need not fear that hypotheses always become prejudiced; rather, we need to see the difference (to be presented in the next paragraph) between scientific and pseudo-scientific (not to mention unscientific) empirical confirmation of hypotheses. This is Whewell's contribution; it was to some extent employed by Bernard; Duhem was aware of it and he even modified it (in a manner Einstein found distasteful). And this is how things stand to date: the theory of test procedures is still very much on the agenda of the philosophy of science and, barring *hoi polloi* who are concerned with inductive probability, whether subjective or objective, there are still the Duhemians, including Thomas S. Kuhn; the Whewellians, including Sir Karl

Popper, and the in-betweens, like W. V. Quine. Perhaps even a present-day Baconian may join the debate and examine test procedures. L. Jonathan Cohen and Adolph Grünbaum declare themselves Baconians and at times they show some concern for test procedures. We cannot judge. Frankly, in our own understanding, if Bacon was firm about anything, it was his (true) claim that every given test procedure is defective and his (erroneous) proposal therefore to avoid all test procedures and operate quasi-randomly. Baconianism and test procedures do not mix. This is why Whewell's rejection of Bacon's views was so monumental: it led him to the study of test procedures which was barred by the Baconian lore.

Whewell claimed that all theory may be confirmable by experience—as Bacon has observed—but not necessarily so: it depends on the test procedures. Given the permission to reinterpret or slightly (and reasonably) modify our hypothesis in the light of experience, Bacon's conclusion holds. Given no permission to modify prior to a decision on the outcome of the test, it does not. Whewell was all for reinterpretation and modification, but only after the test results are assessed; once the test results are assessed, modification and reinterpretation are in order and they invite new tests. Whewell (carefully) studied the history of Newtonian optics and showed a history of a systematic misreading of hypothesis plus test leading to modification plus new test leading to new modifications and so on. For each modification was refuted and invited a new modification, yet since each new modification tallied with the previous results of tests it was considered quite clearly empirically confirmed.

Whewell therefore proposed that only if test-statements are strictly deduced from hypothesis and then compared with new empirical results, is the test scientific. Duhem agreed. Yet Duhem denied Whewell's idea that once there is a discord between test-statement deduced from hypothesis and observation-report, then the hypothesis is not scientific. On the contrary, said Duhem, the tests do not determine truth-values but only domains of applicability. And then, if the reinterpretation widens or broadens the domain of application of the theory, it should be welcomed, but, when necessary, a narrowing interpretation must likewise be admitted.

We shall not discuss here the important significance and influence of Duhem's work. We have argued that his modification of Whewell's theory of test procedure is still the competitor to that theory. We shall conclude by comparing Duhem's history with that of Whewell. For they are both monumental and illustrate the role of bold conjecture in the growth of science. Yet, whereas Whewell attempted to illustrate the growth of science from the Copernican period to date, Duhem attempted to apply almost the same techniques to the predecessors of Copernicus, so as to vindicate Bellarmine's hostility to the modernists and his demand to ensure the slow progress of science—not only for the needs of the Church of Rome, but also in order to accord with Duhem's instrumentalist modification of Whewell's realist history, which requires historical continuity through the ages.

It is easiest to see the impact—direct or not, conscious or not—of Whewell's history on that of Duhem when one approaches the latter critically. The literature which Duhem was the first to study as history of science lay neglected or studied by scholars utterly indifferent to modern science until he aired it. There is no doubt that what he extracted out of it were the scientific nuggets, even when studying the most scientific of texts. Unlike the students of Euclid or of Ptolemy, the historians of science who study the scientific views of the writers of the high Middle Ages—and the great pioneer among them was surely Pierre Duhem—have to approach their material with a sieve and find a scientific nugget tucked here and a passing remark of scientific import there in the midst of much high theology, scholastic metaphysics, and sheer superstition. Query: what was the sieve? Answer: Whewell's theory of test-procedure, usually in Duhem's modification. It is the theory of test procedure which the historian of science employs, which gives his work its major characteristics. And we have, to date, only Whewell's theory, with or without Duhem's modification.

This, then, is the proper place to conclude this admittedly preliminary examination of Whewell's impact on posterity, in which many areas are left untouched, some only mentioned, a few only broadly surveyed, and many questions left unasked. We look to see more studies, and, we hope, they will quote the present one lavishly, rather than consigning it to the dusty archives as a mere predecessor.

Notes on Contributors

Joseph Agassi is Professor of Philosophy at Tel Aviv University and York University, Canada. He has authored over 400 publications on various topics in history, philosophy, and sociology of science. His principal works include: *Towards an Historiography of Science* (1963), *Faraday as Natural Philosopher* (1971), *Science in Flux* (1975), *Science and Society: Studies in the Sociology of Science* (1981), *Technology: Philosophical and Social Aspects* (1985), and *The Gentle Art of Philosophical Polemics* (1988).

Harvey W. Becher is Associate Professor of History at Northern Arizona University. He has published on Cambridge mathematics in *Historical Studies in the Physical Sciences* and *Historia Mathematica*, on the social and intellectual milieu of Cambridge mathematics honours graduates in *Victorian Studies*, and on Cambridge science in *The British Journal for the History of Science*. He is working on a prosopographical study of Cambridge mathematics honours graduates of the second half of the nineteenth century.

John Hedley Brooke is Senior Lecturer in History of Science at the University of Lancaster, where he has also been Senior Tutor and Principal of Bowland College. He has published extensively on the history of organic chemistry and on the historical relations of science and religion. His book, *Science and Religion: Some Historical Perspectives*, is shortly to be published by Cambridge University Press. Since May 1988 he has been editor of *The British Journal for the History of Science*.

Gerd Buchdahl is emeritus Reader in History and Philosophy of Science at the University of Cambridge and a Fellow of Darwin College, Cambridge. He has published widely in the field of the philosophy of science, in particular in relation to Kant. He is the author of *Metaphysics and the Philosophy of Science: The Classical Origins, Descartes to Kant* (1969).

Geoffrey N. Cantor is Senior Lecturer in History and Philosophy of Science in the Philosophy Department at the University of Leeds. He is author of *Optics after Newton* (1983), co-editor of *The Companion to the History of Modern Science* (1989), and is currently writing a biography of Michael Faraday.

Menachem Fisch is Lecturer at the Institute for the History and Philosophy of Science and Ideas at Tel Aviv University. He has published widely on various topics in the history and philosophy of science. He is author of *William Whewell: Philosopher of Science* (forthcoming from Oxford University Press).

Jonathan Hodge is a member of the Division of the History and Philosophy of Science in the Department of Philosophy at the University of Leeds. He is editor (with G. N. Cantor) of *Conceptions of Ether* (1981), and is currently preparing monographs on Lyell and Darwin.

Michael Ruse is Professor of Philosophy and Zoology at the University of Guelph. He is a Fellow of the Royal Society of Canada, and author of a number of books on Darwinism and related topics, including: *Darwinism Defended: A Guide to the Evolution Controversies* (1982), *Taking Darwin Seriously: A Naturalistic Approach to Philosophy* (1986), *Homosexuality: A Philosophical Inquiry* (1988), and *Philosophy of Biology Today* (1988).

Simon Schaffer is University Lecturer in History and Philosophy of Science at the University of Cambridge. He is co-author (with Steven Shapin) of *Leviathan and the Air Pump* (1985) and co-editor (with David Gooding and Trevor Pinch) of *The Uses of Experiment* (1989).

John Wettersten is currently engaged in research at the University of Mannheim. His publications include *Learning from Error* (with William Berkson), 'On Two Non-Justificationist Moves' (*Synthese*), 'On Adventurous and Conservative Styles of Scientific Research' (*Minerva*), and 'The Historiography of Scientific Psychology' (*Journal of the History of Behavioral Sciences*).

Perry Williams works at the Cambridge Wellcome Unit for the History of Medicine. He has published on various aspects of nineteenth-century history, including psychical research, women's higher education, nursing and sanitary reform. He is now preparing, with Andrew Cunningham, an edited collection *Medicine and the Laboratory* and a popular book *The Invention of Science*.

David B. Wilson is Professor of History and Mechanical Engineering at Iowa State University. Among his numerous publications are *Kelvin and Stokes: A Comparative Study in Victorian Physics* (1987) and 'Herschel and Whewell's Version of Newtonianism', *Journal of the History of Ideas* (1974).

Richard Yeo is Senior Lecturer in the Division of Humanities at Griffith University, Brisbane. He has published extensively on various aspects of nineteenth-century science, including Baconianism, natural theology, scientific authority, images of Newton, and on the work of William Whewell. He has edited (with John Schuster) *The Politics and Rhetoric of Scientific Method* (1986).

Bibliography

Aarsleff, H. (1967) *The Study of Language in England 1780–1860* (Princeton).

—— (1971) 'Locke's Reputation in Nineteenth-Century England', *Monist 55*, 392–422.

—— (1982), *From Locke to Saussure: Essays on the Study of Language and Intellectual History* (Minneapolis).

Abrams, M. H. (1958) *The Mirror and the Lamp: Romantic Theory and the Critical Tradition* (New York).

—— (1973) *Natural Supernaturalism: Tradition and Revolution in Romantic Literature*, 2nd edition (New York).

Agassi, J. (1971) 'Sir John Herschel's Philosophy of Success', *Historical Studies in the Physical Sciences*, 1–36, reprinted in J. Agassi (1981) *Science and Society* (Dordrecht and Boston), 388–420.

Airy, G. B. (1845) 'Tides and Waves', *Encyclopaedia Metropolitana* (London), v. 241–400.

—— (1896) *Autobiography of Sir George Biddell Airy* (Cambridge).

Altick, R. D. (1957) *The English Common Reader: Social History and the Mass Reading Public* (Chicago, reprinted Phoenix Books, 1963).

Alvarez, L. W. *et al.* (1980) 'Extraterrestrial Cause for the Cretaceous-Tertiary Extinction', *Science 208*, 1095–108.

Annan, N. (1955) 'The Intellectual Aristocracy', in J. H. Plumb, (ed.) *Studies in Social History: A Tribute to G. M. Trevelyan* (London), 241–87.

Anonymous (1817) 'Metaphysics' *Encyclopaedia Britannica*, 5th edn., 20 vols. (Edinburgh), xiii. 555–685.

—— (1821) 'Normandy—Architecture of the Middle Ages', *Quarterly Review 25*, 113–47.

—— (1830), 'Whewell's Notation of Political Economy: A Fragment', *Fraser's Magazine 1*, 303–7.

—— (1831) 'The Cambridge System of Education', *Quarterly Journal of Education 2*, 242–50.

—— (1834) 'The Cambridge Controversy: Admission of Dissenters to Degrees', *Quarterly Review 52*, 466–87.

—— (1841) 'Whewell's Philosophy of the Inductive Sciences', *University Magazine 18*, 555–72.

—— (1842) 'Principles of Gothic Architecture', *Quarterly Review 69*, 11–149.

—— (1847) 'An Historical and Critical View of *Speculative Philosophy of Europe in the Nineteenth Century*. By J. D. Morell, A. M. Pickering: 1846', *Westminster Review 48*, 6–23.

—— (1847) 'On Whewell's Inductive Sciences', *United Services Magazine*, March, 3–4.

—— (1855) 'Contemporary Literature', *Westminster Review 65*, 221–312.

—— (1866) Obituary, 'The Reverend William Whewell', *Athenaeum*, March, 333–4.

—— (1875–89) 'Whewell', *Encyclopaedia Britannica*, 9th edn., 25 vols. (Edinburgh), xxiv. 539.

—— (1881) 'The Life of William Whewell', *Nature*, 16 June, 137–9.

Anschütz, R. P. (1953) *The Philosophy of J. S. Mill* (Oxford).

Anscombe, G. E. M. (trans.) (1953) *Wittgenstein's Philosophical Investigations* (Oxford).

Atkins, J., and Shan-lan Li (trans.) (1867) *Ch'ung Hsüeh* (Shanghai). Translation of Whewell, (1819).

Ayer, A. J. (1946) *Language, Truth and Logic*, 2nd edn. (New York).

Babbage, C. (1838) *Ninth Bridgewater Treatise: A Fragment* (London).

—— (1864) *Passages from the Life of a Philosopher* (London).

Bacon, F. (1883) *The Works of Francis Bacon*, ed. James Spedding (London).

Bain, A. (1882) *James Mill: A Biography* (London).

Barlow, N. (ed.) (1958) *The Autobiography of Charles Darwin* (New York).

Barrett, P. H., *et al.* (eds.) (1987) *Darwin's Notebooks, 1836–1844* (Cambridge).

Barrow, L. (1986) *Independent Spirits: Spiritualism and English Plebeians 1850–1910* (London).

Barton, R. (1976) 'The X Club: Science, Religion, and Social Change in Victorian England', Ph.D. thesis (University of Pennsylvania).

Becher, H. W. (1971) 'William Whewell and Cambridge Mathematics', Ph.D. dissertation (University of Missouri).

—— (1975) 'William Whewell and the Preservation of a "Liberal Education" in an Age of Challenge', *Rocky Mountain Social Science Journal 12*, 59–70.

—— (1980) 'Woodhouse, Babbage, Peacock, and Modern Algebra', *Historia Mathematica* 7, 389–400.

—— (1981) 'William Whewell and Cambridge Mathematics', *Historical Studies in the Physical Sciences 11*, 1–48.

—— (1986) 'Voluntary Science in Nineteenth-Century Cambridge University to the 1850s', *British Journal for the History of Science 19*, 57–87.

Beer, G. (1985) *Darwin's Plots: Evolutionary Narrative in Darwin, George Eliot and Nineteenth-Century Fiction* (London).

Belsey, A. (1974) 'Interpreting Whewell', *Studies in History and Philosophy of Science 5*, 49–58.

Benjamin, A. E., G. N. Cantor, and J. R. R. Christie (eds.) (1987) *The Figural and the Literal: Problems of Language in the History of Science and Philosophy, 1630–1800* (Manchester).

Bentham, J. (1924) *The Handbook of Political Fallacies*, ed. H. A. Larrabee (New York).

Berkeley, G. (1710) *Principles of Human Knowledge* (London).

—— (1734) 'The Analyst', in *The Works of George Berkeley*, (London), iv. 53–102.

Berman, M. (1978) *Social Change and Scientific Organization: The Royal Institution 1799–1844* (London).

Berry, L. G., and B. Mason (n.d.) *Mineralogy* (San Francisco).

Birks, T. R. (1834) *Oration on the Analogy of Mathematical and Moral Certainty* (Cambridge), reprinted in T. R. Birks, *The Treasure of Wisdom, or Thoughts on the Connection between Natural Science and Revealed Truth, with an Essay on the Analogy between Mathematical and Moral Certainty* (London, 1855).

Blake, R. M., C. J. Ducasse, and E. H. Madden (1960) *Theories of Scientific Method* (Seattle and London).

Blakey, R. (1850) *History of the Philosophy of Mind: Embracing the Opinions of All Writers on Mental Science from the Earliest Period to the Present Time*, 4 vols. (London).

Blanché, R. (1935) *Le Rationalisme de Whewell* (Paris).

Bowden, J. W. (1834) 'The British Association', *Oxford University Magazine 1*, 401–12.

—— (1839) 'The British Association for the Advancement of Science', *British Critic 25*, 1–48.

Boyer, C. B. (1949) *History of the Calculus* (New York).

Braithwaite, R. B. (1953) *Scientific Explanation* (Cambridge).

Brent, R. (1987) *Liberal Anglican Politics: Whiggery, Religion, and Reform 1830–1841* (Oxford).

Brewster, D. (1837) 'On the History of the Inductive Sciences', *Edinburgh Review 66*, 110–51.

—— (1842) 'Whewell's Philosophy of the Inductive Sciences', *Edinburgh Review 74*, 139–161.

—— (1854) *More Worlds Than One* (London).

Brooke, J. H. (1974) *New Interactions between Theology and Natural Science* (Milton Keynes).

—— (1977) 'Natural Theology and the Plurality of Worlds: Observations on the Brewster–Whewell Debate', *Annals of Science 34*, 221–86.

—— (1977*a*) 'Richard Owen, William Whewell and the *Vestiges*', *British Journal for the History of Science 10*, 132–45.

—— (1979) 'The Natural Theology of the Geologists: Some Theological Strata' in L. J. Jordanova and R. S. Porter (eds.), *Images of the Earth* (Chalfont St Giles), 39–64.

—— (1985) 'Why did the English Mix their Science and their Religion?' in S. Rossi (ed.), *Science and Imagination in Eighteenth-Century British Culture* (Milan), 57–78.

—— (1986) 'Joseph Priestley (1733–1804) and William Whewell (1794–1866): Apologists and Historians of Science. A Tale of Two Stereotypes' in R. Anderson and C. Lawrence (eds.), *Science, Medicine, and Dissent, Joseph Priestley (1733–1804)* (London).

Brougham, H. (1825) 'New University in London', *Edinburgh Review 42*, 346–67.

Buchdahl, G. (1963) *Induction and Necessity in the Philosophy of Aristotle* (London).

—— (1969) *Metaphysics and the Philosophy of Science: The Classical Origins: Descartes to Kant* (Oxford).

—— (1970) 'History of Science and Criteria of Choice' in R. Stuewer (ed.), *Historical and Philosophical Perspectives of Science* (Minneapolis).

—— (1974) 'The Conception of Lawlikeness in Kant's Philosophy of Science' in L. W. Beck (ed.), *Kant's Theory of Knowledge* (Dordrecht).

—— (1982) 'Reduction-Realization: a Key to the Structure of Kant's Thought', in J. N. Mohanty and R. W. Shahan (eds.), *Essays on Kant's Critique of Pure Reason* (Norman, Okla.).

—— (1983) 'Styles of Scientific Thinking', in F. Bevilacqua and P. J. Kennedy (eds.), *Proceedings of the International Conference on Using History of Physics in Innovatory Physics Education* (Pavia).

—— (1986) 'Metaphysical and Internal Realism: The Relations between Ontology and Methodology in Kant's Philosophy of Science', in R. Barcan Marcus *et al.* (eds.), *Proceedings of the Seventh International Congress of Logic, Methodology, and Philosophy of Science* (Salzburg), 623–41.

—— (1987) 'Philosophy of Science: Its Historical Roots', in *Les Relations mutuelles entre la philosophie des sciences et l'histoire des sciences* (Bruxelles).

—— and L. L. Laudan (eds.), (1967) *The Historical and Philosophical Works of William Whewell* (London.

Burke, J. G. (1966) *Origins of the Science of Crystals* (Berkeley).

Burkhardt, F., and S. Smith (eds.) (1985–7) *The Correspondence of Charles Darwin* (Cambridge), 3 vols. published.

Burrow, J. W. (1967) 'The Uses of Philology in Victorian England' in Robert Robson (ed.), *Ideas and Institutions in Victorian Britain* (London).

—— (1981) *A Liberal Descent: Victorian Historians and the English Past* (Cambridge).

Butts, R. E. (1965) 'Necessary Truth in Whewell's Theory of Science', *American Philosophical Quarterly 2*, 161–81.

—— (1965*a*) 'On Walsh's Reading of Whewell's View of Necessity', *Philosophy of Science 32*, 175–81.

—— (1967) 'Professor Marcucci on Whewell's Idealism', *Philosophy of Science 34*, 175–83.

—— (1968) *William Whewell's Theory of Scientific Method* (Pittsburgh).

Bynum, W. F. *et al.* (eds.) (1981) *Dictionary of the History of Science* (London).

Cairns, D. (1956) 'Thomas Chalmers' *Astronomical Discourses*: A Study in Natural Theology', *Scottish Journal of Theology* 9, 410–21.

Campbell, L., and W. Garnett (1884) *The Life of James Clerk Maxwell*, 2nd edn. (London).

Cannon, S. F. (1978) *Science in Culture: The Early Victorian Period* (New York).

Cannon, W. F. (1960) 'The Problem of Miracles in the 1830s', *Victorian Studies 4*, 5–32.

—— (1960*a*) 'The Uniformitarian–Catastrophist Debate', *Isis 51*, 38–55.

—— (1961) 'The Impact of Uniformitarianism: Two Letters from John Herschel to Charles Lyell 1836–1837', *Proceedings of the American Philosophical Society 105*, 301–14.

—— (1961*a*) 'The Bases of Darwin's Achievement: A Re-evaluation', *Victorian Studies* 5, 109–34.

—— (1964) 'William Whewell: Contributions to Science and Learning', *Notes and Records of the Royal Society 19*, 176–91.

—— (1976) 'The Whewell–Darwin Controversy', *Journal of The Geological Society of London 132*, 377–84.

Cantor, G. (1975) 'The Reception of the Wave Theory of Light in Britain', *Historical Studies in the Physical Sciences 4*, 109–32.

—— (1981) 'The Theological Significance of Ethers', in G. N. Cantor and M. J. S. Hodge (eds.), *Conceptions of Ether: Studies in the History of Ether Theories 1740–1900* (Cambridge).

—— (1983) *Optics after Newton: Theories of Light in Britain and Ireland, 1704–1840* (Manchester).

—— (1985) 'Reading the Book of Nature: The Relation between Faraday's Religion and Faraday's Science' in D. Gooding and F. James (eds), *Faraday Rediscovered* (London).

Caravale, G. A. (1985) *The Legacy of Ricardo* (Oxford).

Cardwell, D. S. L. (1957) *The Organisation of Science in England* (London).

Carlyle, H. (1882) 'William Whewell', *Macmillan's Magazine 45*, 138–44.

Carlyle, T. (1829) 'Signs of the Times', *Edinburgh Review 49*, 439–58.

Carpenter, W. B. (1838) 'Physiology an Inductive Science', *British and Foreign Medical Review 5*, 317–42.

Cartwright, N. (1983) *How the Laws of Physics Lie* (London and New York).

Chadwick, O. (1971) *The Victorian Church: Part I, 1829–1859*, 3rd edn. (London).

Challis, J. (1863) 'On Newton's Foundation of All Philosophy', *Philosophical Magazine 4*, 280–96.

—— (1869) *Notes on the Principles of Pure and Applied Calculation and Applications of Mathematical Principles to Theories of the Physical Forces* (Cambridge and London).

—— (1873) *An Essay on the Mathematical Principles of Physics, with Reference to the Study of Physical Science by Candidates for Mathematical Honours in the University of Cambridge* (Cambridge and London).

Chalmers, T. (1817) *Astronomical Discourses* (Glasgow).

—— (1833) *On the Power, Wisdom, and Goodness of God as Manifested in the Adaptation of External Nature to the Moral and Intellectual Constitution of Man*, 2 vols. (London).

Chambers, R. (1844) *Vestiges of the Natural History of Creation* (London, reprinted Leicester, 1969).

—— (1846) *Explanations: A Sequel to Vestiges of the Natural History of Creation* (London).

—— (1858) *Vestiges of the Natural History of Creation with a Sequel* (New York).

Checkland, S. G. (1951) 'The Advent of Academic Economics in England', *Manchester School of Economic and Social Studies 19*, 43–70.

Clark, J. W. (1889–92) 'The Foundations and Early Years of the Society', *Proceedings of the Cambridge Philosophical Society 12*, i–xlviii.

—— and T. M. Hughes (1890) *Life and Letters of the Reverend Adam Sedgwick*, 2 vols. (Cambridge).

Clark, W. G. (1866) 'William Whewell', *Macmillan's Magazine 13*, 545–52.

Clayton, J. P. (1986) 'Gottesbeweise II and III', *Theologische Realenzyklopädie 13*, 724–84.

Cochrane, J. L. (1970) 'The First Mathematical Ricardian Model', *History of Political Economy 2*, 419–31.

Cohen, I. B. (1976) 'The Eighteenth-Century Origins of the Concept of Scientific Revolution', *Journal of the History of Ideas 37*, 257–88.

Coleridge, S. T. (1818) *The Friend*, ed. B. Rooke 2 vols. (London and Princeton, 1969).

—— (1830) *On the Constitution of the Church and State*, ed. J. Colmer (London and Princeton, 1976).

—— (1840) *Aids to Reflection*, 4th edn., ed. H. N. Coleridge, (reprinted 1971, Port Washington, NY).

—— (1850) *A Preliminary Treatise on Method*, 3rd edn. (London).

Collini, S., D. Winch, and J. Burrow (1983) *That Noble Science of Politics: A Study in Nineteenth-Century Intellectual History* (Cambridge).

Comte, A. (1830, 1835) *Cours de philosophie positive* (Paris).

Cooter, R. (1984) *The Cultural Meaning of Popular Science: Phrenology and the Organizaton of Consent in Nineteenth-Century Britain* (Cambridge).

Cornell, J. F. (1986) 'Newton of the Grassblade? Darwin and the Problems of Organic Teleology', *Isis 77*, 405–21.

Corrigan, T. J. (1980) '*Biographia Literaria* and the Language of Science', *Journal of the History of Ideas 41*, 399–419.

Corsi, P. (1987) 'The Heritage of Dugald Steward: Oxford Philosophy and the Method of Political Economy, 1809–1832', *Nuncius 2*, 89–144.

—— (1988) *Science and Religion: Baden Powell and the Anglican Debate 1800–1860* (Cambridge).

Crowe, M. J. (1986) *The Extraterrestrial Life Debate 1750–1900* (Cambridge).

Cumming, D. (1985) 'John MacCulloch, High Priest, Blackguard and Thief, Reassessed', in A. Wheeler and J. Price (eds.), *From Linnaeus to Darwin* (London), 77–88.

Cumming, J. (1823) 'On the Development of Electro-Magnetism by Heat', *Transactions of the Cambridge Philosophical Socety* 2, 47–76.

Curtis, R. C. (1986) 'Are Methodologies Theories of Scientific Rationality?', *The British Journal for the History of Science 37*, 135–61.

—— (1987) 'Darwin as an Epistemologist', *Annals of Science 44*, 379–408.

Darwin, C. (1974) 'Autobiography', in G. de Beer (ed.), *Autobiographies of Charles Darwin and Thomas Henry Huxley* (Oxford).

Darwin, F. (ed.) (1877) *The Life and Letters of Charles Darwin*, 3 vols. (London).

Daston, L. J. (1986) 'The Physicalist Tradition in Early Nineteenth-Century French Geometry', *Studies in History and Philosophy of Science 17*, 269–95.

Davies, G. (1857) *On the Proposed Conditions of the Tenure of Fellowships of Trinity College* (Cambridge).

Deacon, M. (1967) *Scientists and the Sea, 1650–1900: A Study of Marine Science* (New York and London).

Deas, H. (1959) 'Crystallography and Crystallographers in England in the Early Nineteenth Century: A Preliminary Survey', *Centaurus 6*, 129–48.

de Beer, G. (ed.) (1960) 'Darwin's Notebooks on Transmutation of Species', *Bulletin of the British Museum (Natural History)* (Historical Series, London).

de la Beche, H. (1834) *Researches in Theoretical Geology* (London).

de Marchi, N. B., and R. P. Sturges (1973) 'Malthus and Ricardo's Inductive Critics: Four Letters to William Whewell', *Economica 40*, 379–93.

Demonferrand, J. F. (1827) *A Manual of Electro-Dynamics*, trans. J. Cumming (Cambridge).

De Morgan, A. (1834) 'Cambridge Differential Notation', *Quarterly Journal of Education*, *8*, 100–110.

Desmond, A. (1982) *Archetypes and Ancestors: Palaeontology in Victorian London 1850–1875* (London).

—— (1985) 'Richard Owen's Reaction to Transmutation in the 1830s', *British Journal for the History of Science 18*, 25–50.

—— (1987) 'Artisan Resistance and Evolution in Britain 1819–1848', *Osiris 3*, 77–110.

—— (1989) *The Politics of Evolution* (Chicago).

Distad, N. M. (1979) *Guessing at Truth: The Life of Julius Charles Hare* (Shepherdstown).

Dixon, R., and S. Mathesuis (1978) *Victorian Architecture* (New York and Toronto).

Dubbey, J. M. (1963) 'The Introduction of the Differential Notation to Great Britain', *Annals of Science 19*, 37–48.

Ducasse, C. J. (1966) 'William Whewell's Philosophy of Scientific Discovery', in Edward H. Madden (ed.), *Theories of Scientific Method: The Renaissance Through the Nineteenth Century* (Seattle).

Duhem, P. (1954) *The Aim and Structure of Physical Theory*, translated from the 2nd edition 1914 (New York).

Durant, J. (1985) *Darwinism and Divinity* (Oxford).

Elkana, Yehuda (ed.) (1984) *William Whewell: Selected Writings on the History of Science* (Chicago).

—— (1984*a*) 'William Whewell, Historian', *Rivista de Storia della scienza 1*, 149–97.

—— (1984*b*) 'Transformations in Realist Philosophy of Science from Victorian Baconianism to the Present Day', in E. Mendelsohn (ed.), *Transformation and Tradition in Science* (Cambridge), 487–511.

Engelhardt, D. von (1979) *Historisches Bewusstsein in der Naturwissenschaft von der Aufklärung bis zum Positivismus* (Munich).

Enros, P. C. (1981) 'Cambridge University and the Adoption of Analytics in Early Nineteenth-Century England', in H. Mehrtens *et al.* (eds.), *Social History of Nineteenth-Century Mathematics* (Boston), 135–48.

—— (1983) 'The Analytical Society 1812–1813', *Historia Mathematica 10*, 24–47.

Faraday, M. (1839–55) *Experimental Researches in Electricity*, 3 vols. (London).

—— (1846) 'Thoughts on Ray Vibrations', *Philosophical Magazine 28*, 345–50.

—— (1849) *Experimental Researches in Electricity*, 2nd edn., (London), i.

—— (1850) 'On the Polar and Other Conditions of Diamagnetic Bodies', *Philosophical Transactions 140*, 171–88.

—— (1971) *Selected Correspondence*, ed. L. Pearce Williams, 2 vols. (Cambridge).

Farquharson, W. G. (1953) 'International Hydrography 1835', *Hydrographic Review 30*, 123–35.

Fisch, M. (1985) 'Necessary and Contingent Truth in William Whewell's Antithetical Theory of Knowledge', *Studies in History and Philosophy of Science 16*:4, 275–314.

—— (1985*a*) 'Whewell's Consilience of Inductions: An Evaluation', *Philosophy of Science 52*:3, 239–55.

—— (1986) 'Antithetical Knowledge: William Whewell's Theory of Science, its Development and Historiographical Implications', Ph.D. dissertation (Tel Aviv University).

—— (1987) 'Il Collegamento dei fatti: la nozione non necesaria de verità empirica di William Whewell', in R. Similli (ed.), *L'epistomologia di Cambridge 1850–1950* (Bologna).

Forbes, D. (1952) *The Liberal Anglican Idea of History* (Cambridge).

Foucault, M. (1973) *The Order of Things* (New York), trans. A. Sheridan from *Les Mots et les choses* (Paris, 1966).

Garland, M. M. (1980) *Cambridge before Darwin: The Ideal of a Liberal Education 1800–1860* (Cambridge).

Gash, N. (1972) *Sir Robert Peel* (London).

Gillespie, N. C. (1979) *Charles Darwin and the Problem of Creation* (Chicago).

Glas, E. (1986) 'On the Dynamics of Mathematical Change in the Case of Monge and the French Revolution', *Studies in History and Philosophy of Science 17*, 249–68.

Goldman, L. (1983) 'The Origins of British "Social Science": Political Economy, Natural Science and Statistics 1830–1835', *Historical Journal 26*, 587–616.

Gooding, D. (1980) 'Faraday, Thomson and the Concept of the Magnetic Field', *British Journal for the History of Science 13*, 91–120.

—— (1981) 'Final Steps to the Field Theory: Faraday's Study of Magnetic Phenomena, 1845–1850', *Historical Studies in Physical Science 11*, 231–75.

—— (1982) 'Empiricism in Practice: Teleology, Economy and Observation in Faraday's Physics', *Isis 73*, 46–67.

—— (1985) '"He Who Proves, Discovers": John Herschel, William Pepys and the Faraday Effect', *Notes and Records of the Royal Society 39*, 229–44.

Goodman, N. (1954) *Fact, Fiction, and Forecast* (London).

Goodwin, H. (1845) *Considerations Respecting the Exercises in the Schools and Other Matters Connected with the Qualifications for the Degree of BA* (London).

Gordon, E. O. (1894) *Life and Correspondence of William Buckland* (London).

Graham, L. W., W. Lepenies, and P. Weingart (eds.), (1983) *Functions and Uses of Disciplinary Histories* (Dordrecht).

Grattan-Guinness, I. (1979) 'Babbage's Mathematics in its Time', *The British Journal for the History of Science 12*, 82–8.

—— (1981) 'Berkeley's Criticism of the Calculus as a Study in the Theory of Limits', *Janus 56*, 215–27.

—— (1985) 'Mathematics and Mathematical Physics from Cambridge 1815–1840' in P. M. Harman (ed.) *Wranglers and Physicists: Studies on Cambridge Mathematical Physics in the Nineteenth Century* (Manchester), 84–111.

—— (1986) 'French *Calcul* and English Fluxions around 1800: Some Comparisons and Contrasts', *Jahrbuch überblicke Mathematik*, 167–78.

Graves, R. P. (1882–9) *Life of Sir William Rowan Hamilton*, 3 vols. (Dublin).

—— (1975) *Life of Sir William Rowan Hamilton*, 3 vols. (New York).

Greene, M. T. (1983) *Geology in the Nineteenth Century* (Ithaca).

Grote, J. (1861) *A Few Words of Criticism: Being an Examination of the Article in the Saturday Review of April 20, 1861, upon Dr Whewell's Platonic Dialogues for English Readers* (Cambridge).

Grove, W. R. (1862) *The Correlation of Physical Forces*, 4th edn. (London).

Guiney, L. I. (1904) *Hurrell Froude: Memoranda and Comments* (London).

Halevy, E. (1928) *The Growth of Philosophic Radicalism* (London).

Hamilton, P. (1983) *Coleridge's Poetics* (Oxford).

Hamilton, W. (1831) 'Universities of England: Oxford', *Edinburgh Review 53*, 384–427.

—— (1831*a*) 'English Universities: Oxford', *Edinburgh Review 54*, 478–504.

—— (1834) 'Admission of Dissenters to Universities', *Edinburgh Review 60*, 202–30.

—— (1836) 'Study of Mathematics: University of Cambridge', *Edinburgh Review 62*, 409–55.

Hankins, T. L. (1980) *Sir William Rowan Hamilton* (Baltimore).

Hanson, N. R. (1958) *Patterns of Discovery* (Cambridge).

Harcourt, W. V. (1831) *British Association Reports* (York).

Harding, A. J. (1985) *Coleridge and the Inspired Word* (Kingston and Montreal).

Hare, J. (1835) 'Samuel Taylor Coleridge and the English Opium Eater', *British Magazine* 7, 15–27.

—— and A. Hare (1873) *Guesses at Truth by Two Brothers* (London).

Harman, P. M. (1982) *Energy, Force and Matter: The Conceptual Development of Nineteenth-Century Physics* (Cambridge).

—— (1985) 'Edinburgh Philosophy and Cambridge Physics: The Natural Philosophy of James Clerk Maxwell', in P. M. Harman (ed.), *Wranglers and Physicists: Studies on Cambridge Mathematical Physics in the Nineteenth Century* (Manchester), 202–24.

Hartshorne, C. (1964) *A Natural Theology of Our Time* (La Salle, Ill.).

Harvie, C. (1976) *The Lights of Liberalism: University Liberals and the Challenge of Democracy* 1860–86 (London).

Hendry, J. (1986) *James Clerk Maxwell and the Theory of the Electromagnetic Field* (Bristol).

Herbert, S. (1985) 'Darwin the Young Geologist', in D. Kohn (ed.), *The Darwinian Heritage* (Princeton).

Herschel, J. (1827) 'Treatise on Light', *Encyclopaedia Metropolitana* (London).

—— (1830) *A Preliminary Discourse on the Study of Natural Philosophy* (London).

—— (1841) 'Whewell on Inductive Sciences', *Quarterly Review 68*, 177–238, reprinted (1857) in *Essays from the Edinburgh and Quarterly Reviews* (London), 142–256.

—— (1851) *Outlines of Astronomy*, 4th edition (London).

—— (1857) *Essays from the Edinburgh and Quarterly Reviews* (London).

—— (1867–8) 'The Reverend William Whewell DD', *Proceedings of the Royal Society of London 135*, 51–61.

Hesse, M. B. (1964) 'Francis Bacon' in D. J. O'Connor (ed.), *A Critical History of Western Philosophy* (Glencoe).

Hilton, Boyd (1988) *The Age of Atonement: The Influence of Evangelicalism on Social and Economic Thought 1795–1865* (Oxford).

Hodge, M. J. S. (1977) 'The Structure and Strategy of Darwin's "Long Argument"', *The British Journal for the History of Science 10*, 237–45.

—— (1985) 'Darwin as a Life-Long Generation Theorist', in D. Kohn (ed.), *The Darwinian Heritage*, (Princeton), 207–44.

—— (1989) 'Darwin's Theory and Darwin's Argument', in M. Ruse (ed.), *What Philosophy of Biology Is: Essays Dedicated to David Hull* (Dordrecht), 163–82.

—— (1990) 'Darwin's Studies at Work: A Re-examination of Three Decisive Years, 1835–7' in T. Levere and W. Shea (eds.), *Nature, Experiment and the Sciences: Essays on Galileo and the History of Science in Honour of Stillman Drake* (Dordrecht), 249–74.

—— and D. Kohn (1985) 'The Immediate Origins of Natural Selection', in D. Kohn (ed.), *The Darwinian Heritage* (Princeton) 185–206.

Hogg, T. J.? (1831) 'The Universities of Oxford and Cambridge', *Westminster Review 15*, 56–69.

Hollander, S. (1983) 'William Whewell and John Stuart Mill on the Methodology of Political Economy', *Studies in History and Philosophy of Science 14*, 127–68.

Hollis, P. (1970) *The Pauper Press: A Study in Working-Class Radicalism of the 1830s* (Oxford).

Hooykaas, R. (1959) *Natural Law and Divine Miracle: A Historical Study of the Principle of Uniformity in Geology, Biology and Theology* (Leiden).

Houghton, W. E. (1966) *The Wellesley Index of Victorian Periodicals, 1824–1900* (Toronto).

Humber, R. D. (1968) *Heversham: The Story of a Westmorland School and Village* (Kendal).

Hume, D. (1888) *A Treatise of Human Nature*, ed. L. A. Selby-Bigge (Oxford).

—— (1902) *An Enquiry Concerning Human Understanding*, ed. L. A. Selby-Bigge (Oxford).

Huxley, T. H. (1854) 'Vestiges of Creation', *British and Foreign Medical Review 13*, 425–39.

—— (1887) 'On the Reception of *The Origin of Species*', in F. Darwin (ed.), *The Life and Letters of Charles Darwin*, 3 vols. (London), 179–204.

—— (1970) 'Science and Pseudoscience', in T. H. Huxley, *Collected Essays*, 9 vols. (New York), v. 90–125.

Hyman, A. (1982) *Charles Babbage, Pioneer of the Computer* (Oxford).

Jackson, H. J. (1983) 'Coleridge, Etymology and Etymologic', *Journal of the History of Ideas 44*, 75–88.

Jaki, S. L. (1978) *Planets and Planetarians* (Edinburgh).

James, F. (1985) ' "The Optical Mode of Investigation": Light and Matter in Faraday's

Natural Philosophy', in D. Gooding and F. James (eds.), *Faraday Rediscovered* (London).

James, L. (1976) *English Popular Literature 1819–1851* (New York).

Jarausch, K. H. (1974) 'The Sources of German Student Unrest 1815–1848', in L. Stone (ed.), *The University in Society*, 2 vols. (Princeton), ii. 533–69.

Jardine, L. (1974) *Francis Bacon: Discovery and the Art of Discourse* (Cambridge).

Johnson, W. E. (1924) *Logic* (Cambridge).

Jones, R. (1831) *An Essay on the Distribution of Wealth and on the Sources of Taxation*, (London).

—— (1859) *Literary Remains*, ed. William Whewell, (London).

Jordanova, L. (ed.) (1986) *Languages of Nature: Critical Essays on Science and Literature* (London).

Kant, I. (1910) *Gesammelte Schriften*, 29 vols. (Berlin).

—— (1933) *The Critique of Pure Reason*, trans. N. Kemp Smith (London).

—— (1965) *First Introduction to the Critique of Judgement*, trans. J. Haden (New York).

Kemble, J. M. (1837) 'British and Foreign Universities', *British and Foreign Review* 5, 168–209.

Kent, C. (1978) *Brains and Numbers: Elitism, Comtism and Democracy in Mid-Victorian England* (Toronto).

Keynes, J. M. (1921) *A Treatise of Probability* (London).

Kneale, W. (1949) *Probability and Induction* (Oxford).

Knight, D. (1970) 'The Physical Sciences and the Romantic Movement', *History of Science* 9, 54–75.

—— (1970*a*) *Atoms and Elements*, 2nd edn. (Cambridge).

Knights, B. (1978) *The Idea of the Clerisy in the Nineteenth Century* (Cambridge).

Kohn, D. (ed.) (1985) *The Darwinian Heritage* (Princeton).

Kuhn, T. S. (1962) *The Structure of Scientific Revolutions* (Chicago).

Lagrange, J. L. (1788) *Traite de mécanique analytique* (Paris).

—— (1797) *Théorie des fonctions analytiques* (Paris).

Laudan, L. (1971) 'William Whewell on the Consilience of Inductions' *Monist* 55, 368–91.

Laudan, R. (1982) 'The Role of Methodology in Lyell's Geology', *Studies in History and Philosophy of Science 13*, 215–50.

—— (1987) *From Mineralogy to Geology: The Foundations of a Science 1650–1830* (Chicago).

Lenoir, T. (1982) *The Strategy of Life: Teleology and Mathematics in Nineteenth-Century German Biology* (Dordrecht).

Lesch, J. (1984) *Science and Medicine in France: The Emergence of Experimental Physiology 1790–1855* (Cambridge, Mass.).

Levere, T. H. (1981) *Poetry Realized in Nature: Samuel Taylor Coleridge and Early Nineteenth-Century Science* (Cambridge).

Levine, G. (ed.) (1987) *One Culture: Essays in Science and Literature* (Madison).

Lowes, J. L. (1927) *The Road to Xanadu: A Study in the Ways of Imagination* (London).

Lunn, F. (1824–5) 'Electricity', *Encyclopaedia Metropolitana* (London), iv. 41–170.

—— (1826) 'Heat', *Encyclopaedia Metropolitana*, iv. 225–340.

—— (1827) 'Chemistry', *Encyclopaedia Metropolitana*, iv. 587–746, 747*–762*.

Lyell, C. (1827) 'State of the Universities', *Quarterly Review 34*, 216–68.

—— (1830–3) *Principles of Geology*, 3 vols. (London).

—— (1845) *Travels in North America*, 2 vols. (London).

Lyell, K. M. (ed.) (1881) *Life, Letters and Journals of Sir Charles Lyell*, 2 vols. (London).

Macaulay, T. B. [?Brougham, Henry] (1825–6) 'The London University', *Edinburgh Review 43*, 315–41.

Mach, E. (1917) *Erkenntnis und Irrtum*, 3rd edn. (Leipzig).

Mackintosh, J. (1862) *Dissertation on the Progress of Ethical Philosophy*, ed. William Whewell, 3rd edn. (Edinburgh).

—— (1872) *On the Progress of Ethical Philosophy Chiefly during the Seventeenth and Eighteenth Centuries*, 4th edn. (Cambridge).

Manier, E. (1978) *The Young Darwin and his Cultural Circle* (Boston and Dordrecht).

Marcucci, S. (1963) *L'idealismo scientifico di William Whewell* (Pisa).

Marston, V. P. (1984) *Science, Methodology, and Religion in the Work of Adam Sedgwick*, Ph.D. thesis (Open University).

Maurice, F. (1874) 'Essay on Deacon Hare's Position in the Church, with Reference to the Parties that Divide it', in J. C. Hare, *The Victory of Truth*, 3rd edn. (London).

—— (1884) *The Life of Frederick Denison Maurice*, 4th edn., 2 vols. (London).

Maxwell, J. Clerk (1876) 'Whewell's Writings and Correspondence', *Nature*, 6 July, 206–8.

—— (1890) *Scientific Papers*, ed. W. D. Niven, 2 vols. (Cambridge).

—— (1927) 'Grove's Correlation of Physical Forces', in W. D. Niven (ed.), *Scientific Papers of James Clerk Maxwell*, 2 vols. (Paris) 400–405.

McCormmach, R. and C. Jungnickel (1986) *The Intellectual Mastery of Nature* (Chicago).

McFarland, G. F. (1964) 'J. C. Hare, Coleridge, de Quincey and German Literature', *Bulletin of John Rylands Library 47*, 165–97.

McKusick, J. C. (1986) *Coleridge's Philosophy of Language* (New Haven).

McMullin, E. (1985) 'The Significance of Newton's *Principia* for Empiricism', in M. J. Osler and P. L. Farber (eds.), *Religion, Science and World-View* (Cambridge).

McNally, D. H. (1982) 'Science and the Divine Order: Law, Idea and Method in William Whewell's Philosophy of Science', Ph.D. dissertation (University of Toronto).

Mendelsohn, E.(ed.) (1984) *Transformation and Tradition in Science* (Cambridge).

Merleau-Ponty, J. (1976) 'Situation et rôle de l'hypothèse cosmogonique chez Laplace', *Revue de l'histoire des sciences 29*, 21–49.

Merz, J. T. (1896–1904) *A History of European Thought in the Nineteenth Century*, 4 vols. (Edinburgh and London).

Middleton, W. K. (1969) *The Invention of Meteorological Instruments* (Baltimore).

Mill, J. S. (1836) 'On the Definition of Political Economy, and on the Method of Philosophical Investigation in that Science', *London and Westminster Review 26*, 1–29.

—— (1843) *A System of Logic Ratiocinative and Inductive: Being a Connected View of the Principles of Evidence and the Methods of Scientific Investigation* (London); 2nd edn. (1846); 7th edn. (1868); 10th edn. (1879).

—— (1852) 'Doctor Whewell on Moral Philosophy', reprinted (1875) in *Dissertations and Discussions*, 4 vols. (London) ii. 450–509.

—— (1865) *An Examination of Sir William Hamilton's Philosophy* (London), 8th edn. (1889).

—— (1873) *Autobiography* (London).

Miller, D. P. (1986) 'The Revival of Physical Sciences in Britain', *Osiris*, 2nd series, *2*, 107–34.

Miller, W. H. (1839) *Treatise on Crystallography* (Cambridge).

Morell, J. D. (1847) *An Historical and Critical View of the Speculative Philosophy of Europe in the Nineteenth Century*, 2nd edn. (London).

Morgan, S. R. (1985) 'William Whewell's *Philosophy of the Inductive Sciences*', M.Phil. thesis (Cambridge University).

Morrell, J. and A. Thackray (1981) *Gentlemen of Science: Early Years of the British Association for the Advancement of Science* (Oxford).

—— (eds.) (1984) *Gentlemen of Science: Early Correspondence of the British Association for the Advancement of Science* (London).

Morris, R. J. (1976) *Cholera 1832* (London).

Morus, I. R. (1988) 'The Sociology of Sparks', *Social Studies of Science 18*, 387–417.

Murphy, R. (1833) *Elementary Principles of the Theories of Electricity, Heat, and Molecular Actions* (Cambridge).

Nagel, E. (1961) *The Structure of Science* (New York).

Nesbitt, G. L. (1934) *Benthamite Reviewing: The First Twelve Years of the Westminster Review* (New York).

Newman, J. H. (1841) *The Tamworth Reading Room: Letters on an Address Delivered by Sir Robert Peel* (London).

Newsome, D. (1974) *Two Classes of Men: Platonism and English Romantic Thought* (London).

Newton, I. (1962) *Principia*, ed. F. Cajori (Berkeley).

Oldroyd, D. (1972) 'Edward Daniel Clarke and his Role in the History of the Blow-pipe', *Annals of Science 29*, 213–35.

Oppenheim, J. (1985) *The Other World: Spiritualism and Psychical Research in England, 1850–1914* (Cambridge).

Owen, R. (1848) *On the Archetype and Homologies of the Vertebrate Skeleton* (London).

Pap, A. (1962) *Introduction to the Philosophy of Science* (Glencoe).

Palter, R. (1970) 'An Approach to the Early History of Astronomy', *Studies in History and Philosophy of Science 1*, 93–133.

Patterson, E. (1983) *Mary Somerville and the Cultivation of Science, 1815–1840* (The Hague).

Peacock, G. (1831) *Observations on the Plans for the New Library, etc., by a Member of the First Syndicate* (Cambridge).

—— (1831*a*) *Remarks on the Replies to the Observations for the New Library, by a Member of the First Syndicate* (Cambridge).

Peacock, Thomas Love (1963) *Complete Novels*, ed. D. Garnett (London).

Peirce, C. S. (1865) 'Lecture on the Theories of Whewell, Mill, and Comte', in Peirce (1982), 205–23.

—— (1869) 'Lectures on British Logicians', in Peirce (1984), 310–45.

—— (1931–58) *Collected Papers*, 8 vols., C. Hartshorne and P. Weiss (eds.) (Cambridge, Mass.).

—— (1982, 1984) *Writings of Charles S. Peirce: Chronological Edition*, 2 vols., M. Fisch (ed.) (Bloomington).

Pevsner, N. (1970) *Robert Willis* (Northampton, Mass.).

Playfair, J. (1807–8) 'Review of Laplace's *Mechanique Celeste*', *Edinburgh Review 11*, 249–84.

—— (1802) *Illustrations of the Huttonian Theory of the Earth* (Edinburgh).

Popper, K. (1959) *The Logic of Scientific Discovery* (London).

Powell, B. (1834) *An Historical View of the Progress of the Physical Sciences* (London).

—— (1838) *The Connexion of Natural and Divine Truths* (London).

—— (1849) *On Necessary and Contingent Truth* (Oxford).

—— (1850) 'Letter to Dr Whewell', *Philosophical Magazine*, series 3, *35*, 235.

—— (1855) *Essays on the Spirit of the Inductive Philosophy, the Unity of Worlds and the Philosophy of Creation* (London).

—— (1859) *The Order of Nature, Considered with Reference to the Claims of Revelation* (London).

Preyer, R. (1957) 'Julius Hare and Coleridgean Criticism', *Journal of Aesthetics and Art Criticism 21*, 449–60.

—— (1981) 'The Romantic Tide Reaches Trinity', in J. Paradis and T. Postlewait (eds.), *Victorian Science and Victorian Values: Literary Perspectives* (New York), 39–68.

Priestley, J. (1772) *The History and Present State of Discoveries Relating to Vision, Light, and Colours* (London).

—— (1775) *The History and Present State of Electricity*, 3rd edn. (London).

Pritchard, A. (1897) *Charles Pritchard. Late Savilian Professor of Astronomy in the University of Oxford. Memoirs of his Life* (London).

Pryme, G. (1870) *Autobiographic Recollections* (Cambridge).

Pugin, A. W. (1841) *Contrasts or a Parallel between the Noble Edifices of the Middle Ages and Corresponding Buildings of the Present Day*, 2nd edn. (London).

Rashid, S. (1979) 'Richard Jones and Baconian Historicism at Cambridge', *Journal of Economic Issues 13*, 159–73.

—— (1980) 'The Growth of Economic Studies at Cambridge 1776–1860', *History of Education Quarterly 20*, 281–94.

Raub, C. Q. (1988) 'Robert Chambers and William Whewell: A Nineteenth-Century Debate over the Origin of Language', *Journal of the History of Ideas 49*, 287–300.

Recker, D. (1987) 'Causal Efficacy: The Structure of Darwin's Argument Strategy in the *Origin of Species*', *Philosophy of Science 54*, 147–75.

Rehbock, P. (1983) *The Philosophical Naturalists: Themes in Early Nineteenth-Century British Biology* (Madison, Wisconsin).

Rescher, N. (1973) *The Primacy of Practice* (Oxford).

—— (1973*a*) *Conceptual Idealism* (Oxford).

Rice, D. F. (1971) 'Natural Theology and the Scottish Philosophy in the Thought of Thomas Chalmers', *Scottish Journal of Theology 24*, 23–46.

Rickman, T. (1835) *An Attempt to Discriminate the Styles of Architecture in England*, 4th edn., 8 vols. (London).

Rimmer, W. G. (1960) *The Marshalls of Leeds: Flax Spinners 1788–1886* (Cambridge).

Robson, R. (1964) 'William Whewell, FRS: 1. Academic Life', *Notes and Records of the Royal Society 19*, 168–76.

—— (1967) 'Trinity College in the Age of Peel' in R. Robson (ed.), *Ideas and Institutions of Victorian Britain* (London), 213–35.

Rorty, R., J. B. Schneewind, and Quentin Skinner (eds.) (1984) *Philosophy in History: Essays on the Historiography of Philosophy* (Cambridge).

Rose, H. J. (1826) *The Tendency of Prevalent Opinions about Knowledge Considered* (Cambridge).

—— (1834) *An Apology for the Study of Divinity* (Cambridge).

—— (1845) 'Preface', *Encyclopaedia Metropolitana*, 26 vols. (London).

Rosman, D. (1984) *Evangelicals and Culture* (London and Canberra).

Ross, S. (1961) 'Faraday Consults the Scholars: The Origins of the Terms of Electrochemistry', *Notes and Records of the Royal Society 16*, 187–220.

—— (1962) ' "Scientist": The Story of a Word', *Annals of Science 18*, 65–85.

Rothblatt, S. (1976) *Tradition and Change in English Liberal Education: An Essay in History and Culture* (London).

—— (1981) *The Revolution of the Dons: Cambridge and Society in Victorian England* (Cambridge).

Rouse Ball, W. W. (1899) *A History of the Study of Mathematics at Cambridge* (Cambridge).

—— (1908) *A Short Account of the History of Mathematics* (New York).

Rudwick, M. J. S. (1971) 'Uniformity and Progression: Reflections on the Structure of Geological Theory in the Age of Lyell', in D. H. D. Roller (ed.), *Perspectives in the History of Science and Technology* (Norman, Okla.), 209–77.

—— (1972) *The Meaning of Fossils* (New York).

—— (1974) 'Poulett Scrope on the Volcanoes of Auvergne: Lyellian Time and Political Economy', *British Journal for the History of Science* 7, 205–42.

—— (1979) 'Transposed Concepts from the Human Sciences in the Early Work of Charles Darwin', in L. J. Jordanova and R. S. Porter (eds.), *Images of Earth: Essays in the History of the Environmental Sciences* (Chalfont St Giles), 67–73.

—— (1985) *The Great Devonian Controversy: The Shaping of Scientific Knowledge among Gentlemanly Specialists* (Chicago).

Rupke, N. (1983) *The Great Chain of History: William Buckland and the English School of Geology (1814–1849)* (Oxford).

Ruse, M. (1973) *The Philosophy of Biology* (London).

—— (1975) 'Britain 1830–1870', *Church History 44*, 505–22.

—— (1975*a*) 'Darwin's Debt to Philosophy', *Studies in History and Philosophy of Science* 6, 159–81.

—— (1976) 'The Scientific Methodology of William Whewell', *Centaurus 20*, 227–57.

—— (1976*a*) 'Charles Lyell and the Philosophy of Science', *The British Journal for the History of Science* 9, 121–31.

—— (1977) 'William Whewell and the Argument from Design', *Monist 60*, 244–68.

—— (1979) *The Darwinian Revolution: Science Red in Tooth and Claw* (Chicago).

—— (1981) 'What Kind of Revolution Occurred in Geology?', in P. Asquith and I. Hacking (eds.), *PSA 1978* (East Lansing, Mich.) 2, 240–73.

—— (1982) *Darwinism Defended* (Reading, Mass.).

Russell, B. (1912) *The Problems of Philosophy* (London).

Sanders, C. R. (1942) *Coleridge and the Broad Church Movement* (Durham, NC).

Sarton, G. (1936) *The Study of the History of Science* (New York).

Schaffer, S. (1989) 'The Nebular Hypothesis and the Science of Progress', in J. R. Moore (ed.), *History, Humanity and Evolution* (Cambridge), 131–64.

Schagrin, M. L. (1973) 'Whewell's Theory of Language', *Studies in History and Philosophy of Science 4*, 231–40.

Scheffler, I. (1967) *Science and Subjectivity* (Indianapolis).

Schneewind, J. B. (1968) 'Whewell's Ethics', *American Philosophical Quarterly Monograph 1*, 108–41.

—— (1974) 'Sidgwick and the Cambridge Moralists', *Monist 58*, 371–404.

—— (1977) *Sidgwick's Ethics and Victorian Moral Philosophy* (Oxford).

Schnuse, C. H. (trans.) (1841) *Elementar-Lehrbuch der Mechanik. Zum Gebrauch für Technische Lehranstalten* (Braunschweig).

Schweber, S. S. (1979) 'The Young Darwin', *Journal for the History of Biology 12*, 175–92.

—— (1980) 'Darwin and the Political Economists', *Journal of the History of Biology 13*, 195–289.

—— (1981) *Aspects of the Life and Thought of Sir John Frederick Herschel* (New York).

Secord, J. A. (1982) 'King of Siluria: Roderick Murchison and the Imperial Theme in Nineteenth-Century British Geology', *Victorian Studies 25*, 413–42.

—— (1986) *Controversy in Victorian Geology: the Cambrian–Silurian Dispute* (Princeton).

Sedgwick, A. (1833) *A Discourse on the Studies of the University*, Victorian Library edn. (New York, 1969).

—— (1834) *A Discourse on the Studies of the University of Cambridge*, 3rd edn. (Cambridge).

Sewell, W. (1841) 'Principles of Gothic Architecture', *Quarterly Review 69*, 111–49.

Shaffer, E. (1974) 'Coleridge and Natural Philosophy: A Review of Recent Literary and Historical Research', *History of Science 12*, 284–98.

Shapin, S., and B. Barnes (1977) 'Science, Nature and Control: Interpreting Mechanics' Institutes', *Social Studies of Science 7*, 31–74.

Shapiro, B. J. (1968) 'Latitudinarianism and Science in Seventeenth-Century England', *Past and Present 40*, 16–41.

—— (1969) *John Wilkins 1614–1672: An Intellectual Biography* (Berkeley and Los Angeles).

Shattock, J., and M. Wolff (eds.) (1982) *The Victorian Periodical Press* (Leicester).

Sidgwick, H. (1876) 'Philosophy at Cambridge', *Mind 1*, 235–46.

Skinner, Q. (1969) 'Meaning and Understanding in the History of Ideas', *History and Theory 8*, 1–53.

Sloan, P. (1985) 'Darwin's Invertebrate Program, 1826–1836: Preconditions of Transformism', in D. Kohn (ed.), *Darwinian Heritage* (Princeton), 71–120.

Smith, A. (1872) 'The Principles which Lead and Direct Philosophical Enquiries: as Illustrated by the History of Astronomy', in A. Smith, *Essays* (London), 325–85.

Smith, C. (1976) ' "Mechanical Philosophy" and the Emergence of Physics in Britain 1800–1850', *Annals of Science 33*, 3–29.

—— (1985) 'Geologists and Mathematicians: The Rise of Physical Geology', in P. M. Harman (ed.), *Wranglers and Physicists: Studies on Cambridge Mathematical Physics in the Nineteenth Century* (Manchester).

Smith, O. (1984) *The Politics of Language 1791–1819* (Oxford).

Smith, S. (1809–10) 'Review of Edgeworth's *Professional Education*', *Edinburgh Review 15*, 40–53.

Somerville, M. (1834) *On the Connexion of the Physical Sciences* (London).

Stair-Douglas, J. (1881) *The Life and Selections from the Correspondence of William Whewell D.D.* (London).

—— (1882), 2nd edn. (London).

Stanton, P. (1968) 'Sources of Pugin's Contrasts', in J. Summerson (ed.), *Concerning Architecture* (London), 120–39.

Stephen, L. (1885–1890) 'Whewell', *Dictionary of National Biography*, 21 vols. (London), xx. 1371.

—— (1900) *The English Utilitarians*, 3 vols. (New York).

Stephens, F. G. (1894) 'Thomas Woolner, R.A.', *The Art Journal*, 80–6.

Stewart, D. (1824) *Elements of the Philosophy of the Human Mind*, 3rd edn. (Edinburgh).

Stewart, R. (1985) *Henry Brougham: His Public Career 1778–1868* (London).

Stichweh, R. (1984) *Zur Entstehung des modernen Systems wissenschaftlicher Disziplinen: Physik in Deutschland 1740–1890* (Frankfurt-am-Main).

Stocking, G. W. (1968) *Race, Culture and Evolution: Essays in the History of Anthropology* (New York).

Stoll, M. R. (1929) *Whewell's Philosophy of Induction* (Lancaster, Pa.).

Thagard, P. R. (1977) 'Discussion: Darwin and Whewell', *Studies in History and Philosophy of Science 8*, 353–6.

Thirlwall, C. (1834) *A Letter to the Rev. Thomas Turton, on the Admission of Dissenters to Academical Degrees* (Cambridge).

—— (1881) *Letters Literary and Theological*, J. J. Stewart Perowne and L. Stokes (eds.), (London).

Thomas, K. (1984) *Man and the Natural World* (Harmondsworth).

Thomas, W. (1979) *The Philosophic Radicals: Nine Studies in Theory and Practice 1817–1841* (Oxford).

Thomson, W. (1882–1911) 'On a Mechanical Representation of Electric, Magnetic, and Galvanic Forces', *Mathematical and Physical Papers*, 6 vols. (Cambridge) i.76–80.

Tillyard, A. I. (1913) *A History of University Reform* (Cambridge).

Todhunter, I. (1876) *William Whewell, D.D. Master of Trinity College Cambridge: An Account of his Writings with Selections from his Literary and Scientific Correspondence*, 2 vols. (London).

Turner, F. M. (1978) 'The Victorian Conflict between Science and Religion: A Professional Dimension', *Isis 69*, 356–76.

—— (1981) *The Greek Heritage in Victorian Britain* (New Haven).

University of London (1912) *The Historical Record* (London).

Walsh, H. T. (1962) 'Whewell on Necessity', *Philosophy of Science, 29*:2, 139–45.

Watson, J. R. (1970) *Picturesque Landscape and English Romantic Poetry* (London).

Webb, R. K. (1955) *The British Working Class Reader 1790–1848* (London).

Whately, R. (1827) *Elements of Logic*, 2nd edn. (London).

Whewell, W. (1819) *An Elementary Treatise on Mechanics* (Cambridge).

—— (1822) 'On the Double Crystals of Fluor Spar', *Transactions of the Cambridge Philosophical Society 1*, 331–42.

—— (1823) *A Treatise on Dynamics* (Cambridge).

—— (1825) 'A General Method of Calculating the Angles Made by Any Planes of Crystals, and the Laws According to which They are Formed', *Philosophical Transactions 115*, 87–130.

—— (1826) 'On the Classification of Crystalline Combinations and the Canons by which their Laws of Derivation may be Investigated', *Transactions of the Cambridge Philosophical Society 2*, 391–425.

—— (1827) 'On the Angle Made by Two Planes, or Two Straight Lines, Referred to Three Oblique Co-ordinates', *Transactions of the Cambridge Philosophical Society 2*, 197–202.

—— (1827*a*) 'Specimen of the Use of Notation in the Analysis of Crystalline Forms', *Edinburgh Journal of Science 6*, 1–6.

—— (1827*b*) 'Reasons for the Selection of a Notation to Designate the Planes of Crystals', *Transactions of the Cambridge Philosophical Society 2*, 427–39.

—— (1828) *Account of Experiments Made at Dolcoath Mine in Cornwall* (Cambridge).
—— (1828*a*) *An Essay on Mineralogical Classification and Nomenclature* (Cambridge).
—— (1828*b*) 'On the Principles of Dynamics, Particularly as Stated by French Writers', *Edinburgh Journal of Science 8*, 27–38.
—— (1830) 'Mathematical Exposition of Some Doctrines of Political Economy', *Transactions of the Cambridge Philosophical Society 3*, 191–230.
—— (1830*a*) *Mathematical Exposition of Some Doctrines of Political Economy* (Cambridge).
—— (1831) 'Lyell's *Principles of Geology* volume 1', *British Critic 9*, 180–206.
—— (1831*a*), 'Modern Science: Inductive Philosophy', *Quarterly Review 45*, 374–407.
—— (1831*b*) 'Mathematical Exposition of Some of the Leading Doctrines in Mr Ricardo's Principles of Political Economy and Taxation', *Transactions of the Cambridge Philosophical Society 4*, 155–98.
—— (1831*c*) 'On the Employment of Notation in Chemistry', *Journal of the Royal Institution 1*, 437–53.
—— (1831*d*) Review of *An Essay on the Distribution of Wealth and the Sources of Taxation* by the Revd Richard Jones, *The British Critic, Quarterly Theological Review and Ecclesiastical Record, 10*, 41–61.
—— (1831*e*) 'On Isomorphism', *The Philosophical Magazine, or Annals of Chemistry, Mathematics, Astronomy, Natural History, and General Science 10*, 401–12.
—— (1831*f*) *Reply to Observations on the Plans for a New Library. By a Member of Both Syndicates* (Cambridge).
—— (1831*g*) *Mathematical Exposition of Some of the Leading Doctrines in Mr Ricardo's Principles of Political Economy and Taxation* (Cambridge).
—— (1832) 'Lyell's *Principles of Geology*, volume 2', *Quarterly Review 93*, 103–32.
—— (1832*a*) 'English Adjectives', *Philological Museum 1*, 359–72.
—— (1832*b*) 'Report on the Recent Progress and Present State of Mineralogy', *Report of the First and Second Meetings of the British Association for the Advancement of Science*, 2nd edn. (London, 1835), 322–65.
—— (1832*c*) *The First Principles of Mechanics: With Historical and Practical Illustrations* (Cambridge).
—— (1832*d*) *An Introduction to Dynamics Containing the Laws of Motion and the First Three Sections of the Principia* (Cambridge).
—— (1832*e*) *On the Free Motion of Points . . .* (Cambridge).
—— (1832*f*) *On the Uses of Definitions* (Cambridge).
—— (1833) 'Address', *Report of the Third Meeting of the British Association for the Advancement of Science* (London, 1834), xi–xxvi.
—— (1833*a*) 'On the Uses of Definitions', *Philological Museum 2*, 263–72.
—— (1833*b*) 'Essay Towards a First Approximation to a Map of Cotidal Lines', *Philosophical Transactions, 123*, 147–236.
—— (1833*c*) *Astronomy and General Physics Considered with Reference to Natural Theology* (London).
—— (1833*d*) *Analytical Statics* (Cambridge).
—— (1833*e*) *An Elementary Treatise on Mechanics*, 4th edn. (Cambridge).
—— (1834) *Astronomy and General Physics Considered with Reference to Natural Theology*, 2nd edn. (London).
—— (1834*a*) 'Mrs Somerville on the Connection of the Sciences', *Quarterly Review 51*, 54–68.

Whewell, W. (1834*b*) 'On the Empirical Laws of the Tides in the Port of London: With Some Reflections on the Theory', *Philosophical Transactions*, *124*, 15–45.

—— (1834*c*) 'On the Nature of the Truth of the Laws of Motion', *Transactions of the Cambridge Philosophical Society* 5, 149–72.

—— (1834*d*) *Remarks on Some Parts of Mr Thirlwall's Two Letters on the Admission of Dissenters to Academical Degrees* (Cambridge).

—— (1834*e*) *On the Motion of Points Constrained and Resisted* (Cambridge).

—— (1834*f*) *On the Connexion of the Physical Sciences* (reprint of 1834*a*).

—— (1834*g*) *Additional Remarks on Some Parts of Mr Thirlwall's Two Letters on the Admission of Dissenters to Academical Degrees* (Cambridge).

—— (1834*h*) 'Remarks on the Recent Statement by Berzelius Respecting the Use of Chemical Formulae', *London and Edinburgh Philosophical Magazine and Journal of Science 4*, 9–10.

—— (1835) *Architectural Notes on German Churches*, new edn. (London).

—— (1835*a*) *Thoughts on the Study of Mathematics as Part of a Liberal Education* (Cambridge), reprinted in (1837).

—— (1835*b*) 'Report on the Recent Progress and Present Condition of the Mathematical Theories of Electricity, Magnetism and Heat', *Report of the British Association for the Advancement of Science Fifth Meeting*, 1–34.

—— (1836) *Newton and Flamsteed: Remarks on an Article in Number 109 of the Quarterly Review* (Cambridge).

—— (1836*a*) 'On the Solar Inequality and on the Diurnal Inequality of the Tides at Liverpool', *Philosophical Transactions 126*, 131–47.

—— (1836*b*) 'On the Empirical Laws of the Tides in the Port of Liverpool', *Philosophical Transactions 126*, 1–15.

—— (1836*c*) 'On the Results of an Extensive System of Tide Observations Made on the Coasts of Europe and America in June 1835', *Philosophical Transactions 126*, 289–341.

—— (1836*d*) *Astronomy and General Physics Considered with Reference to Natural Theology*, 3rd edn. (Cambridge).

—— (1837) *On the Principles of English University Education* (London).

—— (1837*a*) *The History of the Inductive Sciences, from the Earliest to the Present Time*, 3 vols. (London).

—— (1837*b*) 'On the Diurnal Inequality of the Height of the Tide, Especially at Plymouth and at Singapore: and on the Mean Level of the Sea', *Philosophical Transactions 127*, 75–85.

—— (1837*c*) 'On the Progress of the Diurnal Inequality Wave along the Coasts of Europe', *Philosophical Transactions 127*, 227–44.

—— (1837*d*) *On the Foundation of Morals: Four Sermons Preached before the University of Cambridge* (Cambridge and London).

—— (1837*e*) *Mechanical Euclid* (Cambridge).

—— (1838) *The Doctrine of Limits with its Applications Namely Conic Sections, the First Three Sections of Newton, the Differential Calculus* (Cambridge).

—— (1838*a*) *Address Delivered at the Anniversary Meeting of the Geological Society of London, on the 16 of February 1838; and the Announcement of the Wollaston Medal and the Donation Fund for the same year* (London).

—— (1839) 'Presidential Address', *Proceedings of the Geological Society of London 3*, 61–93.

—— (1839*a*) *Astronomy and General Physics Considered with Reference to Natural Theology*, 4th edn. (London).

—— (1840) *The Philosophy of the Inductive Sciences, Founded upon their History*, 2 vols. (London).

—— (1840*a*) *The Isle of Sirens* (Cambridge).

—— (1841) 'Discussion of the Leith Tide Observations', in *Report of British Association for the Advancement of Science, Tenth Meeting* (London), 33–5.

—— (1841*a*) 'Demonstration that All Matter is Heavy', *Transactions of the Cambridge Philosophical Society* 7, 197–207.

—— (1841*b*) *An Elementary Treatise on Mechanics*, 6th edn. (London).

—— (1841*c*) *The Mechanics of Engineering* (London).

—— (1841*d*) *Two Introductory Lectures on Moral Philosophy* (Cambridge).

—— (1842) *Architectural Notes on German Churches: With Notes Written during an Architectural Tour in Picardy and Normandy, to Which are Added, Notes on the Churches of the Rhine, by M. F. Lassaulx, Architectural Inspector to the King of Prussia*, 3rd edn. (London).

—— (trans.) (1842*a*) *The Knight of Toggenburg* by Schiller (Shelford).

—— (1842*b*) 'Presidential Address', *Report of the British Association for the Advancement of Science, Eleventh Meeting* (London), xxvii–xxxv.

—— (1844) 'On the Fundamental Antithesis of Philosophy', *Transactions of the Cambridge Philosophical Society* 7, part 2, 170–81.

—— (1845) *Of a Liberal Education in General, and with Particular Reference to the Leading Studies of the University of Cambridge* (London).

—— (1845*a*) *Indications of the Creator* (London and Philadelphia).

—— (1845*b*) *The Elements of Morality, Including Polity*, 2 vols. (London).

—— (1846) *Indications of the Creator*, 2nd edn. (London and Philadelphia).

—— (1846*a*) *Lectures on Systematic Morality* (London).

—— (1846*b*) 'Letters on English Hexameters', *Blackwood's Magazine 60*, 19–21.

—— (1847) *Philosophy of the Inductive Sciences: Founded upon their History*, 2nd edn., 2 vols. (London).

—— (1847*a*) *Sunday Thoughts and Other Verses* (Cambridge).

—— (1847*b*) 'Dialogue on English Hexameters', *Fraser's Magazine 36*, 665–70.

—— (1847*c*) *Verse Translations from the German, including Lenore, Schiller's Song of the Bell* (London).

—— (1848) 'On the Tides of the Pacific and on the Diurnal Inequality', *Philosophical Transactions 138*, 1–29.

—— (1848*a*) *Butler's Three Sermons on Human Nature* (Cambridge and London).

—— (1848*b*) 'Second Memoir on the Fundamental Antithesis of Philosophy of Induction', *Transactions of the Cambridge Philosophical Society 8*, 614–20.

—— (1848*c*) 'Evangeline, a Tale of Acadie', *Fraser's Magazine 37*, 295–8.

—— (1848*d*) *Second Memoir on the Fundamental Antithesis of Philosophy* (Cambridge).

—— (1848*e*) *The Elements of Morality, Including Polity*, 2nd edn., 2 vols. (London).

—— (1849) 'On Mr Macaulay's Praise of Superficial Knowledge', *Fraser's Magazine 40*, 171–5.

—— (1849*a*) *A Sermon Preached before the University of Cambridge on the Day of General Thanksgiving* (Cambridge).

—— (1849*b*) 'Sacred Latin Poetry', *Fraser's Magazine 39*, 527–32.

—— (1849*c*) *Of Induction* (London).

—— (1850) 'Mathematical Exposition of some Doctrines of Political Economy: Second Memoir', *Transactions of the Cambridge Philosophical Society 9*, 128–49.

Whewell, W. (1850*a*) Review of *The Lamps of Architecture, Fraser's Magazine, 41*, 151–9.
—— (1850*b*) 'Goethe's Herman and Dorothea', *Fraser's Magazine, 41*, 33–40.
—— (1850*c*) 'Mathematical Exposition of some Doctrines of Political Economy: Third Memoir', *Transactions of the Cambridge Philosophical Society* 9, 1–7.
—— (1850*d*) *Of a Liberal Education in General . . . 2. Discussions and Changes 1840–1850* (London).
—— (1851) *The General Bearing of the Great Exhibition of the Progress of Art and Science* (London).
—— (1851*a*) 'On the Transformation of Hypotheses in the History of Science', *Transactions of the Cambridge Philosophical Society* 9, 139–46.
—— (1852) 'Letter to Edwin Guest', *Proceedings of the Philological Society* 5, 133–42.
—— (1852*a*) *Lectures on the History of Moral Philosophy in England* (London).
—— (1852*b*) 'A Letter to the Author of the *Prolegomena Logica*', in the reprint of the 2nd edn. of *The History of the Inductive Sciences from the Earliest to the Present Time*, 3 vols. (London).
—— (1852*c*) *Of a Liberal Education in General . . . 3. The Revised Statutes 1851–1852* (London).
—— (1853) *Of the Plurality of Worlds: An Essay* (London).
—— (1853*a*) 'English Hexameters', *North British Review 19*, 129–50.
—— (1854) *The Influence of the History of Science upon Intellectual Education* (London).
—— (1854*a*) *Of the Plurality of Worlds: An Essay*, 3rd edn. (London).
—— (1854*b*) *Notes on the Oxford University Bill in Reference to the Colleges at Cambridge* (Cambridge).
—— (1855) 'A Few Words on the *Edinburgh Review* Article on the Plurality of Worlds'.
—— (1855*a*) *Remarks on the Proposed Reform of the University of Cambridge* (Cambridge).
—— (1855*b*) *Additional Remarks on the Proposed Reform of the University of Cambridge* (Cambridge).
—— (ed.) (1855*c*) *Bishop Butler's Ethical Discourses; to Which are Added some Remains Hitherto Unpublished Papers as a Textbook in Moral Philosophy, with a Syllabus* (Philadelphia).
—— (1856) *On the Philosophy of Discovery: Chapters Historical and Critical* (London).
—— (1857) *History of the Inductive Sciences from the Earliest to the Present Time*, 3rd edn., 3 vols. (London), reprinted 1867 (London).
—— (1857*a*) *Remarks on Proposed Changes in the College Statutes* (Cambridge).
—— (1858) *Novum Organum Renovatum* (London).
—— (1858*a*) *The History of Scientific Ideas*, 2 vols. (London).
—— (1858*b*) 'Of the Platonic Theory of Ideas', *Transactions of the Cambridge Philosophical Society, 10*, 94–104.
—— (ed.) (1859) *Literary Remains, Consisting of Lectures and Tracts on Political Economy, of the late Rev Richard Jones* (London).
—— (1860) *On the Philosophy of Discovery* (London).
—— (trans.) (1861) Plato's *Republic* (Cambridge).
—— (1862) *Six Lectures on Political Economy* (Cambridge).
—— (1862*a*) *Lectures on the History of Moral Philosophy*, new edn. (Cambridge).
—— (1862*b*) 'English Hexameters: Mr Dart's Translation of *The Iliad*', *Macmillan's Magazine* 5, 487–96.
—— (1862*c*) 'New Hexameter Translations of *The Iliad*', *Macmillan's Magazine* 6, 297–304.

—— (1864) *Astronomy and General Physics Considered with Reference to Natural Theology*, 7th edn. (London).

—— (1864*a*) *The Elements of Morality including Polity*, 4th edn. (London).

—— (1865–6) 'Comte and Positivism', *Macmillan's Magazine 13*, 353–62.

—— (1866) 'Grote's *Plato*', *Fraser's Magazine 73*, 411–23.

—— (1884) *History of the Inductive Sciences, From the Earliest to the Present Time*, 3rd American edn., 2 vols. (New York).

—— (1967) *The Historical and Philosophical Works of William Whewell*, ed. G. Buchdahl and L. Laudan (London).

White, J. F. (1962) *The Cambridge Movement: The Ecclesiologists and the Gothic Revival* (Cambridge).

Williams, L. Pearce (1971) *Michael Faraday* (New York).

Williams, R. (1976) *Keywords: A Vocabulary of Culture and Society* (London).

Willis, R. (1830) *Remarks on the Architecture of the Middle Ages* (Cambridge).

Wilson, D. B. (1974) 'Herschel and Whewell's version of Newtonianism', *Journal of the History of Ideas 35*, 79–97.

—— (1984) 'A Physicist's Alternative to Materialism: The Religious Thought of George Gabriel Stokes', *Victorian Studies 28*, 69–96.

—— (1985) 'The Educational Matrix: A Comparative Study of Physics Education in Early-Victorian Cambridge, Edinburgh, and Glasgow Universities', in P. M. Harman (ed.), *Wranglers and Physicists: Studies on Cambridge Mathematical Physics in the Nineteenth Century* (Manchester).

—— (1987) *Kelvin and Stokes: A Comparative Study in Victorian Physics* (Bristol).

Wilson, J. (1834) 'Admission of Dissenters to Degrees in the English Universities', *Blackwood's Magazine 35*, 716–780.

Wilson, L. G. (1972) *Charles Lyell: The Years to 1841* (New Haven).

Winsor, M. P. (1976) *Jellyfish and the Order of Life: Issues in Nineteenth-Century Science* (New Haven).

Winstanley, D. A. (1940) *Early Victorian Cambridge* (Cambridge), reprinted 1977.

—— (1947) *Later Victorian Cambridge* (Cambridge), reprinted 1977.

Wise, M. Norton (1982) 'The Maxwell Literature and British Dynamical Theory', *Historical Studies in the Physical Sciences 13*, 175–205.

Wood, P. B. (1980) 'Methodology and Apologetics: Thomas Sprat's *History of the Royal Society*', *British Journal for the History of Science 8*, 1–26.

Wordsworth, W. (1978–88) *Letters: The Later Years*, A. Hill and E. de Selincourt (eds.), 4 vols. (Oxford).

Wright, G. H. von (1957) *The Logical Problem of Induction*, 2nd edn. (Oxford).

Wright, J. F. M. (1827) *Alma Mater or Seven Years at the University of Cambridge* (London).

Yeo, R. (1977) 'Natural Theology and the Philosophy of Knowledge in Britain, 1819–1869' Ph.D. thesis (University of Sydney).

—— (1979) 'William Whewell, Natural Theology and the Philosophy of Science in mid-Nineteenth Century Britain', *Annals of Science 36*, 493–512.

—— (1981) 'Scientific Method and the Image of Science, 1831–1891', in R. M. Macleod and P. Collins (eds.), *The Parliament of Science* (Northwood), 65–88.

—— (1984) 'Science and Intellectual Authority in mid-Nineteenth Century Britain: Robert Chambers and *Vestiges of the Natural History of Creation*', *Victorian Studies 28*, 5–31.

Yeo, R. (1985) 'An Idol of the Marketplace: Baconianism in Nineteenth-Century Britain', *History of Science 23*, 251–98.

—— (1986) 'Scientific Method and the Rhetoric of Science in Britain, 1830–1917' in J. Schuster and R. Yeo (eds.), *The Politics and Rhetoric of Scientific Method: Historical Studies* (Boston and Dordrecht).

—— (1987) 'William Whewell on the History of Science', *Metascience 5*, 25–40.

Young, R. M. (1980) 'Natural Theology, Victorian Periodicals and the Fragmentation of a Common Context', in C. Chant and J. Fauvel (eds.), *Darwin to Einstein: Historical Studies on Science and Belief* (Harlow), 69–106.

—— (1985) *Darwin's Metaphor: Nature's Place in Victorian Culture* (Cambridge).

Index

This is an amalgamated index of names and subjects. References to Whewell's works are listed by title gathered under his name. All other entries on Whewell are listed elsewhere under the appropriate topic. The index refers to the body of the text and also to substantial discussions in footnotes. There are no entries for the names of the contributors to this book.